Topics in Geobiology

Volume 42

Series Editors
Neil H. Landman
American Museum of Natural History, New York, NY, USA

Peter J. Harries
North Carolina State University, Raleigh, NC, USA

The **Topics in Geobiology** series covers the broad discipline of geobiology that is devoted to documenting life history of the Earth. A critical theme inherent in addressing this issue and one that is at the heart of the series is the interplay between the history of life and the changing environment. The series aims for high quality, scholarly volumes of original research as well as broad reviews.

Geobiology remains a vibrant as well as a rapidly advancing and dynamic field. Given this field's multidiscipline nature, it treats a broad spectrum of geologic, biologic, and geochemical themes all focused on documenting and understanding the fossil record and what it reveals about the evolutionary history of life. The Topics in Geobiology series was initiated to delve into how these numerous facets have influenced and controlled life on Earth.

Recent volumes have showcased specific taxonomic groups, major themes in the discipline, as well as approaches to improving our understanding of how life has evolved.

Taxonomic volumes focus on the biology and paleobiology of organisms – their ecology and mode of life – and, in addition, the fossil record – their phylogeny and evolutionary patterns – as well as their distribution in time and space.

Theme-based volumes, such as predator-prey relationships, biomineralization, paleobiogeography, and approaches to high-resolution stratigraphy, cover specific topics and how important elements are manifested in a wide range of organisms and how those dynamics have changed through the evolutionary history of life.

Comments or suggestions for future volumes are welcomed.

Neil H. Landman Department of Paleontology American Museum of Natural History New York, USA E-mail: landman@amnh.org Peter J. Harries Department of Marine, Earth and Atmospheric Sciences North Carolina State University Raleigh, USA E-mail: pjharrie@ncsu.edu

More information about this series at http://www.springer.com/series/6623

Thomas Defler

History of Terrestrial Mammals in South America

How South American Mammalian Fauna Changed from the Mesozoic to Recent Times

Springer

Thomas Defler
Department of Biology
National University of Colombia, Bogota
Bogota, Colombia

ISSN 0275-0120
Topics in Geobiology
ISBN 978-3-319-98448-3 ISBN 978-3-319-98449-0 (eBook)
https://doi.org/10.1007/978-3-319-98449-0

Library of Congress Control Number: 2018953014

This Springer imprint is published by the registered company Springer Nature Switzerland AG
The registered company address is: Gewerbestrasse 11, 6330 Cham, Switzerland

I dedicate this book to all students of neotropical mammals.

Preface

I wrote this book during a 10–15-year period of teaching a course in the evolution of South American tetrapod mammals at the Universidad Nacional de Colombia in Bogotá. While the book is primarily intended for the lay reader and the student of South American mammals, I hope that it might also be of service to many professional zoologists and paleontologists, who have been unable to keep abreast of the flood of new discoveries and yet wish to learn something of the more significant published results. How far I have succeeded in a most difficult task must be left to the judgment of such readers. I am not a paleontologist but rather a vertebrate zoologist who has specialized in modern neotropical primates. Since I have rather broad interests in vertebrate zoology and primatology, I elected to teach a course on the evolution of mammals after transferring from another campus to the Bogotá Campus, and frankly, I wanted to learn something more about this field of South American mammals. The course was well-attended throughout the years, and I dedicated myself continuously to study the literature. There is now a lot of literature! For a generalist like myself, I have acquired a broad education on the evolution of terrestrial mammals in South America, but I cannot claim to have dominated the hundreds of technical articles now available. Nevertheless, I have learned to appreciate the vast story that has resulted in the mammalian fauna of today, and I am surely an aficionado on the outline of the story of diversity that has gone before.

Who cannot be fascinated by the evolution of the multitude of species that filled South America during earlier epochs of the Cenozoic, such as the specialized carnivorous Sparassodonta metatherians and the hundreds of species of ungulates that we can now associate with the evolution of perissodactyl and artiodactyl ungulates from the northern continents? What primatologist does not thrill to the evolutionary history of the platyrrhine primates and their last African origins, lost in the veil of time and their improbable rafting from Africa to South America? As well, the rafting of tiny rodents from Africa and their evolution into giant 1000 kg semiaquatic mammals provide us with so much evolutionary color that we pine for more detail and we dream of being able to see these fascinating beasts. I thrilled when I discovered the South American platypus in the literature and the narrow peninsula that connected South America to a warmer Antarctic, all of which provided a highway for South American marsupials to Australia.

Since we can no longer see these animals, we have to rely on reconstructions. When I read GG Simpson's 1980 book on the evolution of South American mammals, I was somewhat disappointed by the poverty of illustrations in that book. At the time there were fewer artists who try to recreate prehistoric fauna, and I guess that the great paleontologist himself illustrated the book, although not too successfully, even though people like the paleoartist Charles Knight were well-known. Reading about lost, extinct fauna is one thing, but an effort to visualize the fauna is an effort that anybody wanting to learn about these animals tends to make, and my wish was to be able to illustrate these animals as richly as possible, just as the dinosaurs have been brought to life by illustrators. I was lucky to discover the Ukrainian artist Roman Uchytel on the Internet, who has specialized in using computer art to depict how these animals might have looked. Computer art fools the eye into thinking that we are viewing a photograph of the animal, and if it is well-done, we are drawn into a prehistoric world that lets us imagine how this world might have been, satisfying the basic urge of all of us to be a witness to such lost worlds. This book, then, publishes many images of Roman that illustrate the animals discussed here, although images of a few other artists also supplement the story that I tell.

Roman Uchytel was born in Ukraine of the Soviet Union and grew up near a zoo, and he spent most of his time there dreaming of becoming a zoologist and sketching animals. Eventually he graduated from art school and university with a love for nature and training as an artist. His knowledge of anatomy helps him to depict all manner of birds and beasts; although he rarely draws a dinosaur, he specializes in rarely depicted ancient mammals and birds, and with computer art he is able to place them in a natural setting that suggests a photograph of an ancient world. Roman, like myself, continues to be fascinated by animals and is dedicated to depicting the ancient ones in now lost worlds. He has considerably enriched this book. You can see more of Roman's work at his Internet site (https://prehistoric-fauna.com).

Bogota, Colombia Thomas Defler

Acknowledgments

I am particularly grateful for the time and space provided to me by the Universidad Nacional de Colombia to continuously teach a course by this name and to continue preparing to manage this great and voluminous theme. I could not have managed the library research necessary without the help of Dr. Scott Raymond and the University of Calgary library services in Canada. Thank you, Scott, for being there when I needed your help and with no thought of recompense. I am very grateful to the many neotropical paleontologists and aficionados of ancient mammals who positively answered my questions and to some who made very helpful suggestions. There were many (aficionados and professionals) who, like myself, are anxious to illustrate to the best of their abilities, the prehistoric mammalian life that lived on this continent, and I thank those who allowed me to reproduce their images in this book. I hope you are satisfied with the result. I am also grateful to the many Colombian students who listened to me over the years and, as a result, learned something about our mammalian history. I am especially grateful to the careful reading of the text and the many spelling corrections made to this book by Claudia Moreno and Enrique Forrero. They have improved the text in a manner that is very important to me and to all who read it.

About the Book

During the last 10 years, as I taught a course at the Universidad Nacional de Colombia with the same title as this book, it came home to me that the information for this fascinating subject was spread out in many journals and that the only attempt to address a history of South American mammals had been G. G. Simpson's 1980 book *Splendid Isolation*, now thoroughly out of date and almost bereft of images. My Internet research had taught me that there now were many images attempting to illustrate the prehistoric fauna of South America, but nobody had attempted to use them to tell the complete story as it is known up till the present. There were hundreds of published articles available, but they were not organized into the great story that is emerging about how South American mammals became what they are today and what history had gone before, though many understand that there was a strange and beautiful mammalian fauna before the two Americas became connected and the invasion of northern fauna occurred that changed everything. What was the fauna like before the American interchange? What were the origins of the several fantastic groups that became extinct when northerners arrived and out-competed the southerners? How did the modern mammalian fauna come into being with such disparate elements as two great rodent groups with two different origins, primates (where none had lived before), strange armadillos, sloths and anteaters with little understanding of their origin, and other mammals such as felines, canines, and deer that clearly had evolutionary connections to the north?

So I set out to write the history of these South American mammals that I was dedicated to studying, hoping that others would be interested in this story as well. I have attempted to write the story on a level that might be interesting to university students and professionals working with mammals and their paleontology. Perhaps those more broadly interested in South American fauna will also want to read this book. This is not a technical book, though some basic biology and paleontology are assumed. Very important are the many illustrations used that will hopefully indicate how these animals might have looked. The major artist used is the Ukrainian Roman Uchytel, who is an expert at illustrating prehistoric mammals (www.prehistoric-fauna.com) and has made many species come alive. I attempt to tell this history in 15 chapters.

More Detailed Description

Based on my university course "Historia de los Mamíferos Terrestres de Sudamérica" and the series of lectures that I developed for this course, I have written a book outlining the evolutionary changes in the fauna of South America, beginning in the Mesozoic and ending at the present time. I have summarized a very large and disparate literature for this evolutionary history. Knowing that 35 years (after the publication of Simpson's book) of paleontological studies in South America are found in many technical journals and books, I have organized much of this literature into a history that tells the story of the changes that South American mammals have gone through since their beginning around 165 million years of so ago.

Besides an introductory chapter and a chapter discussing the ancient pre-Cenozoic fauna, the most detailed story, of course, took place during the Cenozoic or "Age of Mammals" of the last 65 million years, since the amount of information in terms of fossils begins to accumulate. So the majority of the chapters deal with this period from Chap. 3, starting with the first mammals of the Paleocene. Then I change focus, highlighting in each chapter the major mammalian groups that have made up the South American mammals: the marsupials, the ungulates, the xenarthrans, the caviomorph rodents, and the platyrrhine primates (Chaps. 3, 4, 5, 6, 7, and 8).

Once again, I change focus to highlight some specific assemblages that illustrate certain points. The La Meseta fossils (Chap. 9) illustrate an Eocene Antarctic group of mammals, and they underline the strong connection that Antarctica had with South America, at least until the two continents became disconnected around 30 million years ago. The La Venta fauna (Chap. 10) illustrates an ecological community of a geologically short period of time ago from about 13 to 12 million years and is the richest description that we have of an ancient ecology. I include a complete chapter on the neotropical mammals that invaded the Caribbean Islands (Chap. 11). The story of the formation of the Amazon River (Chap. 12) is important, inasmuch as it was obviously the center of evolution for so many mammals, although we do not as yet have the richness of information that exists from Patagonia (Chaps. 9, 10, and 11).

Concluding the history, I describe the inter-American interchange of mammalian fauna and how South America changed due to the mammalian invasions from the north. Now we appreciate that these invasions did not occur only during three to four million years after a final land connection of the Americas became complete but in fact the invasions began as early as 9 or 10 million years ago when the first gomphotheres, tapirs, camels, peccaries, and raccoons managed to arrive, probably swimming, to South America. The story gets very interesting, since the real dispersal abilities of mammals come to the fore as well as a new understanding about the formation of the Central American gap.

A penultimate chapter describes the Pleistocene fauna and the megafauna that populated South America briefly before it became extinct. Within this context of course is included a consideration of what role newly arrived human beings might have played in this great "dying off." The book ends with a brief consideration of the modern mammalian neotropical fauna.

Chapter Outline

The book, as written, comprises 15 chapters of which titles follow below with short chapter descriptions.

Chapter 1: Introduction

Chapter 1 includes brief descriptions of the roles played by various early collectors and describers of ancient bones. It also briefly describes the tools that are necessary to interpret the evolutionary history of South American mammals including the problem of calculating geological time, stratification, radiometric dating, paleomagnetism, the establishment of SALMAs (South American Land Mammal Ages), the geological time scale (and table), plate tectonics and the distribution of fauna, the role of Alfred Wegener, continental drift, and molecular phylogenetic research.

Chapter 2: Ancient Mammals of Gondwanan South America

Chapter 2 includes a description of what is known of South American mammals from the Mesozoic Era, which begins (according to present information) in late Jurassic around 168–161 million years ago and comprises australosphenid mammals (relatives of the platypus) and also the now extinct triconodont mammals. The description then moves through the ear Cretaceous with increasing numbers of fossils available. This fauna was dominated in South America by dryolestid mammals, which were closely related to modern placental mammals. During these times, there are a couple spectacular fossils known, like *Vincelestes neuquenianus* and *Cronopio dentiacutus*, both of which are illustrated.

Chapter 3: Early Cenozoic Mammals in South America

Chapter 3 includes a description of the earliest (Paleocene) mammalian assemblages known for the South American Cenozoic. These include complete discussions of the Tiupampan, Peligran, Itaboraian, and Riochican South American Land Mammal Age (SALMA), and the Riochican overlapping into the Eocene. These assemblages are very instructive, as they contain the most ancient groups, the marsupials, and native ungulates. These apparently arrived in South America from North America, perhaps in the latter part of the Cretaceous before the beginning of the Cenozoic. These faunas also contain other elements from North America and from the ancient Mesozoic mammalian fauna, which then become extinct in South America. Several subgroups of marsupials and ungulates make their first appearance, as well as the first xenarthrans. A South American platypus demonstrates the ancient connection of South America to Australia, via Antarctica.

Chapter 4: Marsupials and Other Metatheres of South America

Marsupials and other metatherians apparently arrived in South America before the beginning of the Paleogene, at the end of the Cretaceous. The earliest relatives of Australian marsupials, the Microbiotheria, appear in the earliest assemblage. The carnivorous Sparassodonta evolve and produce the largest marsupial predator known, the 600 kg *Proborhyaena gigantea*, as well as the jaguar-sized saber-toothed marsupial *Thylacosmilus*. Many of these poorly known marsupial predators are well-illustrated by Roman Uchytel.

Chapter 5: The Native Ungulates of South America

The ancient, native ungulates were another fascinating group, which sadly has become totally extinct, but which we now understand were related to the northern Perissodactyla or odd-toed ungulates. These animals apparently evolved from northern condylarths (primitive ungulates) that somehow made it to South America. Evolution produced five orders of quite bizarre ungulates, some more like rodents than like ungulates, but others, toward the end of the Neogene in the Pleistocene, had become large, rhinoceros-like *Toxodon* and camel-like *Macrauchenia*. Although there is much to learn about the many species of South American meridiungulates (native ungulates), many forms are known. Some of the latest forms are known to have been hunted by early humans, who finally arrived in South America. This chapter is well-illustrated especially by images of Roman Uchytel.

Chapter 6: The Xenarthrans: Armadillos, Glyptodonts, Anteaters, and Sloths

From the time that the first xenarthrans appeared as early armadillos in the late Paleocene Itaborai, the group diversified into strange and wonderful forms. Besides the Dasypodidae, a group called the Glyptodontidae arose and diversified; some were the size (and shape) of a Volkswagen bug and were harmless grazers and were very common on the grassy savannas of South America. Also, the sloth lineage appeared with the last species reaching the greatest size of any southern mammal. *Megatherium americanum* one of the largest mammals known, equivalent to an elephant. Finally a short history of the little-known anteaters is given. A discussion of the possible origins of the Xenarthra is suggested, and the group is presented as one of the most ancient lineages of modern mammals.

Chapter 7: The Caviomorphs: First South American Rodents

The first rodents did not arrive in South America until the mid-Eocene, at about 41 million years ago. This recent discovery makes the history of the caviomorphs extremely interesting, since the earliest known rodents are now known to be from tropical forest and not from dry, savanna-like habitats as previously believed. The group is ancient and is clearly related to the African phiomorph rodents. In this chapter and in Chap. 8, I enjoy describing the probable mode of dispersion of caviomorphs and primates from Africa to South America, since so many have difficulties accepting rafting over the Atlantic Ocean. The history of caviomorphs in South America also includes giant species that appeared during the latter part of the Neogene, culminating in the 1000 kg *Josephoartigasia* of the Río de la Plata (River Plate). This chapter also has some original illustrations by Roman Uchytel and by others.

Chapter 8: The Platyrrhine Monkeys

This chapter presents the partial evolutionary history known of the platyrrhine primates through their known fossils. Recently new evidence for the earliest known primates comes to us from Peru, so that we now have tropical evidence of *Perupithecus* and others, superceding the many higher latitude primates known from the southern cone. The previous earliest known primates (*Branisella*, *Szalatavus*) are known also from a totally different habitat and 20 million years later. Here again new evidence proves the important role of tropical forest in the evolution of South American primates. Additionally the newly discovered Eocene primates have similarities to early African fossils. Again, in this chapter, I describe how primates might have (and probably did) arrived in South America and just what the conditions would have had to be for the success of such a precarious voyage.

Chapter 9: An Antarctic Eocene Mammalian Community

This short chapter describes an Antarctic mammalian community and illustrates the faunal connection that existed between Antarctic and South America, since all of these ancient Eocene Antarctic mammals had living relatives in South America. The community existed during the early-middle to late Eocene when the earth was very warm, the warmest of the entire Cenozoic. But since marsupials dispersed from South America to Australia in the early Paleocene, it is obvious that Antarctica was covered by forest until the arrival of global plunging temperatures at the end of the Eocene and early Oligocene. Many plants that have been identified from the Antarctic Peninsula are commonly seen on the southern tip of South America even today.

Chapter 10: La Venta: A Miocene Colombian Mammalian Community

The chapter describes the well-known Colombian La Venta fauna. This is the most detailed tropical faunal assemblage known for South America. Although it is located on the upper Magdalena River, at the time that it existed, it was peripheral to the great Amazonian wetlands to the east: unlike today, there was no Eastern Cordillera barrier. This fauna is represented by about 72 species of mammals from the richest deposits (the Monkey Beds) dated from about 11.8 to 13.5 million years ago, so it really represents a tropical community from a very narrow time window. The La Venta habitat was an open riparian-savanna with gallery forests, so mammals from several different conditions illustrate forest and savanna mammals. Compared to previous chapters, this is a fairly modern fauna, yet no elements of the north are yet to be found, and the mix is rather different than that found in the high latitudes further south. It is notable that many species and genera of mammals from La Venta have also been found in tropical central Peru, suggesting that a band of similar habitat west of the Amazonian wetlands was continuous from northern Colombia to southern Peru for that period.

Chapter 11: Mammalian Invasion of the Caribbean Islands

Chapter 11 describes the South American fauna that populated the Caribbean Islands and how this fauna might have arrived. Fossils tell us a story that, because of recent extinctions, probably was at least in part caused by human arrival. Particularly interesting to me are the data about primates that lived on the Caribbean Islands, several species of which lasted until the last few hundred years. All of these primates seem to have descended from one group of South America which arrived as far north as Cuba.

Chapter 12: The Genesis of the Modern Amazon River Basin and Its Role in Mammalian Evolution

This chapter discusses the origin of the Amazon basin and its role in mammalian evolution. Information is still sparse, due to taphonomic difficulties in a moist tropical, acid environment, but lately especially new finds of earliest rodent and primate fossils have added to our knowledge of how the tropical regions have played a large role in South American mammalian evolution. The process by which the great river was originally established is also a theme discussed, and what the conditions must have been to allow invading mammals from the north to arrive to the southernmost parts of the continent is considered.

Chapter 13: The Great American Biotic Interchange

This is the story of the revolutionary changes to South American mammals that occurred when it became possible for mammals from North America to pass to South America. This probably began as early as 8–10 million years ago when a proboscid, camelid, tayasuid, tapirid, and procyonid arrived in South America. Later at about three million years ago, the invasion became a flood when (apparently) the terrestrial connection between the two continents became complete. Of course some South American fauna went north, as well, including the terrestrial sloths and the glyptodonts, but they became extinct after modest success. Other southern elements persist yet in Central America, including primates and caviomorph rodents.

Chapter 14: Pleistocene Mammal Communities and Their Extinction

During the Pleistocene, the diversity of mammals in South America became extremely elevated. It seems that hyperdiversity reached the highest-known in the world, and there has been nowhere else where the 37 megamammals (weighing up to 1000 kg) were to be found until they all became extinct, the last just 8000–9000 years ago. Of course ecological factors played a huge role in leveling this out-of-balance fauna, but the intriguing question has always been about the role that human beings had to help these mammals to extinction.

Chapter 15: The Modern Mammals of South America

This short chapter discusses the modern mammalian fauna in South America and shows the comparative balance of mammal orders. The rodents make up well over half of the entire terrestrial mammalian biota (without counting the bats). The total terrestrial mammalian fauna numbers about 1500 species, and South America is the most diverse continent in the world. Despite the extensive Pleistocene extinctions, the fauna is amazing in its diversity. Many illustrations are provided for each group.

Contents

List of Figures

List of Tables

About the Author

Dr. Thomas Defler was born in the United States (Denver, Colorado) on November 26, 1941, and did university studies in Albert Ludwigs University of Freiburg, Germany (1 year); University of Miami, Florida (1 year); University of Colorado (7 years); and University of Colorado at Denver (2 years). He holds a B.A.(biology), M.A (botany), and Ph.D. (zoology). Working as a laboratory assistant in a Denver primate laboratory, he became interested in primates and determined to relocate to a tropical primate habitat country. He moved to Colombia in January 1976 as a Peace Corps Volunteer and worked for 5 years as a PC primatologist with the Colombian government agency INDERENA (natural resources) doing primate research for Colombia in a remote national park called El Tuparro National Park that abuts the Orinoco River. In 1981, Dr. Defler moved south to the Colombian Amazon in Vaupés where he established a tropical research station in undisturbed and virgin rainforest in an isolated and wild part of the Colombian Amazon near the Brazilian border. He directed this station for 17 years, studying primates and other mammals and receiving Colombian students.

In 1997, Dr. Defler became a professor at the Colombian National University in Leticia. He was obligated to leave his research station in 1998 by the Colombian guerilla group the FARC. He lived sporadically for about 7.5 years in Leticia, Colombia, as a National University professor, and meanwhile established another research station in another remote part of Amazonian rainforest in southern Colombia near the Amacayacu National Park (15 km from the Brazilian border). In 2006, he was transferred by request to the main campus in Bogotá where he taught until retiring in December, 2016.

Of the several courses that Dr. Defler has taught (including evolution), the course "History of South American Land Mammals" (Historia de los Mamíferos Terrestres de Sudamérica) set him on a new path of studying the evolution of South American mammals, though he continues also to work as a primatologist and to write and to publish as a scientist.

Chapter 1
Introduction

South American fauna and its evolution is a theme of fascinating interest for many. There are many reasons to be enthralled by the study of South American mammals since there are many very strange and wonderful endemic forms of animals that are found only on the continent of S.A. and in some instances overlap into Central and North America (edentates, hystricomorphic rodents, platyrrhine primates) and many strange forms have become extinct and are known only from South America.

The literature on the evolution of South American land mammals has grown enormously since the 1980 publication of G. G. Simpson's book, *The Curious History of South American Land Mammals*. Many new fossils have come to light, and a better understanding of the geological history of the continent has obligated new interpretations of the history of South American mammals. Then, too, as interesting as G. G. Simpson's book is, it is poorly illustrated, frustrating the interested reader who likes to picture how these ancient animals looked. In the past 20 years, talented illustrators have produced a range of images that greatly help us to picture what these extinct mammals might have looked like, and many of these illustrations are included in this book, hopefully converting it into a visual tool that will assist readers' imaginations about what these ancient mammals might have appeared.

Of course we can only indirectly know anything about these extinct animals' behavior and ecology, and in most cases, the details of these mammals' lives will remain a mystery. However, archeological and paleontological sites have improved our understanding of some of the more recent animals, like the magnificent *Smilodon* saber-toothed cat social behavior and the ecological preferences of South American mastodons, to name only two, but it is certain that paleoecological studies will eventually reveal more details of these ancient mammals' lives. Surprisingly, detailed studies have revealed particulars of the most ancient-known South American mammalian communities from the Late Jurassic Cañadon Asfalto Formation of Argentina's Chubut Province, so it seems probable that much information will be discovered in the future about more recent mammalian communities (Rougier et al. 2007).

T. Defler, *History of Terrestrial Mammals in South America*, Topics in Geobiology 42, https://doi.org/10.1007/978-3-319-98449-0_1

Other important recent contributions in the past 20 years have been a host of discoveries in Argentina and the extraordinary syntheses of R. Pascual and colleagues, who give us a more concrete understanding of the ancient Gondwana precursors of ancient pre-Cenozoic South American fauna. Recent discoveries in Argentina and in Antarctica have made it clear that rather than being the totally isolated continent of legend, South America enjoyed a connection with Antarctica and through that continent with Australia up until about 30–50 million years ago (by 37 Ma the breach of the South Tasman Rise between Tasmania and Antarctica certainly put an end to any mammalian overland dispersal; Australia had definitively disconnected from Antarctica) (Raven and Axelrod 1972; Goin et al. 2016). Fossils and molecular studies have illustrated the ancient relationship between Australian and South American fauna. Other relationships to other parts of *Gondwana* are still not easily understood, since these connections have been broken by tectonic forces.

1.1 Early Studies of South American Mammals

The first well-known and recognized fossil of a South American mammal was collected at Luján, Argentina, although other comments by early Spanish explorers had been published, but not recognized for what they were. The Luján find was an almost complete skeleton of a gigantic *Megatherium americanum*, the largest sloth known, whose remains were found because of stream erosion on the banks of a small tributary of the Luján river, Argentina around 1788. Supposedly the fossil was discovered by a Dominican Friar, Manuel Torres in 1788. This specimen was sent by the Viceroy de la colonia Nicolás del Campo to Madrid where it was placed in the Gabinete Real de Historia Natural (a small natural history museum) founded in 1772 by King Charles III .

1.1.1 Juan Bautista Bru Y Ramón

The fossil bones came into the hands of Juan Bautista Bru y Ramón, who had been educated as an "anatomical painter" (pintor anatómico) and was later offered the position of "*pintor y disecador*," perhaps best translated as *museum preparator* in the Real Gabinete de Historia Natural de Madrid. Juan Bautista Bru de Ramón is best known for his attempt to mount this important skeleton (Fig. 1.1), which he did, gaining fame as the first time such a reconstruction had been attempted. The description that this man published of the fossil, including illustrations (*Colección de láminas que representan los animales y monstros del Real Gabinete de Historia Natural de Madrid*, 1784, 1786) was the only description known until it was popularized by Cuvier in 1796. Although the posture of the giant sloth was somewhat erroneous (Bru de Ramón switched the fore- and hind limbs and positioned the animal with

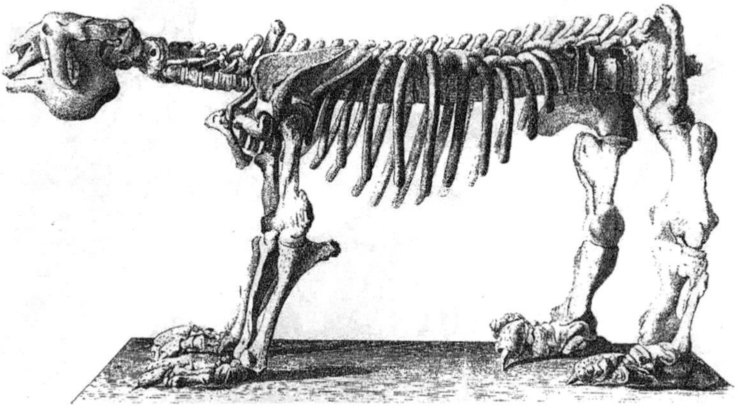

Fig. 1.1 Juan Bautista Bru de Ramón's mount of *Megatherium*, the first attempt to mount the skeleton of a fossil vertebrate. Unfortunately, besides the strange posture, he confused the front and hind feet in his mounting

four feet on the ground), it was obvious to all that this great South American mammal was something new and strange (Pasquali and Tonni 2008; Anonymous 2014c; López Piñero 1988, 2014).

1.1.2 George Cuvier

The young George Cuvier (1769–1832) (Fig. 1.2) had apparently read Bru de Ramón's earlier account and received the proofs of the illustration sent to him by Philippe-Rose Roume, an official of French West India, traveling to Paris by way of Spain. How Roume managed to talk Bru de Ramón out of the proofs of the illustration that he was about to publish, nobody can say. But Cuvier (1796) published his own account (without having seen the fossil), where he named the animal *Megatherium americanum*, thus gaining priority for himself and condemning Bru de Ramón to scientific obscurity. This was Cuvier's first publication on fossil animals, based on an already published description in Spanish (*Notice sur le squelette d'une très-grande espéce du Quadrupède inconnue jusqu'à, trouvé au Paraguay, et deposé au Cabinet d'Histoire Naturelle de Madrid,* Magasin encylopédique, 2me anné, vol. 1, 1796). In his publication, he gave recognition to Bru de Ramón as the person who had painted the plates and had assembled the fossil.

Of course, despite the arguable lack of ethics of this first publication, Cuvier did much to deserve his reputation as the first comparative anatomist and paleontologist in history in underlining comparative techniques of interpretation that continue today in the description of many other fossil animals, including the first fossil primate ever described. To be fair, although there is no evidence, perhaps Bru de

Fig. 1.2 Young George Cuvier, "Father of Paleontology," 1769–1832. (From *Règne animal*—1829 vol. I)

Ramón meant for Cuvier to publish the material that he sent to Paris, since Paris was the scientific center of the world of that time.

Cuvier became a real superstar of French science of the time, though he was a lifelong "catastrophist" believing that these ancient animals became extinct because of catastrophes, especially the biblical diluvium. For all of his life, he was adamantly opposed to the idea of evolution.

In 1812 Cuvier published his famous *Ossemens Fossiles*, a compendium of various previous studies, comparing in Volume IV the *Megatherium* of South America and *Megalonix* (a North American giant sloth first described by Thomas Jefferson, the third president of the United States, but with a much more complete description published by a colleague of the American Philosophical Society, Caspar Wistar) to the living sloths *Bradypus* and *Choloepus*. This publication first established the phylogenetic relationships of the four mammals as members of the same group, the sloths. In this same work, Cuvier clarified the status of all proboscidian teeth found in South America as belonging to one group, later named gomphotherian mastodonts of the family Gomphotheriidae and not to the true elephants (Elephantidae) or the Mammutidae (mastodonts of North and Central America). The differences among the various proboscideans were not yet appreciated in Cuvier's time, but the North and South American proboscidians were recognized as not belonging to the elephants (Elephantidae), the only living proboscidians, distinguished by very different molars (Prado et al. 2003).

1.1.3 Alexander von Humboldt

Most of the proboscidian specimens that Cuvier discussed in his famous publication had been found by Alexander von Humboldt (Fig. 1.3), who deserves to be recognized as discovering this group of proboscidians. Von Humboldt, student of all of South American natural history during his entire life, took many specimens of proboscidian teeth back to Europe that he discovered in widely divergent sites, such as near Santa Fé de Bogotá (Campo de los Gigantes); Imbabura, Ecuador; Tarija, Bolivia; and Concepción, Chile. These Humboldt turned over to Cuvier who described and illustrated them in his publications. Other naturalists also delivered material to Cuvier that they had collected from the Americas, and this was described and illustrated by this great French scientist (Prado et al. 2003).

1.1.4 Charles Darwin and Richard Owen

Much interest in the English-speaking world in the history of South American mammals was generated by the collections of Charles Darwin (1809–1882) (Fig. 1.4) on his voyage as naturalist on the famous ship the Beagle. Darwin's interest in South American fossils was generated in part by the publication of d'Orbigny's *Voyage dans l'Amerique Méridionale* (see below) and Cuvier's publications (Darwin 1989).

Fig. 1.3 Alexander von Humboldt, 1769–1859. (Portrait painted by Friedrich Georg Weitsch)

Fig. 1.4 Charles Darwin (1809–1882). (Taken about 1784 by Leonard Darwin)

On September, 1832, exploring the area of Punta Alta along the coast and in Buenos Aires Province, Argentina, Darwin found many fossil teeth and bones. Later he returned to this area to collect more and especially he collected around Monte Hermoso, the type locality for one of the land mammal ages for South America (Montehermoso, 6.8–4.0 Ma, during the Miocene epoch). Later Darwin traveled northwest of Buenos Aires to a place mentioned by another English traveler as fossiliferous, Saladillo. Here Darwin found fossil teeth that later turned out to fit into a toothless skull from the same locality. The fossil was named *Toxodon* by Richard Owen, the British anatomist, who later was entrusted with the descriptions of the many fossils collected by Darwin. In all, Darwin visited several Argentinian areas for collections: Bajada de Santa Fé near Paraná; Banda Oriental (Uruguay); Punta Alta and Monte Hermoso near Bahía Blanca; and San Julián in Patagonia (Vizcaíno et al. 2009b).

The collections allowed Owen to make the first description of *Toxodon platensis* and of *Macrauchenia patachonica*, two now well-known Pleistocene ungulate species of South America, although at the time Owen did not recognize either as belonging to ungulate groups. Owen described as well three sloths *Mylodon darwinii, Scelidotherium leptocephalum,* and *Megatherium* as well as various glyptodonts, the horse *Equus*, and the rodent *Ctenomys*. Perhaps the most important find was that of the horse *Equus*, as up to that time it was thought that *Equus* had not evolved in the Americas. But the discovery of a tooth very close to modern horses showed differently. These ancient mammals are some of the strongest evidence for evolution, and they were appreciated as evidence by Darwin (Owen 1839; Vizcaíno et al. 2009a).

1.1.5 Alcide Dessalines dÓrbigny

The Frenchman *Alcide Dessalines dÓrbigny* (1802–1857) (Fig. 1.5) obtained the position of traveling naturalist for the Muséum d'Histoire Naturelle while still in his twenties. His first assignment was to go to South America to make collections for the museum. Many of these collections he described in his publication *Voyage dans l'Amérique Méridionale* which Darwin had read, noting locations where fossils had been located by d'Orbigny. D'Orbigny collected about 18,000 fossils, many of which were mollusks (d'Orbigny, 1847).

1.1.6 Florentino Ameghino and Carlos Ameghino

Although Florentino Ameghino (1853–1911) has been lionized in Argentina and elsewhere, the contribution of Florentino's brother Carlos Ameghino (1865–1936) was completely key to the success that Florentino had during his life. These two brothers (Fig. 1.6) produced extraordinary work in describing ancient Argentinian mammals, and they represent the best tradition of the possibilities of collaboration. Florentino was a public personage all of his life, dealing with people and publishing his results, often to an exclusion of appreciation for the role of Carlos in obtaining the fossils in the first place.

Fig. 1.5 Alcide Dessalines dÓrbigny (1802–1857)

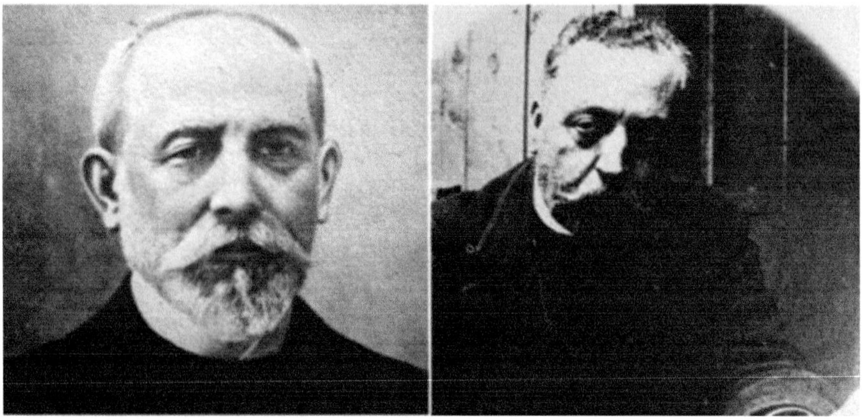

Fig. 1.6 Florentino Ameghino (left) 1853–1911 and Carlos Amegino (right) 1865–1935. (Verlag Naturhistorisches Museum Wien; Buffetaut 2013)

Florentino was the only one of the two who received some formal education in the first municipal school of Luján during 1862–1867, where the two boys had been born. Later, as a 15-year-old boy, he taught for a while in Mercedes, at the same time collecting local fossils.

In 1878, Florentino traveled to Paris to attend the Exposition Universelle de 1878 (the third World's Fair for Paris), taking several of his fossils to display and sell, to help finance his trip. It was at the exposition that Florentino began making important contacts that continued for the rest of his life. Among his contacts, interested in his fossils and his collecting were Edward Cope, Paul Gervais, and Albert Gaudry. After the exposition, Florentino worked with Paul Gervais in writing the book *Los Mamíferos Fósiles de la América del Sud*, published in 1880 (Gervais and Ameghino 1880), and another tome called *La Antiguédad del Hombre en La Plata* published in 1881, before returning to Argentina in 1882 with a new French wife. During his informal education, Florentino learned to read, write, and speak in French and Italian. Later he learned also to read in English, German, and Portuguese.

On his return to Argentina, Florentino was destitute, but burning to work with fossils and to publish his results. It was at this time that he opened his famous bookshop called "El Glyptodon" which at one time or another either he or his younger brother Juan managed, financing his and Carlos Ameghino's research for the next years. Juan's role in supporting the voluminous research published under Florentino's pen is generally unrecognized.

In 1884, Florentino was appointed professor of natural history at the University of Córdoba; however, in 1886, he moved to La Plata as the assistant director of the Museo de Historia Natural de La Plata under the director F. P. Moreno with whom he quickly had a falling out. He left the museum the following year in 1887 and subsisted with his brothers on the proceeds of another bookshop.

Carlos Ameghino was apparently a self-effacing younger brother and preferred the wilderness of Patagonia to the intellectual life aspired to by Florentino. While Carlos spent most of his time collecting fossils, Florentino spent most of his time studying them. Additionally, the youngest brother (of the three) Juan spent most of his time running the bookshop to support the other two brothers' endeavors.

Carlos obtained much less formal education than Florentino, briefly studying in the Mercedes school where Florentino taught. He set off for Patagonia for the first time in 1887 as an official collector for the Museo de La Plata where Florentino had been employed as assistant director. The connection with the Museo de La Plata only lasted until 1889 when Florentino resigned his position, resulting in the necessity to finance further expeditions with the bookshop in Buenos Aires (Simpson 1984).

Later in life, Florentino received much criticism for his interpretations of various fossils:

1. Because of his insistence that many of his finds were much older in age than they were (Ameghino 1881).
2. Because of his belief that the Borhyaenidae marsupials were related to ancient dogs (Fig. 1.7).
3. Relating the ancient primate *Homunculus* to human evolution.
4. He named fossils as new whenever he found minor variations (a typological approach instead of the recognition of the importance of a population approach with its variability).

Fig. 1.7 The Borhyaenidae show many features of canines, leading Florentino to consider erroneously this group as ancestral to the dog family. (By Roman Uchytel)

5. His belief that many fossil groups were ancestors of other groups, and the overestimations of age led him to the conclusion that almost all the groups of mammals (including humans) originated in Argentina and later colonized the rest of the world.

Florentino was apparently a volatile man and fought with many other paleontologists over interpretations of Argentinian fossils and their antiquity.

1.1.7 Santiago Roth

The Swiss-Argentinian Santiago Roth (1850–1924) (Fig. 1.8), who arrived as an immigrant in Argentina in 1866, was self-educated and particularly interested in natural history and geology and by 1870 was assembling collections of fossils and plants. He was soon influenced by the director of the Museo Público de Buenos Aires, Carlos Germán Conrado Burmeister (1807–1892), who encouraged him to collect fossils, many of which were sent to museums in Europe. Later, Burmeister also became a critic of Florentino, and they remained enemies until Burmeister's death, perhaps influenced by Burmeister's biblical anti-evolutionary approach to fossil interpretation, as opposed to Florentino's open adoption of evolutionary thought (Simpson 1984).

Fig. 1.8 Santiago Roth (1850–1924)

In 1881, Roth discovered a human skeleton underneath the carapace of a glyptodont, this being one of the important early pieces of evidence for the existence of paleoindians alongside the extinct South American megafauna (Giacchino and Gurovich 2001). Also Roth realized that the primitive ungulates being found in Argentina belonged mostly to one great order, the Notoungulata, a discovery that has stood the test of time. From 1887 onward, the two scientists Roth and Ameghino were in conflict as scientific rivals. Most seriously this rivalry resulted in the hiding of the source of many fossils which to this day have not been identified with a collecting site. Nevertheless, as a result of their intense rivalry, Roth shares with the Ameghinos the honor of having been the most intrepid explorers of Patagonia and other parts of Argentina.

1.1.8 John Bell Hatcher

John Bell Hatcher (1861–1899) (Fig. 1.9) was the first North American paleontologist to collect in Argentina. Working under the auspices of Princeton University, he made three expeditions to Patagonia (1886, 1887, 1898) where he worked with few resources to amass an impressive collection of fossils. He worked under very difficult conditions, even during the austral winter, yet his collection was remarkable. The fossils were all from the Santacruzian SALMA (17.5–16.3 Ma). After delivering his fossils to his superior, William Berryman Scott, he resigned his position reportedly feeling unappreciated. He spent his few later years working on dinosaurs at the Carnegie Museum when he died at the early age of 42 years of typhoid fever.

Fig. 1.9 John Bell Hatcher (1861–1904)

Before his death, he wrote a very extensive seven-volume account of his expeditions, *Reports of the Princeton University Expeditions to Patagonia, 1896–1899,* whose narrative was later abridged into the book, *Bone Hunters in Patagonia,* by the editors of Ox Bow Press (Hatcher 1985; Simpson 1984).

1.1.9 William Berryman Scott

William Berryman Scott (1858–1947) (Fig. 1.10) was a North American paleontologist associated with Princeton University who spent most of his career studying fossils from the western United States. Since he had received the excellent collection of Argentinean fossils from John Bell Hatcher because of Hatcher's resignation from Princeton, it was Scott's responsibility to analyze and publish the results.

Because part of the collections were fossil marine mollusks and outside of Scott's expertise, these were assigned to the German-American Arnold Edward Ortmann (1863–1927), a felicitous decision, as he was a specialist in mollusks and crustaceans and was able to conclude, making comparisons with European mollusks, that the marine fauna was Early Miocene (Santacrucian, 17.5–16.3 Ma). This placed most of the mammals discovered as later (younger), except for the amazing Pyrotherium fauna, first discovered by Carlos Ameghino and placed at the Eocene-Oligocene boundary. This definitively proved a more recent age for the Argentinean fauna than was accepted by Florentino Ameghino, who had dated the fossils (Simpson 1984).

Fig. 1.10 John Berryman Scott (1858–1947)

1.1.10 André Tournouër

André Tournouër (1871–1929) was a well-off young Frenchman who immigrated to the Argentine Republic to farm. But after contact with Albert Gaudry, he was convinced to collect fossils in Patagonia for the Museum of Natural History in Paris and to do this in imitation of his father who had done paleontological work. Since his father had died in 1882 when André was only 11 years old, he was eager to do something in homage of his father, who had worked in geology and paleontology and had published some of his scientific work. While working as a collector for the Paris Museum, André made five expeditions to Patagonia to collect fossils for the museum. Beginning in 1898, he threw himself into his adopted work and collected much important material.

Being the end of the nineteenth century, working in isolated sites in Patagonia was not easy. However, due to resources available to him, he mounted ambitious expeditions. One of his expeditions that he described in an article in 1922 involved 40 mules and the men to manage them, necessitating a great deal of supplies for the men and mules and clearly surpassing the few resources that were available to Carlos Ameghino in his many years of collecting in Patagonia.

It seems surprising to many that André related to the two Ameghinos very well and that the brothers helped him, disclosing many collection sites and enabling Tournouër to make an outstanding collection of the Pyrotherium (Fig. 1.11) fauna.

Fig. 1.11 *Pyrotherium* sp. lived in what is now Argentina, during the Early Oligocene. Its body was 3 m long and 1.50 m tall at the shoulders. Its had robust legs and a short proboscis, and flat, forward facing tusks (two in the upper jaw, one in the lower one). It has sometimes been seen as a descendent of the Xenungulata. (By Roman Uchytel)

Probably part of Andrés success with the Ameghinos was due to Florentino's earlier visit to France and the many European contacts that he had made, particularly writing a book with Paul Gervais. When Carlos Ameghino first met Tournouër, the young Frenchman was wandering about Patagonia without much collecting success. Carlos seems to have taken him under his wing, telling him specifically where he might make the best collections and thus guaranteeing his success.

Gaudry enjoyed the collecting successes of Tournouër who was not really interested in studying his collection, so publications of results came from others. Tournouër published very little, but he did manage to publish on the stratigraphy of the regions he visited and a short note on the feet of the *Astrapotherium*, of which he was able to collect many specimens for the Paris Museum. He never considered himself a paleontologist and spent the rest of his life on other pursuits in France (Buffetaut 2013; Simpson 1984).

1.1.11 Jean Albert Gaudry

Jean Albert Gaudry (1827–1908) (Fig. 1.12) received Tournouër's collections, studied them, and published the results. This resulted in Gaudry's emphasizing the isolation of South America from North America and Africa for the entire Cenozoic, although we now appreciate that a connection was maintained through Antarctica with Australia, intermittently from 40 to 35 Ma until about 30 million years ago, when the two continents rifted apart forming a marine barrier (Raven and Axelrod

Fig. 1.12 Jean Albert Gaudry (1827–1908)

1972; Oglesby 1991; Goin et al. 2016). Gaudry concluded that the fauna that his young helper had collected was Casamayoran SALMA (54–48 Ma), corresponding to the Eocene. This early fauna was not so distinct from mammals from the rest of the world compared to later evolved South American mammals. Gaudry also recognized the possibility of a connection of Patagonia with Australia, though there was no geological evidence at that time (Simpson 1984).

1.1.12 G. G. Simpson

G. G. Simpson (1902–1984) (Fig. 1.13) was a North American paleontologist who specialized on the mammals of the Mesozoic and Cenozoic. He was one of the key figures along with Theodosius Dobzhansky and Ernst Mayr to work on the synthetic theory of evolution. His books *Time and Mode in Evolution* and *The Meaning of Evolution* were very important in the adoption of the synthetic theory, whereby genetics, biogeography, and paleontology were united into modern evolutionary theory.

Simpson worked at the American Museum of Natural History and participated in various expeditions, especially to Patagonia (1930–1931, 1933–1934). He became close friends and an admirer of Carlos and Florentino Ameghino. He wrote his first book to give a general account of the evolutionary history of South American mammals, *Splendid Isolation: The Curious History of South American Mammals* (Simpson 1944, 1967, 1980, 1984).

Fig. 1.13 G. G. Simpson. (Permission of the American Museum of Natural History) 1902–1984

There are many more paleontologists who have dedicated their lives to the study of the South American biota and mammalian evolution in particular, but they have become too numerous to list. It is particularly important to recognize that since the Ameghinos, Argentina has become a leading center for these studies (Vucetich et al. 2007; Simpson 1984).

1.2 The Problem of Assigning Time

1.2.1 Religious Traditions

It is impossible to discuss the history of biota without considering time and acquiring an understanding of geological time, that is, units of time in terms of millions of years. Assigning a time-line is always a problem for anybody considering the evolution of the biota, for how can we calculate the vast periods of time that have passed since life first appeared on the planet? Ideas of time have always been influenced by religious views, and especially in the west where much of the geological time scale was developed, we have had to come to terms with popular religious beliefs that the earth was only a few thousand years old. The Anglican Bishop Ussher (1581–1656) gained fame by calculating the first day of creation as Sunday, 23 October 4004 BC, based on an intricate correlation of Middle Eastern and Mediterranean histories and Holy Writ. This date and that species were fixed in time were commonly accepted beliefs, and, it was believed, they changed only when great catastrophes occurred, such as the great flood of the Bible, all commonly accepted belief (Groves 1996; Anthoni 2004).

1.2.2 Stratification

As more and more fossils appeared, the popular view became very difficult to accept, and other theories were advanced. One of the most influential early theories of the late eighteenth century was offered by the German geologist Abraham Werner, who divided the rocks of the earth's crust into four types, primary, secondary, tertiary, and quaternary, each type said to be formed during a different part of earth's history. This system was used into the twentieth century as, gradually, an appreciation of many strata of fossils permitted a finer division of time and an appreciation of the use of stratification which teaches us that geological layers are usually found in stratified layers such that the layers lower down are older and the layers above are younger. Fossils in the stratified layers are thus assigned the relation of being older or younger than other layers.

But stratification does not allow an assignment of absolute time, which is needed for a more finely divided consideration of evolving species and communities, and it was not until the end of the nineteenth century that any technique was available to calculate absolute ages using radioactivity. Unfortunately, most fossils are not found in rocks that allow this type of dating, although sometimes a rock layer above or below the sedimentary rock in which any fossils are found can be dated, establishing limits to the particular fossils' age.

1.2.3 Radiometric Dating

The discovery of radiation opened the door to the determination of absolute time of formation of certain radioactive substances, called radiometric or radioactive dating. Radioactivity is what happens when the nucleus of an unstable atom loses energy by emitting ionizing radiation. Such materials are radioactive. Since radioactive substances decay in at a constant rate, it has been possible to calculate the time necessary for a substance (an isotope or a radioactive form of an atom) to lose half its radioactivity. This is called the "half-life" of that isotope, and the constant loss of radioactivity for many elementary isotopes can be measured, so that the time from their formation can be calculated. The solidification of many of these isotopes dates from the time when they were formed under the great heat and pressure of the earth's magma, so that a volcanic eruption usually leaves direct evidence of the time of its eruption, via the radioactive isotopes produced.

For even finer considerations of paleofauna, a system gradually became developed on each continent independently that identified "land mammal ages," since it became clear that faunas of different ages tended to be characterized by one or more species of mammals unique for that period. An examination of many different faunas allowed workers to build up a scale based on the evolution and succession of mammal species. Since these land mammal ages are not applicable among all the continents because of huge differences in the fauna, systems have been built up for each continent, and comparisons between continents are difficult and must resort to absolute ages calculated by radioisotopes (Anonymous 2014a).

1.2.4 Paleomagnetism

In the early part of the twentieth century, geologists discovered that volcanic rocks were magnetized in varied directions from the earth's magnetic field. With time, it was understood that the earth's magnetic poles periodically became reversed or wandered from one point to another. While the mechanism for magnetic pole reversal is not well-understood, it became clear during the International Geophysical Year that these changes in the magnetic pole could be read from volcanic and sedimentary rocks containing ferrous minerals, after they had become solidified. A parallel record could be read on both sides of the mid-Atlantic ridge, since seafloor spreading was also discovered emanating from the ridge. Since the 1960s, the record of these pole reversals has been mapped for upward of 250 million years, and absolute ages have been assigned to many, using radiometric dating (Jovane et al. 2013).

1.2.5 Magnetostratigraphy

This technique uses the collection of magnetically oriented strata of sediments at intervals throughout a stratum plus volcanic rocks analyzed for orientation of their magnetic minerals when the lava was laid down, all of which reveals the polar orientation of the earth's magnetic field at the time of the deposits. The results of such samples are then analyzed to be able to compare the different magnetic orientations to the known Global Magnetic Polarity Time Scale, previously constructed with assigned time intervals, allowing geochronological time to be assigned to different rock units. Known dates to these magnetic orientations have been assigned to at least 155 million years before present (Fig. 1.14).

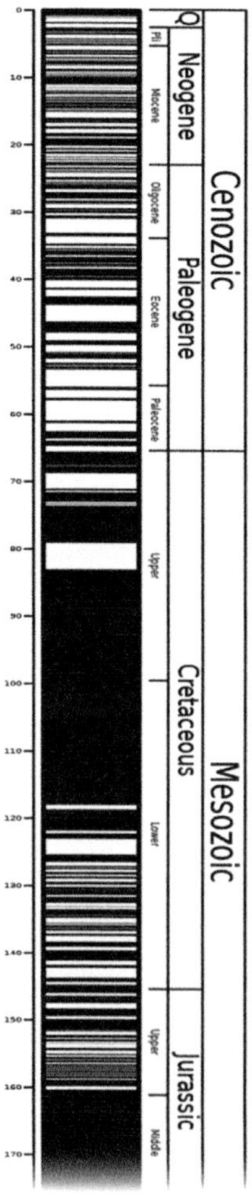

Fig. 1.14 Geomagnetic polarity 0–169 Ma since the Middle Jurassic. Dark areas denote periods where the polarity matches today's polarity, while light areas denote periods where that polarity is reversed. (From Wikipedia "Geomagnetic Reversal", public domain)

1.2.6 South American Land Mammal Ages (SALMAs)

The South American Land Mammal Age (SALMA) system was begun by the Ameghino brothers and was named either by names of genera or by the localities where the fossils were first found. For example, Lujanian SALMA is named for the city where the Ameghinos grew up and begun hunting fossils, Luján, about 67 km from Buenos Aires. The Ensenadan and Uquian SALMAs were Pleistocene faunas also closely related in time and space to the Lujanian and named for places where they were first found. Over the years, this system has been built up to include around 22 SALMAs for the entire Cenozoic Era with many additions and some subtractions, all allowing a more precise consideration of the evolution of mammals (Cione and Tonni 1995; Vucetich et al. 2007) (Table 1.1).

Because absolute ages were still unavailable during the early development of the SALMA system, the Ameghinos tended to overestimate the ages of their finds. Florentino considered that most of his SALMAs were Cretaceous in age, almost as if they were biased for earlier ages. This caused consternation and criticism from

Table 1.1 South American Land Mammal Ages (Krause et al. 2017; Woodburne et al. 2014; Clyde et al. 2014; Flynn and Swisher 1995)

Epoch	South American Land Mammal Age
Pleistocene	Lujanian: lower boundary 0.8 Ma. Upper boundary 0.011 Ma Ensenadan: lower boundary 1.2 Ma. Upper boundary 0.8 Ma Uquian: lower boundary 3 Ma. Upper boundary 1.2 Ma
Pliocene	Chapadmalalan: lower boundary 4 Ma. Upper boundary 3 Ma Montehermosan: lower boundary 6.8 Ma. Upper boundary 4 Ma Huayquerian: lower boundary 9 Ma. Upper boundary 6.8 Ma
Miocene	Chasicoan: lower boundary 10 Ma. Upper boundary 9 Ma Mayoan: lower boundary 11.8 Ma. Upper boundary 10 Ma Laventan: lower boundary 13.8 Ma. Upper boundary 11.8 Ma Colloncuran: lower boundary 15.5 Ma. Upper boundary 13.8 Ma Friasian: lower boundary 16.3 Ma. Upper boundary 15.5 Ma Santacrucian: lower boundary 17.5 Ma. Upper boundary 16.3 Ma
Oligocene	Colhuehuapian: lower boundary 21.0 Ma. Upper boundary 17.5 Ma Hiatus Deseadan: lower boundary 29.0 Ma. Upper boundary 24.5 Ma (Flynn and Swisher 1995)
Eocene	Tinguiririican: lower boundary 36.0 Ma. Upper boundary 29 Ma Divisaderan: lower boundary 42.0 Ma. Upper boundary 36.0 Ma Mustersan: lower boundary 48.0 Ma. Upper boundary 42.0 Ma Casamayoran: lower boundary 54.0 Ma. Upper boundary 48.0 Ma Riochican: lower boundary 46.7 Ma. Upper boundary 42.0 Ma (Krause et al. 2017) or 51.4–42.4 Ma (Krause et al. 2017) Itaboraian: lower boundary 53.0 Ma. Upper boundary 50.0 Ma (Woodburne et al. 2014) or lower boundary 56 Ma. Upper boundary 51.4 Ma. (Krause et al. 2017)
Paleocene	Peligran: lower boundary 62.5 Ma. Upper boundary 59.0 Ma Tiupampan: lower boundary 64.5 Ma. Upper boundary 62.5 Ma

many other paleontologists. There are still many problems dating these SALMAs because of a lack of paleomagnetic or radiometric dates and the existence of several large temporal gaps representing a lack of fossils for those times. The refinement of dates for each SALMA continues.

In many cases, Florentino was mistaken about the phylogenetic relationships of the mammals that he discovered with Carlos with other mammals on other continents. Much of this error was due to the phenomenon of evolutionary convergence, a distinct problem for many paleontologists who sometimes notice great similarities between organisms and conclude a relationship. Homoplasy (similarity in species of different ancestry) may occur if the organisms are living in similar ecosystems and are those exposed to similar selective forces, such that both organisms reflect a distinctive way of life. The Ameghinos did not appreciate that their discoveries were absolutely unique to South America.

More recent evidence explains the evolution of mammalian groups in South America:

1. The adoption of plate tectonics as a mechanism to explain movements of continents does much to explain South American isolation from Africa and North America but also the connection that South America did enjoy with Australia and other elements of the Gondwana fauna by way of Antarctica.
2. Plate tectonics helps explain the changes brought about in the Caribbean that resulted in closer and closer sea connections via islands and finally land connections that permitted interchanges of fauna, particularly the immigration of marsupials and the first placental mammals to South America.
3. These geological processes that resulted in the eventual connection of the two Americas began 12 million years ago when submarine connections and volcanic island permitted the first biotic interchanges.
4. The absence of tribosphenic mammals and the total extinction of endemic non-tribosphenic mammals is an exceptional trait of Patagonian evolution during the Gondwanan episode.

Up to the Early Paleocene, very few Gondwanan mammals survived in South America (with the exception of some native gondwanatherians and an endemic dryolestoid). Another gondwanatherian survived until the Late Eocene in Antarctica. These Gondwanan survivors lived alongside the more modern immigrants. The association could have begun before the end of the Cretaceous, though it has not been demonstrated.

1.3 The Geologic Time Scale

With much effort, the development of dating techniques, and their application to many, many rocks and fossils, a geologic scale has been developed over the last two centuries and continues to be refined. It is one of those great successes of modern science that we can now appreciate the vast expanses of time and how they were related to the evolution of the earth's biota (Table 1.2; Repcheck 2008).

Table 1.2 The geologic time scale (US Department of the Interior/US Geological Survey. URL: http://geomaps.wr.usgs.gov/gmeg/index.htm)

EON	Era	Period		Millions of years ago
		Quaternary	Recent	Began 11,700 years
			Pleistocene	2.58 million years
	Cenozoic	Neogene	Pliocene	5.333
			Miocene	23.03
		Paleogene	Oligocene	33.9
			Eocene	56
			Paleocene	66
		Cretaceous	Late	100.5
			Early	145
	Mesozoic	Jurassic	Late	163.5
			Middle	174.1
			Early	201.3
		Triassic	Late	237
			Middle	242
			Early	251.902
		Permian		298.9
Phanerozoic		Pennsylvanian		323.2
		Mississippian		358.9
	Paleozoic	Devonian		419.2
		Silurian		443.8
		Ordovician		485.4
		Cambrian		541
Proterozoic	Late Archean Middle Proterozoic Early Proterozoic			2500
Archean	Late Archean Middle Archean Early Archean			4000
	Pre-Archean			

1.4 Plate Tectonics and the Distribution of Fauna

1.4.1 Alfred Wegener and Continental Drift

The German Alfred Wegener (1880–1930) was the first to present evidence that the continents actually moved about on the surface of the earth. His studies showed him many instances of close fits of coast lines (especially Africa and South America), as well as large-scale geological features, fossils, and living biota on both sides of wide oceans. He became convinced that the continents were not fixed in place and wrote a book, first edition in 1915, *Die Entstehung der Kontinente und Ozeane* (*The Origin of Continents and Oceans*), outlining his theory of continental drift. It was his thesis that about 300 million years ago, the continents had formed a single large land mass, which he named "ein Urkontinent" (usually referred to as "Pangaea" these days). Pangaea then split apart, and the pieces drifted off to the positions of today. An English translation of the book did not appear until 1925 (Wegener 1929, 1966) (Figs. 1.15 and 1.16).

The reaction to the theory was almost universally hostile, since it was very difficult for people to imagine the continents as moving. Also, he could not propose a convincing argument for how the continents might move. His best guess was that centrifugal and tidal forces pushed the continents through the earth's crust. But this was not convincing, so the majority of geologists continued to believe in a static earth, with a few exceptions, at least into the 1960s when evidence (discovery of paleomagnetism, seafloor spreading, earthquake zones above active continental plates, called Wadati-Benioff zones) had begun to accumulate that Wegener had been right, at least as far as the wandering continents goes (Anonymous 2014b).

As a young biology student at a university in the United States around 1965–1966, I well remember studying biogeography where nothing was mentioned about continental movements and all were explained in terms of land bridges and rafting. This contrasted with a geology course that I monitored at Albert Ludwigs University in Germany in 1960–1961 where the modern theory of plate tectonics was being taught. At the time, Wagner's ideas were more receptive to a German audience than to an American one. Could there have been a prejudice against Germans, as well, at that time (Oreskes 1999)?

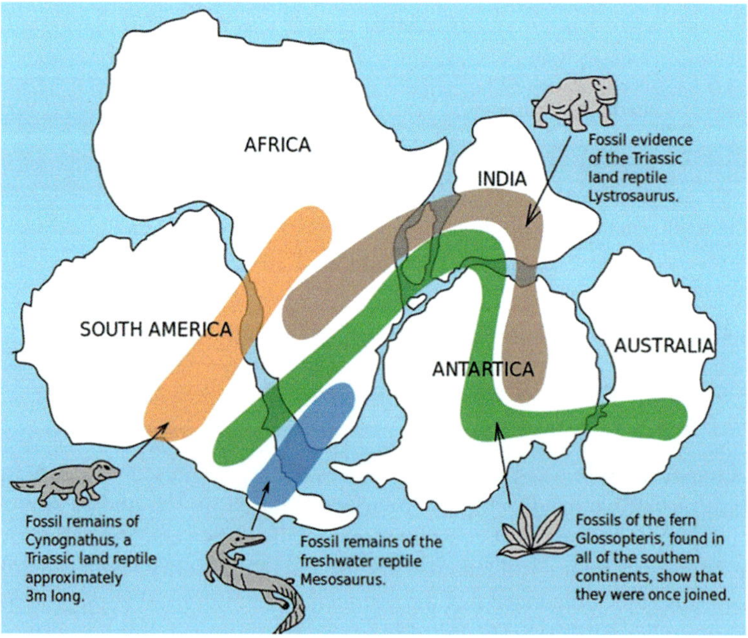

Fig. 1.15 Fossil evidence for continental drift. (By the US Geological Survey)

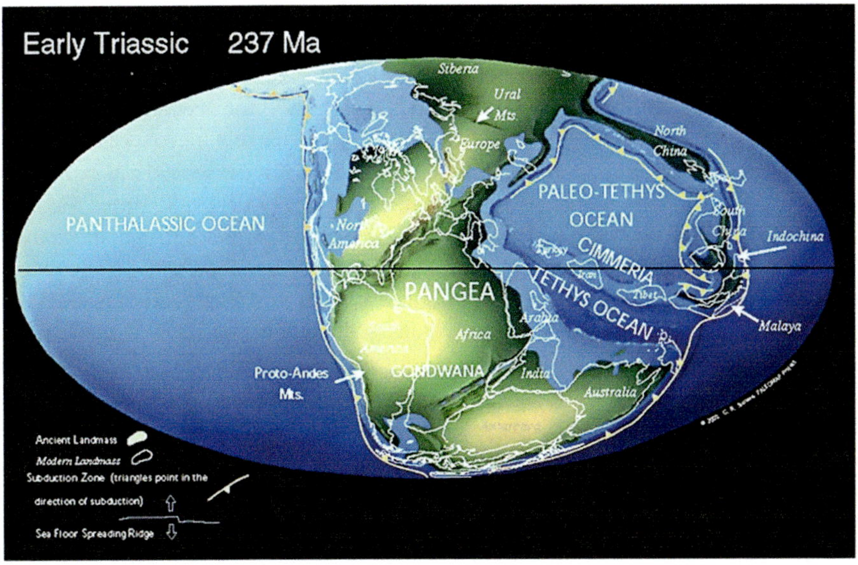

Fig. 1.16 Pangaea and the position of the continents in the Early Triassic. (By C. Scotese Paleomap Project)

1.4.2 Plate Tectonics

Plate tectonics is a modern theory explaining large-scale movements of the continents. The theory was developed from the earlier idea of Alfred Wegener of continental drift. We have discovered that the rigid cooler outer skin or lithosphere of the earth is broken up into many tectonic plates (eight major and many minor) which ride atop the more plastic and hotter asthenosphere, located just below the lithosphere and on which the continental plates float. There are two types of lithosphere, continental lithosphere that is associated with continental crust and oceanic lithosphere associated with oceanic crust. Oceanic lithosphere is denser and less thick than the continental crust but heavier so that it rides on the asthenosphere lower than the continental crust and thus fills up with water. Portions of the continental crust are ancient, in excess of three billion years, but oceanic crust is constantly being created from mid-oceanic ridges and pushes out at each side; this is thought to be part of the basis for continental movements, though convection currents from variations in the mantle are also implicated in the motion. As oceanic crust pushes against continental crusts, it sinks under the continental crusts (because it is heavier) causing upthrust of mountain ranges and volcanic and earthquake activity. Most earthquake and volcanic activity is found near the area of contact between the two types of lithosphere.

Acceptance of Wegener's theory (at least in Europe) had to wait until 1957 when variable magnetic field direction in rocks of differing ages was discovered from magma upwelling, and an immense mid-oceanic ridge was detected in the Atlantic Ocean created by this same magma. The solidified rock was soon shown to have mirrored magnetic field reversals on each side of the ridge. Thus, a mechanism for continental movement was discovered and the reticence for accepting the theory broke down. In 1960, Harry Hess advanced the idea that the earth's crust moved laterally away from long, volcanically active oceanic ridges, which had been discovered during the International Geophysical Year in 1957–1958. Bruce Heezen discovered the Great Global Rift that runs along the Mid-Atlantic Ridge, the source of seafloor spreading, and this helped establish Alfred Wegener's earlier (but generally dismissed at the time) concept of continental drift as scientifically respectable. This triggered a revolution in the earth sciences. Plate motions have now been calculated at 10–40 mm/a, at about the speed of a growing fingernail. Oceanic lithosphere is typically about 100 km thick, while continental lithosphere is around 200 km (Le Pichon 1968).

1.5 Molecular Phylogenetic Research

A scientific revolution has been occurring since the development of techniques allowing comparison of molecules from different organisms (Gilbert et al. 2005). The first impacts of this type of research were published with the results of research on birds by Charles Gald Sibley and Jon Edward Ahlquist, who using DNA-DNA hybridization presented a new and highly influential avian phylogeny. Their important publications, (1983, 1990) *Phylogeny and Classification of Birds* (written with Ahlquist) and (1990) *Distribution and Taxonomy of Birds of the World* (with Burt Monroe) are among the most cited of all ornithological works and establish a new taxonomy.

These and other techniques have been extended to primates, mammals in general, human evolution, and plant evolution and lately have been adapted to extinct taxa when some DNA or RNA could be recovered (Pääbo 1989, 2014). A thorough discussion of molecular techniques as applied to mammalian phylogeny is not the scope of this book, although some data is reported here (Willserslev and Cooper 2005).

Very recent results from the sequencing of collagen, a structural protein comprising two separate chains, (coded by genes on separate chromosomes), provides useful systematic information from Pleistocene fossils of the taxa *Toxodon* (Notoungulata) and *Macrauchenia* (Litopterna) and have indicated that at least these two orders are indeed laurasiatheres, and more specifically, they are closely related to odd-toed ungulates (Welker et al. 2015). This research is being extended to other taxa where remnants of collagen are to be found and is superior to fossil DNA in that it remains stable for a longer period of time.

References

Ameghino F (1881) La Antigüedad del Hombre en La Plata. La Cultura de la Argentina, Buenos Aires

Anonymous (2014a) Radiometric dating. Wikipedia http://en.wikipedia.org/wiki/Radiometric_dating. Accessed 17 Jan 2015

Anonymous (2014b) Geomagnetic reversal. Wikipedia http://en.wikipedia.org/wiki/Geomagnetic_reversal. Accessed 17 Jan 2015

Anonymous (2014c) Bru de Ramón JB. Wikipedia http://www.biologia-en-internet.com/fteixido/s-xviii/juan-bautista-bru-de-ramon-1740-1799/. Accessed 17 Jan 2015

Anthoni JF (2004) Geologic time table: the development of life on earth. http://www.seafriends.org.nz/books/geotime.htm. Accessed 17 Jan 2015

Buffetaut E (2013) André Tournouër, paléontologue méconnu et cryptozoologue sceptique. Le Dinoblog, n° 1, sur Dinosauria (musée), 25 March 2013 (consulté le 27 March 2013); http://www.dinosauria.org/blog/2013/03/25/andre-tournouer-paleontologue-meconnu-et-cryptozoologuesceptique-episode-1

Cione AL, Tonni EP (1995) Chronostratigraphy and "Land-Mammal Ages" in the Cenozoic of southern South America: principles, practices, and the "Uquian" problem. J Paleontol 69(1):135–159

Cuvier G (1796) Notice sur le squelette d'une très-grande espèce du quadrupède inconnue jusqu'à, trouvé au Paraguay, et déposé au Cabinet d'Histoire Naturelle de Madrid. *Magasin encyclop*, 2ème année 1:310

Cuvier G (1812) Recherche sur les ossements fósiles de cuadrúpedas. Chez Deterville, Paris

Clyde WC, Wilf P, Iglesias A, Slingerland RL et al (2014) New age constraints for the Salamanca formation and lower Río Chico Group in the western San Jorge Basin, Patagonia, Argentina: implications for cretaceous-Paleogene extinction recovery and land mammal age correlations. GSA Bull 126(3-4):289–306. https://doi.org/10.1130/B30915.1

d'Orbigny AD (1847) Voyage dans l'Amerique Meridional (Le Brésil, la Republique Orientale de l'Uruguay, la Republique Argentine, la Patagonie, la République du Chili, la République de Bolivia, la République du Pérou), 1826, 1827, 1828, 1829, 1830, 1831, 1832 y 1833. Libraire de la Mociété géologique de France, Paris

Darwin C (1989) The voyage of the Beagle. Penguin; Abridged Ed edition, 29 June 1989

Flynn JJ, Swisher III CC (1995) Cenozoic South American Land Mammal Ages: Correlation to global geochronologies. SEPM Special Publication no. 54, ISBN 1-56576-024-7

Gervais H, Ameghino F (1880) Les mammifères fossils de l'Amérique du Sud – Los mamíferos fósiles de la América del Sud. Igon Hermanos, Buenos Aires/Paris

Giacchino A, Gurovich Y (2001) Homenaje al Doctor Santiago Roth a 150 años de su natalidad. http://www.academia.edu/430640/HOMENAJE_AL_DOCTOR_SANTIAGO_ROTH_A_150_ANOS_DE_SU_NATALICIO?

Gilbert MT, Bandelt H-J, Hofreiter M, Barnes I (2005) Assessing ancient DNA studies. TREE 20(10):541–544. https://doi.org/10.1016/j.tree.2005.07.005

Goin FJ, Woodburne MO, Chornogubsky L (2016) Dispersal of vertebrates from between the Americas, Antarctica, and Australia in the late Cretaceous and early Cenozoic. In: Goin FJ, Woodburn MO, Zimiez AN, Martin GM, Chornogubsky L (eds) A brief history of South American metatherians: evolutionary contexts and intercontinental dispersals. Springer, New York, pp 77–124

Groves C (1996) From Ussher to slusher, from Archbishop to Gish: or not in a million years…. Archaeol Oceania 31:145–151

Hatcher JB (1985) Bone hunters in Patagonia. Ox Bow Press, Woodbridge

Jovane L, Hinnov L, Housen BA, Herrero-Barvera E (2013) Magnetic methods and the timing of geological processes. Geol Soc, London, Special Publications 373:1–12

Krause JM, Clyde WC, Ibañez-Mejía M, Schmitz MD, Barnum T, Bellosi ES, Wilf P (2017) New age constraints for early Paleogene strata of central Patagonia, Argentina: implications for the timing of South American land mammal ages. GSA Bull 129(7/8):886–903. https://doi.org/10.1130/B31561.1

Le Pichon X (1968) Sea-floor spreading and continental drift. J Geophys Res 73(12):3661–3697

López Piñero JM (1988) Juan Bautista Bru (1740–1799) and the description of the genus *Megatherium*. J Hist Biol 21(1):147–163

López Piñero JM (2014) Bru de Ramón (1740–1799). http://www.mcnbiografias.com/app-bio/do/show?key=bru-de-ramon-juan-bautista. Accessed 4 Mar 2014

Oglesby RJ (1991) Joining Australia to Antarctica: GCM implications for the Cenozoic record of Antarctic glaciation. Clim Dynam 6:13–22

Oreskes N (1999) The rejection of continental drift: theory and method in American earth science. Oxford University Press, Oxford

Owen R (1839) The zoology of the voyage of H.M.S. Beagle under the command of Captain Fitzroy, R.N. During the years 1832 to 1836. Fossil mammalia. Smith, Elder and Co, London

Pääbo S (1989) DNA: extraction, characterization, molecular cloning, and enzymatic amplification. Proc Natl Acad Sci U S A 86:1939–1943

Pääbo S (2014) Neanderthal man: in search of lost genomes. Basic Books, New York

Pasquali RC, Tonni EP (2008) Los hallazgos de mamíferos fósiles durante el período colonial en el actual territorio de Argentina. Los Geólogos y La Geología En La Historia Argentina, SCG 24:1–6

Prado JL, Alberdi MT, Sánchez B, Azanza B (2003) Diversity of the Pleistocene gomphotheres (Gomphotheriidae, Proboscidea) from South America. In: Reumer JWF, De Vos J, Mol D (eds) Advances in mammoth research: proceedings of the second international mammoth conference, Rotterdam, 16–20 May 1999. Deinsea 9:347–363

Raven PH, Axelrod DI (1972) Plate tectonics and Australasian paleobiogeography. Science 176(4042):1379–1386

Repcheck J (2008) The man who found time: James Hutton and the discovery of the earth's antiquity. Perseus Book Group, New York

Rougier GW, Martinelli AG, Forasiepi AM, Novacek MJ (2007) New Jurassic mammals from Patagonia, Argentina: a reappraisal of australosphenidan morphology and interrelationships. Am Mus Novit 3566:54 pp

Simpson GG (1944) Tiempo and mode in evolution. Princeton University Press, Princeton

Simpson GG (1967) The meaning of evolution. Yale University Press, New Haven

Simpson GG (1980) Splendid isolation: the curious history of South American mammals. The Murray Printing Co., Westford

Simpson GG (1984) Discoverers of the lost world. Yale University Press, New Haven

Vizcaíno SF, Fariña RA, Fernicola JC (2009a) Young Darwin and the ecology and extinction of Pleistocene South American fossil mammals. Rev Asoc Geol Argent 64(1):160–169

Vizcaíno SF, Vizcaíno SF, DeLulias G (2009b) The fossil mammals collected by Charles Darwin in South America during his travels on board the HMS Beagle. RAGA – Revista de la Asociación Geológica Argentina 64:147–159

Vucetich MG, Reguero MA, Bond M, Candela AM, Carlini AA et al (2007) Mamíferos continentales del Paleógeno argentino: las investigaciones de los últimos cincuenta años. Ameghiniana Publicación Especial 11:239–255

Wegener A (1929) Die entstehung der kontinente und ozeane. Friedr. Vieweg & Sohn, Braunschweig. https://archive.org/details/dieentstehungderoowege

Wegener A (1966) The origin of continents and oceans. Dover Publications, New York

Welker R, Collins MJ, Thomas JA, Wadsley M et al (2015) Ancient proteins resolve the evolutionary history of Darwin's South American ungulates. Nature 522:81. https://doi.org/10.1098/rspb.nature14294

Willserslev E, Cooper A (2005) Review paper, ancient DNA. Proc R Soc B Biol Sci 272:3–16. https://doi.org/10.1098/rspb.2004.2813

Woodburne MC, Goin J, Raigemborn MS, Heizle M, Javier HM, Gelfo JN, Oliveira EV (2014) Revised timing of the South American early Paleogene land mammal ages. J S Am Earth Sci 54:109–119

Chapter 2
Ancient Mammals of Gondwanan South America

2.1 Introduction

The traditional view of southern mammals is that the original South American fauna began to develop from five great clades, the marsupials, the ungulates, the xenarthrans, the caviomorph rodents, and the primates, and that at least four of these clades came from elsewhere, probably North America and Africa. This is certainly true when considering the Cenozoic fauna, the so-called Age of Mammals, but lately it has become very clear that another, more ancient mammalian fauna lived in South America during Mesozoic times. These were part of a Gondwanan fauna, and biogeographically they form a Gondwanan period in the history of South American fauna, called the Gondwanan Episode by the well-known Argentinean paleontologist Rolando Pascual (Pascual and Ortiz-Jaureguizar 2007). Others divide the history of radiations of South American mammalian fauna into as many as five "phases," including early and late Gondwanan phases, early and late South American phases, and an Inter-American phase (Goin et al. 2012a, b). I prefer to look at the history of neotropical mammals as comprising three histories, the Gondwanan Episode or Phase, the South American Episode or Phase, and the Great American Biotic Interchange (Webb 2006).

It is universally agreed that the mammals got their start as early as the Triassic; the oldest mammal identified at this time is generally accepted as *Adelobasileus* (Fig. 2.1) from the Late Triassic of Texas at 225 millions of years, the most primitive mammal known. The fossil itself is of the posterior base of a skull, showing typical mammalian characteristics (Lucas and Luo 1993). At that time in the geological history of the world, there was probably a continuous connection of North America to South America by way of Europe and Africa (Fig. 2.2).

T. Defler, *History of Terrestrial Mammals in South America*, Topics in Geobiology 42, https://doi.org/10.1007/978-3-319-98449-0_2

Fig. 2.1 *Adelobasileus* the most primitive mammal known from the Triassic of 225 million years of Texas. (By N. Tamura)

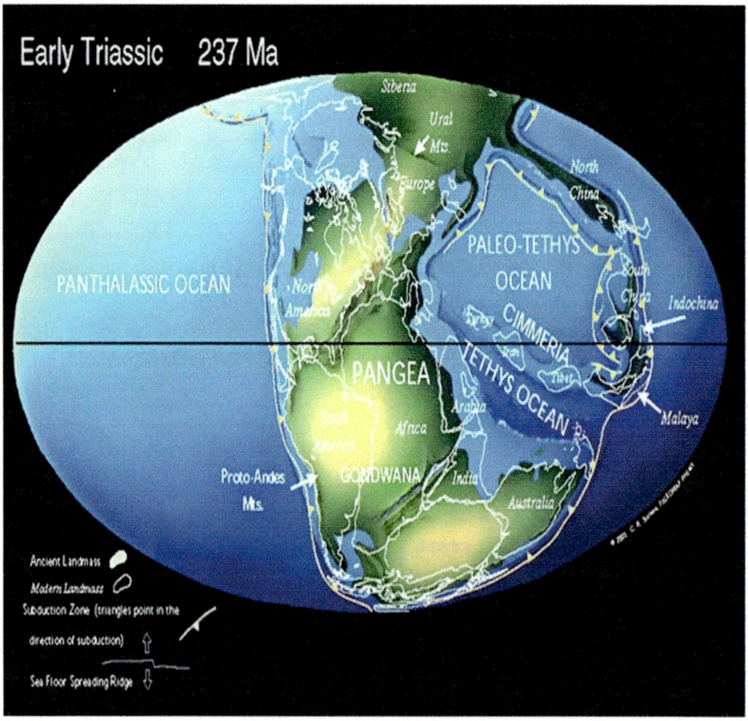

Fig. 2.2 Relation of continents about 200 million years ago, during the time of the first mammals. (By C. Scotese, Paleomap Project)

2.2 Three Histories?

When South America separated from Africa about 120 million years ago, connections were still maintained between South America and Antarctica until about 30 million years ago (Cunningham et al. 1995). Through this connection, there were also connections to Madagascar, India, and Australia.

The early Gondwanan fauna was characterized by both tribosphenic forms (molars with three main cusps arranged in a triangle) and pre- and non-tribosphenic forms. Many animals were clearly and logically related to earlier Pangaea fauna (Pangaea is the name of all continents when they were united as one) as well as forms of Gondwana origin. There were also endemic forms, clearly evolved in South America, judging by the lack of evidence from other Gondwanan continents (Fig. 2.3).

Most of this fauna had disappeared by Paleocene times, but it was characterized generally by endemic (to South America) mammals of Gondwanan origin from

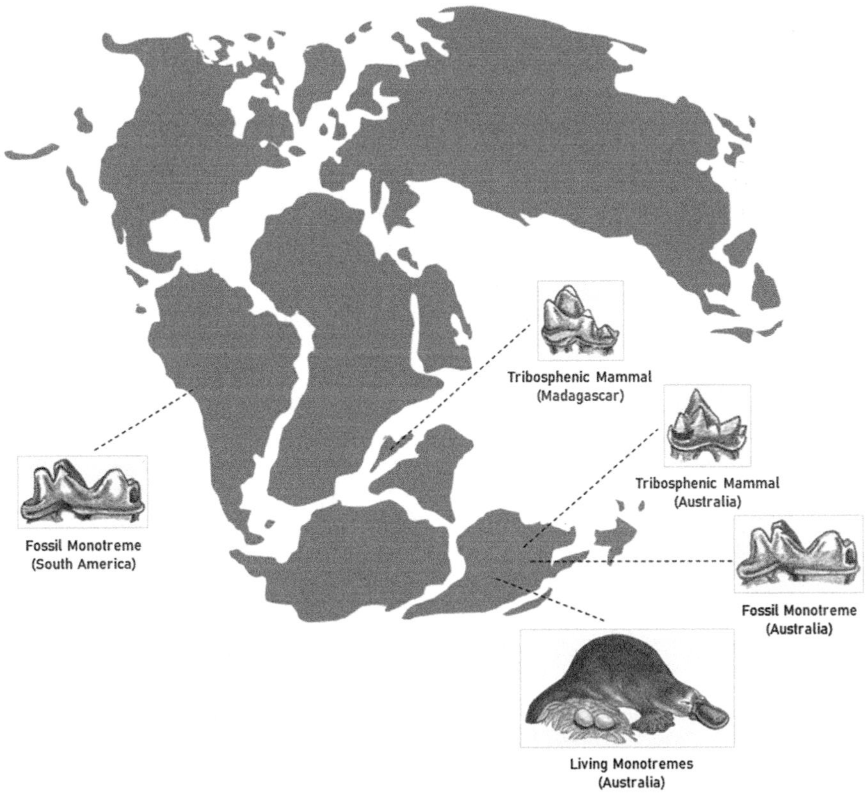

Fig. 2.3 Origin of the first southern tribosphenic australosphenidan mammal was discovered in Argentina in 2001 and other tribosphenic mammals discovered in parts of Gondwana from the Middle Jurassic. (By N. Calvo-Roa, Applied Biodiversity Foundation)

Mesozoic lineages, that is, Mesozoic elements that were the last of the Pangean mammals that were both non-tribosphenic and tribosphenic. It is important to mention that modern mammals all derive from tribosphenic-toothed mammals, a molar tooth architecture considered revolutionary in the evolution of mammals and from which dietary specialized teeth evolved. The basic tribosphenic tooth possesses both cutting edges and grinding surfaces with a three-cuspid arrangement. This molar design is considered one of the most important evolutionary achievements in mammalian history (Rose 2006).

Presently known, South American Mesozoic mammals come from Australosphenida, Trichonodonta, Dryolestoidea, Gondwanatheria, Multituberculata, and a basal stem therian (Forasiepi et al. 2012). Tribosphenic molars apparently evolved at least twice. All northern mammals (marsupials and placentals) evolved from the earliest tribosphenic mammals in the north. Another early clan of mammals, the "australosphenidans" from which the monotremes probably evolved, has been recognized far to the south in Australia, Madagascar, and South America (Figs. 2.3 and 2.4). These South American mammals represent three groups: (1) diverse and varied endemics, (2) mammals obviously from old lines stemming from Laurasia (e.g., the Gondwanatheria as vicariant Multituberculata), and (3) Pangaeic relicts. During these pre-Cenozoic times, South America still maintained a connection with Antarctica (which for long periods of time was not ice-covered and was covered by forest) which lasted up until about 30 million years ago (Dettmann 1989).

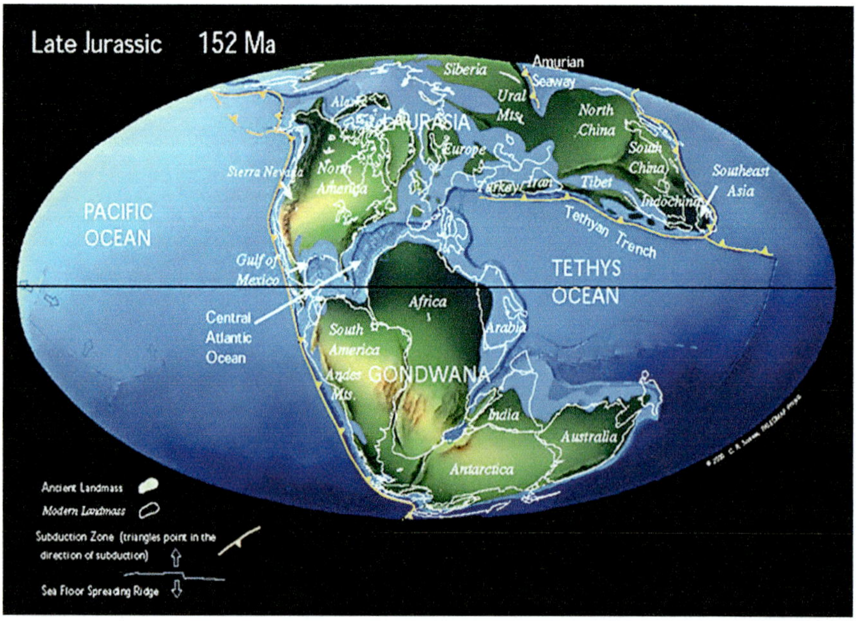

Fig. 2.4 Continental arrangement at the end of the Jurassic (152 Ma). (By Christopher Scotese, Paleomap Project)

2.3 The Australosphenida: Southern Tribosphenic Mammals and Triconodonta – The Earliest Known Mammals in South America

A triconodont molar corresponding to the family Amphilestidae was found in the lower levels of *Cañadon asfalto* in Chubut Province, Argentina, making it the oldest South American mammal, dated from the Early Jurassic (Toartian age, 174.1–182.7 Ma). Named *Condorodon spanios*, this species placed in the Amphilestidae, a family best known from the northern continents. It is the first of the family to be identified in South America (Gaetano and Rougier 2012).

However, the earliest South American mammalian community known so far is from Late Jurassic times, also discovered in Chubut, Argentina, in the *Cañadon asfalto* of about 168–161 Ma. The formation contained three different mammals: two australosphenid mammals *Asfaltomylos patagonicus* and *Henosferos molus* and a triconodont mammal *Argentoconodon fariasorum* (Rauhut et al. 2002; Rougier et al. 2007a, b).

2.3.1 *The Australosphenids (Fig. 2.4)*

The australosphenids represent southern tribosphenic mammals, representatives of which have also been found in Madagascar and Australia. This is generally recognized as an ancient Gondwanan group distinguished from the ancient northern (Laurasia) mammal group Boreosphenida (Luo et al. 2002). These South American mammals are probably related to the monotremes (include duck-bill platypus and echidna) and their early ancestors (Rougier et al. 2007a). The lower mandible and dentition of *Asfaltomylos patagonicus* (Fig. 2.5a, b) was discovered in 2001 in Patagonia. This small, shrew-sized Australosphenodon mammal had tribosphenic teeth, a characteristic that is generally applied to the modern mammals. "The discovery of this new mammal indicates that the Australosphenida diversified and were widespread in Gondwana well before the end of the Jurassic, and that mammalian faunas from the Southern Hemisphere already showed a marked distinction from their northern counterparts by the Middle to Late Jurassic" (Rauhut et al. 2002). The establishment of australosphenids implies that the tribosphenic molar evolved independently from the northern therian mammals that also had a tribosphenic molar (Luo et al. 2001; Martin and Rauhut 2014). However, there is some controversy whether this interpretation is correct or whether, in fact, the "australosphenids" belong to the northern placentals and were unrelated to monotreme evolution, although an analysis by Martin and Rauhut (2014) of the mandible and dentition of *Asfltomylos* appears to confirm that *Asfaltomylos* has fundamental differences from northern placentals.

Fig. 2.5 (**a**) A recreation of *Asfaltomylos patagonicus* and (**b**) jaw bone and tribosphenic teeth of a Mesozoic tribosphenic mammal (related to monotremes, line equals a scale of 1 mm) discovered in Patagonia. (By M. Joelle Giraud, Grupo Evolución y Ecología de Mamíferos Neotropicales, Universidad Nacional de Colombia).

2.3.2 *The Triconodonts*

The third mammalian tooth from this Middle to Upper Jurassic community in Patagonia has been identified as a new triconodont genus and species *Argentoconodon fariasorum* and shares similarities with poorly known triconodonts from the Jurassic of North America and Morocco (Rougier et al. 2007b). The age is probably Callovian-Oxfordian which is from the Middle Jurassic Callovian corresponding to an age of 164.7 ± 4.0 Ma–161.2 ± 4.0 Ma to about 161.2 ± 4 Ma–155.7 ± 4 Ma (million years ago) (Oxfordian) (Gaetano and Rougier 2011). Comparison of post-cranial characteristics of *Argentoconodon* to another triconodont sister clade *Volaticotherium* of China, which had been interpreted as a gliding mammal, suggests the possibility that *Argentoconodon* could also have been a glider (Gaetano and Rougier 2011).

Triconodonts have three major crown cusps arranged from the front (anteriorly) to the back (distally). They were relatively widespread especially in the Holarctic communities during the Jurassic and Early Cretaceous and may represent a paraphyly (all the descendants of the last common ancestor of the group's members minus a small number of monophyletic groups of descendants) grouping (Gaetano and Rougier 2011). The triconodonts were a sister or stem clade to modern therian mammals.

These three Early Jurassic mammals were found in association with many other tetrapods and rich plant life. The dominant plants were conifers (Araucariaceae) as well as ferns and equisetaleans, a group including the last surviving genus, the modern-day *Equisetum* (horsetails or snake grass). Other tetrapods from this community were crocodilians, turtles, frogs, and dinosaurs, most identified at least to genus. This fossil record represents "the most completely known biota from the continental Middle to Late Jurassic of South America and one of the most complete of the entire world" (Escapa et al. 2008). Additional possible triconodont mammals from South America have recently been reviewed (Gaetano et al. 2013). South America remained in contact with the northern continents throughout the Jurassic. The lacustrine *Cañadon asfalto* Basin existed in the early part of the Jurassic in northern Patagonia with the flora mentioned above typical of a moist temperate coniferous forest of Jurassic times.

Another ancient mammal from this period *Ameghinichnus patagonicus* (Fig. 2.6) has been described as an ichnospecies, i.e. a species identified based upon footprints or other traces that are not part of a skeleton. An ichnogenus is a group of trace fossils that is given a name because of the similarity of the traces that suggest they were made by closely related species of organisms. Ichnological taxonomy applies the principles of biological nomenclature to non-biological material and is governed by the *International Code of Zoological Nomenclature*. Above the level of genus, the Code indicates names should be used formally only to the family level; at higher levels all names are informal. Names of ichnogenera are conventionally written italicized and with a capital initial; ichnogenus is abbreviated as igen. *Ameghinichnus patagonicus* is only recognized as a small mammal or mammal-like form, but its higher taxonomy cannot be established at this time (Rougier et al. 2010). This species is based only on an ample register of tracks that indicate that the mammal was a rapid jumper and that there may have been more than one species. The tracks were found associated with dinosaurs.

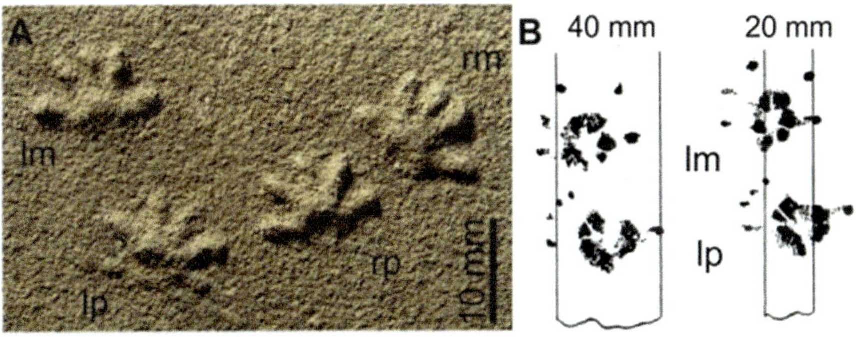

Fig. 2.6 Photograph of ichnites of *Ameghinichnus patagonicus*, fossil footprints in sand, used here to study the gait of the animal (A. *lm = left manus; lp = left pes; rm = right manus; rp = right pes*). (Kümmell 2014)

2.4 Cretaceous Mammals in South America

The South American mammalian fauna during the Cretaceous was dominated by the Dryolestida, plus a few triconodonts, gondwanatherians, and a possible docodont (Rauhut et al. 2002). The Dryolestida belong to an ancient clade, the Dryolestoidea, and are frequently referred to as eupantotherians. The "eupantotheres" ("" indicating paraphyletic origin) are generally seen as being very close to the modern mammal line, but they appeared too late to have been ancestral to marsupials and placentals. The group was common in South America up to the end of the Mesozoic era and even survived to the beginning of the Cenozoic, evolving endemic South American forms (Rougier et al. 2011).

2.4.1 *Vincelestes neuquenianus*

The Early Cretaceous *Vincelestes neuquenianus* is the sister lineage of therians (marsupials and placental mammals) (Fig. 2.7). Although not the direct ancestor of therian mammals, *Vincelestes* is exceptionally important because it gives us an idea of what the ancestral therian might have looked, and an indication of when the therians originated and began to diversify. *Vincelestes neuquenianus* (Fig. 2.8) is from the Lower Cretaceous La Amarga Formation of southern Neuquén Province, Argentina. At least six individual specimens are known, all of which were recovered from a single locality (Rougier et al. 2007a, 2010). The species seems to be a stem clade (sister clade) to modern mammals (i.e. placentals and marsupials) (Kielan-Jaworowska et al. 2004; Kielan-Jaworowska 2013). *Vincelestes* shows numerous anatomical features that are transitional between more primitive extinct mammals

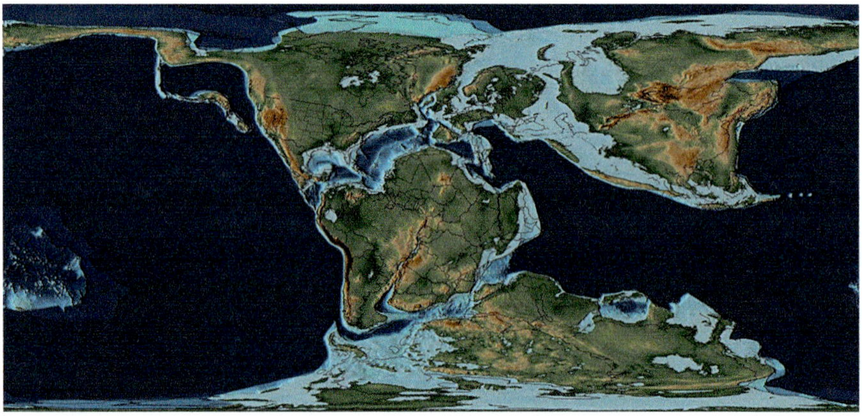

Fig. 2.7 Position of continental plates during the Early Cretaceous (135 Ma). (By C. Scortese, Paleomap Project)

Fig. 2.8 *Vincelestes neuquenianus.* (By Roman Uchytel)

and extant therians. For example, the cheek teeth of *Vincelestes* have a "reversed triangle occlusal pattern" that closely approximates the fully tribosphenic pattern of therians, an innovation that enables them to process food more thoroughly. Many features of the internal cranial anatomy of *Vincelestes* that cannot be seen on visual inspection of the skull are readily discernible in CT slices (computed tomography scan) (Macrini et al. 2007).

Vincelestes was originally identified as a eupantothere (includes Dryolestoidea), based on information then available. But a more recent comparison of *Vincelestes* lower jaw with other Laurasian and Gondwanan taxa using 90 characters suggests that *Vincelestes* is associated with other Gondwanan mammals and has nothing to do with northern taxa (Bonaparte 2008) and represents an ancient southern (Gondwanan) evolution. Time will tell which of these interpretations of *Vincelestes* is correct.

2.4.2 Cretaceous Mammal Diversity

During the Late Cretaceous (Fig. 2.9), a much greater diversity of Mesozoic mammals is known. There have been at least 11 genera of mammals recognized for the Late Cretaceous of South America: *Austrotriconodon* (Triconodonta), *Bondesius* (Symmetrodonta), *Groebertherium*, *Brandonia*, *Leonardus*, *Mesungulatum*, *Paraungulatum*, *Quirogatherium*, *Reigitherium* (?), *Casamiquelia*, *Alamitherium* and *Rougiertherium* (all Dryolestoidea); *Ferugliotherium* and *Gondwanatherium* (Multituberculata or Gondwanatheria, the designation is controversial) (de la Fuente et al. 2007; Rougier et al. 2009). Bonaparte (1986a, b, 1990, 1992, 2002) identifies

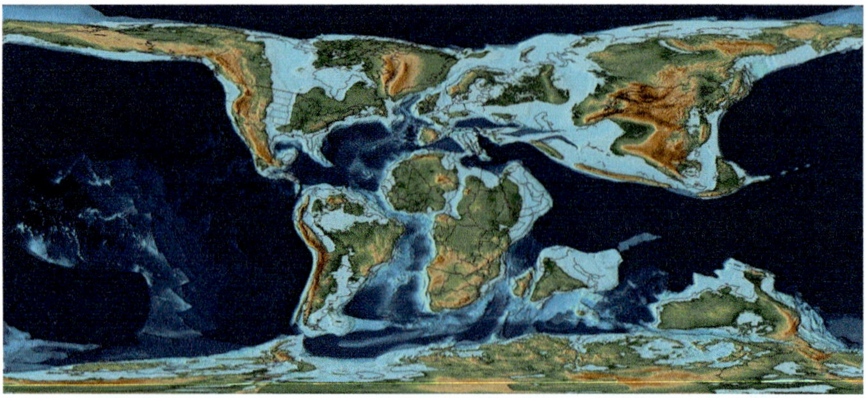

Fig. 2.9 Continental positions in the Late Cretaceous (100.5–66 Ma) at about 90 Ma. (By Christopher Scotese, Paleomap Project)

18 genera of triconodonts, symmetrodonts, dryolestoids, multituberculates, and gondwantherians (*Barberenia, Trapalcotherium, Coloniatherium, Argentoditis*, Mammalia indet. 3 sp.) (Forasiepi et al. 2012; Rougier et al. 2009) although Rougier et al. (2010) felt that the fauna might be overestimated and that triconodonts and symmetrodonts were not present. Finally, a Late Cretaceous eutherian from Brazil must be added to the list, which is supported by various workers (Bertini et al. 1993; Rougier et al. 2010).

Apparently the dryolestoids had become extinct on all continents except for South America where the Dryolestidae were apparently the most abundant mammal in the latter part of the Cretaceous, even lasting into the Cenozoic as *Peligrotherium* (Kielan-Jaworowska et al. 2004; Gelfo and Pascual 2001; Forasiepi et al. 2012). At the beginning of the South American episode (Early Paleocene), there were still two relictual Gondwanan taxa, a gondwanatherian and a **dryolestoid** *Peligrotherium* as well as an Ornithorynchidae (*Monotrematum sudamericanum*) in Patagonia. Another gondwanatherian relict was also found in the Antarctic Eocene deposits on Elsmore Island off the Antarctic Peninsula.

The discovery of *Reigitherium* from the Late Cretaceous of Patagonia was identified first as a dryolestid when it was first described by Bonaparte (1990) and then as a docodont (Pascual et al. 2000), a group typically of Laurasia and not Gondwana. But new material makes it more likely that this animal was a dryolestoid, another primitive Mesozoic mammal that has been put forward as containing the ancestors of modern therian mammals (Rougier et al. 2011).

A highly specialized dryolestoid *Cronopio dentiacutus* (Fig. 2.10) was discovered in Rio Negro Province of Argentina from the Late Cretaceous. This animal was medium-sized with an extremely elongated rostrum (snout) and very long canines with evidence of highly specialized chewing. The skull shows a mixture of primitive mammalian features and highly specialized traits, underlining the endemic dryolestoid fauna found at this period of the Mesozoic Era (Rougier et al. 2011).

Fig. 2.10 *Cronopio dentiacutus* was a highly specialized dryolestoid with a very narrow snout. (By Guillermo Rougier). What type of specialization these animals had is conjectural. They do, however, remind us of a certain acorn-pushing mammal from the film "Ice Age"

2.4.3 *The Gondwanatheres*

The gondwanatheres were a rare and bizarre group (suborder?) restricted to the Late Cretaceous–Early Paleocene of SA, Middle Eocene of Antarctica, Upper Cretaceous of Madagascar and India, and perhaps from Africa (Krause et al. 1997). Four genera have now been described for South America: the *Gondwanatherium* and *Ferugliotherium* of upper Cretaceous Patagonia, *Sudamerica* from Patagonian Paleocene (which may have actually belonged to a multituberculates, according to some), and *Greniodon* of the late Early Eocene (45–47 Ma, Tejedor et al. 2009; Bonaparte 1986; Scillato-Yané and Pascual 1985; Krause and Bonaparte 1993; Goin et al. 2012a, b). *Ferugliotherium* has been placed in the family Ferugliotheriidae along with *Argentodites* and *Trapalcotherium* from the end of the Cretaceous, considered by some to be a multituberculate and by others as a gondwanathere unrelated to the multituberculates (Rougier et al. 2009). These may have been the earliest grazers known, judging by the high-crowned molars that distinguish them. Some believe that this group may have been the ancestors of the xenarthrans; in fact, it seems to be the only candidate group (Bonaparte 1986).

The multituberculates were an ancient and very successful early rodent-like mammals in Laurasia and especially in North America. As a group, they appear in the Jurassic and were able to last into the Eocene. Multituberculates were very scarce in South America. Two genera of these animals have been recently identified from teeth in Patagonia with a Late Cretaceous age, *Argentodites* and an unnamed species and genus (Kielan-Jaworowska et al. 2007). The unnamed genus was previously identified as a gondwanatheran, derived from the multituberculates

Feruglíotherium (Krause and Bonaparte 1993; Pascual et al. 1999), but reconsiderations place it firmly within the multituberculates (Bonaparte 1986a, b; Kielan-Jaworowska et al. 2004). However, others have seen the gonwanatheres as ***incertae sedis*** (placement uncertain), not connecting the group to other mammals (Pascual et al. 1999). The relationship of multituberculates to other mammals is controversial and unresolved (Luo et al. 2002).

2.4.4 Connections with Australia

Most surprising was the discovery in Argentina by Rosendo Pascual et al. (1992) and associates of two platypus molars (*Monotrematum sudamericanum*) (Fig. 2.11) in Early Tertiary Paleocene (Peligran SALMA) of southern Argentina. This astonishing find is the only monotreme known outside of Australia, and it indicates the earlier widespread geographic distribution of platypus, obviously having also existed in Antarctica in order to exist in South America (Woodburne and Case 1996). This find also supports the idea that the southern part of Patagonia had a distinct biotic history from the rest of the continent (Pascual et al. 1992). It might be thought by some that two molars might not cinch the case for a South American platypus, but the species was obviously closely related to the early Australian platypus *Obdurodon dicksoni* which had teeth, unlike the modern platypus. Both *Obdurodon* and *Monotrematum* were about 50% larger than the present-day platypus (Fig. 2.11).

A reanalysis of old material collected in the nineteenth century in Patagonia concluded that the Early Miocene *Necrolestes patagonensis* was, in fact, a dryolestoid meridiolestid, the most recent non-therian so far found (Rougier et al. 2012). Although a recent phylogenetic analysis using 137 morphological characters among 44 taxa suggested that *Necrolestes* was a dryolestoid cladotherians but were

Fig. 2.11 *Obdurodon sp.* with three species existing in Australia during the Miocene 15 and 25 Ma. They probably kept their teeth until adulthood, unlike the modern (*Ornithorhynchus anatinus*) which has horny plates in place of teeth. Since the South American *Monotrematum sudamericanum* existed around 61 Ma, it may be an ancestor to the later platypus. The two molars that are known show close affinities to *Obdurodon dicksoni*. (By M. Joelle Giraud, Grupo de Evolución y Ecología de Mamíferos Neotropicales)

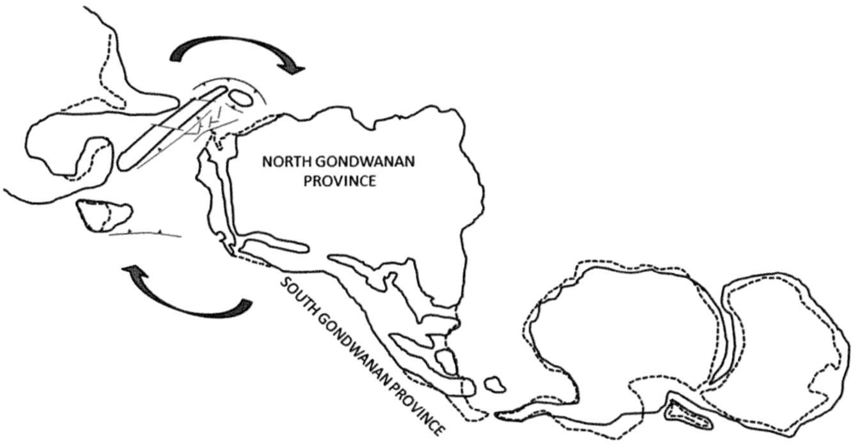

Fig. 2.12 Two ancient biogeographic provinces, North Gondwanan Province and South Gondwanan Province. Two paleobiogeographical provinces dated between 85 and 63 Ma reflect the climax of the Gondwanan Episode with a strong African influence in the Northern Gondwanan Province and a strong Australian (and Antarctic?) influence in the Southern Gondwanan Province, reflected by lungfish and turtles. (By D. Casallas-Pabón, Applied Biodiversity Foundation) (Pascual 2006)

rather Symmetrodonta in the superfamily Spalacotherioidea (Averianov et al. 2013). These Miocene mammals together with the known Paleocene australosphenida monotremes (*Monotrematum*), the gondwanatherians (*Sudamerica*) and other meridiolestidans (*Peligrotherium*) make it clear that survivors of non-therian Mesozoic mammals are a distinctive part of the Cenozoic mammal fauna (Rougier et al. 2012).

Between 85 and 63 million years ago, South America possessed a shallow sea that separated two biogeographical provinces North Gondwana (Brazil and adjacent lands) and South Gondwana (Argentina or Patagonia and adjacent lands) (Fig. 2.12). Tectonic changes in the proto-Caribbean caused many volcanic islands and facilitated the immigration of tetrapods from North America, including the mammals that formed a part of the fauna of the South American Episode, including early ungulates and early marsupials, but also including the earliest Cretaceous eutherians, Paleocene docodonts and other mammalian and nonmammalian fauna clearly connected with the north.

The biogeographic provinces of North Gondwana and South Gondwana are clearly defined by the turtles and the dipnoid (lungfishes) fishes that lived in each province during this epoch, a fauna related to Africa (in the north) and to Australia (in the south) and by the survival of archaic lineages of mammals in Patagonia in the Paleocene (Pascual 2006; Rougier et al. 2010). South American biotic history is characterized by long periods of isolation alternating with brief connections and so explains the relatively few episodes that are generally sharply and clearly separable (Pascual and Ortiz-Jaureguizar 1992). Goin et al. (2016) summarize that the Late

Cretaceous of Argentina comprises 17 identified non-tribosphenic genera consisting of a symmetredont, 1 sudamericid, 1 ferugliotheriid gondwanatherian, 1 multituberculate, and 13 dryolestids.

References

Averianov AO, Martin T, Lopatin AV (2013) A new phylogeny for basal Trechnotheria and Cladotheria and affinities of South American endemic Late Cretaceous mammals. Naturwissenschaften 100:311–326

Bertini RJ, Marshall LG, Gayet M, Brito P (1993) Vertebrate faunas from the Adamantina and Marilia formations (upper Baurú Group, Late Cretaceous, Brazil) in their stratigraphic and paleobiographic context. Neues Jahrb Geol P-A 188:71–101

Bonaparte JF (1986a) A new and unusual Late Cretaceous mammal from Patagonia. J Vertebr Paleontol 6(3):264–270

Bonaparte JF (1986b) Sobre *Mesungulatum houssayi* y nuevos mamíferos Cretácicos de Patagonia, Argentino. IV Congreso Aregentino de Paleontología y Estratigrafía, Actas 2:63–95

Bonaparte JF (1990) New Late Cretaceous mammals from the Los Alamitos Formation, northern Patagonia. Natl Geogr Res 6:63–93

Bonaparte JF (1992) Una nueva especie de Triconodonta (Mammalia), de la Formación Los Alamitos, Provincia de Río Negro y comentarios sobre su fauna de mamíferos. Ameghiniana 29:99–110

Bonaparte JF (2002) New dryolestid (Theria) from the Late Cretaceous of Los Alamitos Formation, Argentina, and paleogeographical comments. Neues Jahrb Geol P-A 224:339–371

Bonaparte JF (2008) On the phyletic relationships of *Vincelestes neuquenianus*. Hist Biol 20(2):81–86

Cunningham WD, Dalziel WD, Lee T-Y, Lawver LA (1995) Southernmost South America-Antarctic peninsula relative plate motions since 84 Ma: implications for the tectonic evolution of the Scotia Arc region. J Geophys Res 100(B5):8257–8266

de la Fuente MS, Salgado L, Albino A, Báez AM, Bonaparte JF, Calvo JO, Chiappe LM, Codorniú LS, Coria RA, Gasparini Z, González R, Novas FE, Pol D (2007) Tetrápodos continentales del Cretácico de la Argentino: una síntesis actualizada. Ameghiniana Publicación Especial 11:137–153

Dettmann ME (1989) Antarctica: Cretaceous cradle of austral temperate rainforests. Geophys Soc London, Special Publications 47:89–105

Escapa IH, Sterli J, Pol D, Nicoli L (2008) Jurassic tetrapods and flora of Cañadon Asfalto formation in Cerro Cóndor area, Chubut province. Rev Asoc Geol Argent 63(4):613–624

Forasiepi AM, Coria RA, Hurum J, Currie P (2012) First dryolestoid (Mammalia, Dryolestoidea, Meridiolestida) from the Coniacian of Patagonia and new evidence on their early radiation in South America. Ameghiniana 49(4):497–504

Gaetano LC, Rougier GW (2011) New materials of *Argentoconodon fariasorum* (Mammaliaformes, Triconodontidae) from the Jurassic of Argentina and its bearing on triconodont phylogeny. J Vertebr Paleontol 31(4):829–843

Gaetano LC, Rougier GW (2012) First Amphilestid from South America: a molariform from the Jurassic Cañadón Asfalto formation, Patagonia, Argentina. J Mamm Evol 19:235. https://doi.org/10.1007/s10914-012-9194-1

Gaetano LC, Marsicano CA, Rougier GW (2013) A revision of the putative Late Cretaceous triconodonts from South America. Cretac Res 46:90–100

Gelfo JN, Pascual R (2001) *Peligotherium tropicalis* (Mammalia, Dryolestida) from the early Paleocene of Patagonia, a survival from a Mesozoic Gondwanan radiation. Geodiversitas 23(3):369–378

Goin FJ, Gelfo JN, Chornogubsky L, Woodburne MO, Marin T (2012a) Origins, radiations, and distribution of South American mammals: from greenhouse to icehouse worlds. In: Patterson BD, Costa LP (eds) Bones, clones, and biomes: the history and geography of recent neotropical mammals. University of Chicago Press, Chicago, pp 20–50

Goin FJ, Tejedor MF, Chornogubsky L, López GM et al (2012b) Persistence of a Mesozoic, non-therian mammalian lineage (Gondwanatheria) in the mid-Paleogene of Patagonia. Naturwissenschaften 99:449–463

Goin FJ, Woodburne MO, Chornogubsky L (2016) Dispersal of vertebrates from between the Americas, Antarctica, and Australia in the Late Cretaceous and early Cenozoic. In: Goin FJ, Woodburn MO, Zimiez AN, Martin GM, Chornogubsky L (eds) A brief history of South American metatherians: evolutionary contexts and intercontinental dispersals. Springer, New York, pp 77–124

Kielan-Jaworoska Z (2013) In pursuit of early mammals. Indiana University Press, Bloomington

Kielan-Jaworowska Z, Cifelli RL, Luo Z-X (2004) Mammals from the age of dinosaurs: origins, evolution, and structure. Columbia University Press, New York

Kielan-Jaworowska Z, Ortiz-Jaureguizar E, Vieytes C, Pascual R, Goin FJ (2007) First ?cimolodontan multituberculate mammal from South America. Acta Palaeontol Pol 52(2):257–262

Krause DW, Bonaparte JF (1993) Superfamily Gondwanatherioidea: a previously unrecognized radiation of multituberculate mammals in South America. Proc Natl Acad Sci U S A 90:9379–9383

Krause DW, Prasad GVR, von Koenigswald W, Sahni A, Grine FE (1997) Cosmopolitanism among Gondwanan Late Cretaceous mammals. Nature 390:504–507

Kümmell S (2014) Range of movement in ray I of manus and pes and the prehensility of the autopodia in the Early Permian to Late Cretaceous non-anomodont synapsida. PLoS One 9(12):e113911. https://doi.org/10.1371/journal.pone.0113911

Lucas SG, Luo Z (1993) *Adelobasileus* from the upper Triassic of West Texas: the oldest mammal. J Vertebr Paleontol 13(3):309–334

Luo Z-X, Cifelli RL, Klelan-Jaworowska Z (2001) Dual origin of tribosphenic mammals. Nature 409:53–57

Luo Z-X, Kielan-Jaworowska Z, Cifelli RI (2002) In quest for a phylogeny of Mesozoic mammals. Acta Palaeontol Pol 47(1):1–78

Macrini TE, Rougier GW, Rowe T (2007) Description of a cranial endocast from the fossil mammal Vincelestes neuquenianus (Theriiformes) and its relevance to the evolution of endocranial characters in therians. Anat Rec 290:875–892

Martin T, Rauhut OW (2014) Mandible and dentition of *Asfaltomylos patagonicus* (Australosphenida, Mammalia) and the evolution of tribosphenic teeth. J Vertebr Paleontol 25(2):414–425

Pascual R (2006) Evolution and geography: the biogeographic history of South American land mammals. Ann Mo Bot Gard 93:209–230

Pascual R, Ortiz-Jaureguizar E (1992) Evolutionary pattern of land mammal faunas during the Late Cretaceous and Paleocene in South America: a comparison with the North American pattern. Ann Zool Fenn 28:245–252

Pascual R, Ortiz-Jaureguizar E (2007) The Gondwanan and South American episodes: two major and unrelated moments in the history of the South American mammals. J Mamm Evol 14(2):75–137

Pascual R, Archer M, Ortiz Jureguizar E, Prado JL, Godthelp H, Hand SJ (1992) First discovery of monotremes in South America. Nature 356:704–706

Pascual R, Goin FJ, Krause KW et al (1999) The first gnathic remains of Sudamerica: implications for gondwanathere relationships. J Vertebr Paleontol 19(2):373–382. https://doi.org/10.1080/02724634.1999.10011148

Pascual R, Goin FJ, Gonzalez P, Ardolino A, Puerta PF (2000) A highly derived docodont from the Patagonian Late Cretaceous. Geodiversitas 22(3):395–414

Rauhut OWM, Martin T, Ortiz-Jaureguizar E, Puerta P (2002) A Jurassic mammal from South America. Nature 416:165–168
Rose KD (2006) The beginning of the age of mammals. The Johns Hopkins University Press, Baltimore
Rougier GW, Garrido A, Gaetano L, Puerta PF, Corbitt C, Novacek MJ (2007a) First Jurassic triconodont from South America. Am Mus Novit 3590:1–17
Rougier GW, Martinelli AG, Forasiepi AM, Novacek MJ (2007b) New Jurassic mammals from Patagonia, Argentina: a reappraisal of australosphenidan morphology and interrelationships. Am Mus Novit 3566:1–54
Rougier GW, Chornogubsky L, Casadio S, Paéz Arango N, Giallombardo A (2009) Mammals from the Allen Formation, Late Cretaceous, Argentina. Cretac Res 30:223–238
Rougier GW, Gaetano L, Bradley D, Colella R, Gómez RO, Arango NP (2010) A review of the Mesozoic mammalian record of South America. In: Calvo J, Porfiri J, Gonzáliez Riga B, Dos Santos D (eds) Paleontologia y dinosaurios desde América Latina. Universidad Nacional del Cuyo, San Rafael, pp 195–213
Rougier GW, Apesteguía S, Gaetano LC (2011) Highly specialized mmmalian skulls from the Late Cretaceous of South America. Nature 479:98. https://doi.org/10.1038/nature10591
Rougier GW, Wible JR, Beck RMD, Apesteguía S (2012) The Miocene mammal *Necrolestes* demonstrates the survival of a Mesozoic nontherian lineage into the late Cenozoic of South America. Proc Natl Acad Sci U S A 109(49):20053–20058. https://doi.org/10.1073/pnas.1212997109
Scillato–Yané GJ, Pascual R (1985) Un peculiar Xenarthra del Paleoceno medio de Patagonia (Argentina). Su importancia en la sistemática de los Paratheria. Ameghiniana 21:173–176
Tejedor MF, Goin FJ, Gelfo JN, López GM, Bond M, Carlini AA, Scillato-Yané GJ, Woodburne MO, Chornogubsky L, Aragón E, Reguero MA, Czaplewski NJ, Vincon S, Martin GM, Ciancio MR (2009) New early Eocene mammalian fauna from western Patagonia, Argentina. Am Mus Novit 3638:1–48
Webb SD (2006) The Great American Biotic Interchange: patterns and processes. Ann Mo Bot Gard 93(2):245–257
Woodburne MO, Case JA (1996) Dispersal, vicariance, and the Late Cretaceous to Early Tertiary land mammal biogeography from South America to Australia. J Mamm Evol 3(2):121–161

Chapter 3
Early Cenozoic Mammals in South America

3.1 Introduction

The earliest Paleocene *therian* mammal community known so far (and earliest for the entire Cenozoic) is the Tiupampan community (SALMA) from Bolivia, dated at about 64–62.5 Ma. So far, South American fauna before this community is unknown except for *Cocatherium lefipanum*. There is no information available for the presence of therian mammals for the latest Cretaceous. The exception for a probable earlier Late Cretaceous presence of modern mammals is a molar of a therian found in La Plata, Argentina, and dated from about 66.043 to 61.7 Ma, *Cocatherium lefipanum,* most likely a polydolopimorphian marsupial (Goin et al. 2006).

The Cretaceous, as far as we know, was dominated by triconodonts, symmetrodonts, dryolestoids, multituberculates, and gondwanatherians, representatives of the Gondwanan Episode of South America (de Muizon and Cifelli 2001). The Tiupampan community itself was a huge break from the Late Cretaceous, since the majority of the community was made up of modern mammals (=Theria) whose ancestors had immigrated from North America, especially the marsupials, the metatherians, and the ancient "ungulates" and these mammals form the great majority of the South American Cenozoic mammal community until the first rodents and primates arrived, maybe 15–20 million years later. This meant there was a lot of evolving going on of these two large groups.

This faunal break between the Cretaceous and Cenozoic Eras was caused by the events in a proto-Central America when a huge asteroid or comet crashed into our planet, calculated with argon/argon dating at 66.043 ± 0.011 Ma (Renne et al. 2013). This caused worldwide ecological disturbances which purportedly wiped out the dinosaurs and other fauna (Pope et al. 1999; Arenillas et al. 2006; Schulte et al. 2010; Cowen 1999). Concurrent with this impact and the eruption of the Deccan

T. Defler, *History of Terrestrial Mammals in South America*, Topics in Geobiology 42, https://doi.org/10.1007/978-3-319-98449-0_3

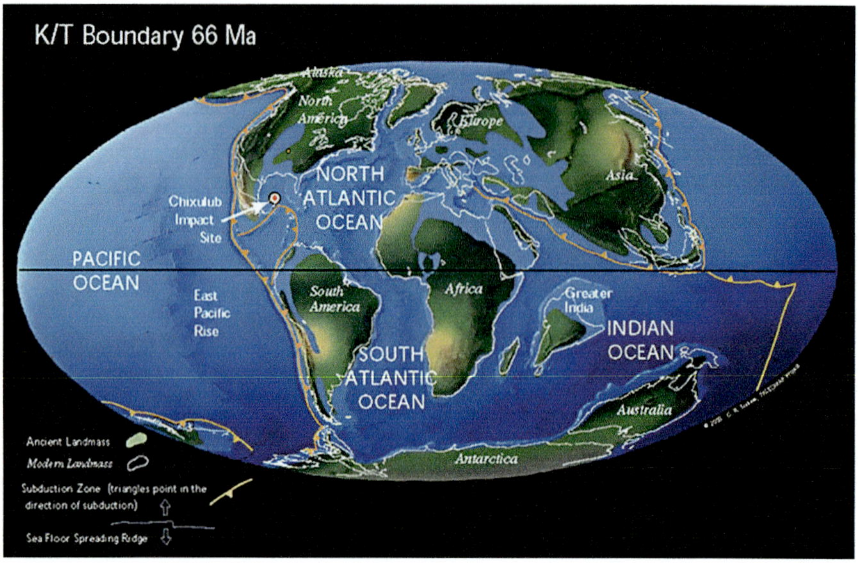

Fig. 3.1 Continental geography just before Chicxulub impact. (By Christopher Scotese, Paleomap Project)

Traps (possibly caused by the impact and enormous outgassing of sulfur dioxide from the traps), great climate changes took place, ocean productivity and currents stopped or were altered, and the food chain was drastically perturbed due to a cessation of photosynthesis (Taylor et al. 2011; Schulte et al. 2010; Keller 2014). As we shall see, the K-T (or K-Pg) extinction event was not so abrupt in South America since the Gondwanan Episode intruded into southern Patagonia with many Gondwanan faunal elements, mixed with the ever-increasing invasion of northern modern therian mammals. All major Cretaceous mammalian lineages (monotremes, multituberculates, marsupials and placentals, dryolestoideans, docodonts, and gondwanatheres) survived the extinction event in South America but mostly in the extreme south and not abundantly. Basically a new episode begins in the evolution of South American mammals at the end of the Cretaceous and beginning of the Paleocene, the South American Episode (Pascual and Ortiz-Jaureguizar 2007; MacLeod et al. 1997) (Fig. 3.1).

Much of the Paleocene in South America is divided into known SALMAs except for the very beginning from 66 million years to the beginning of the Tiupampan at 64.5 million years. These SALMAs represent discontinuities in strata and mammal sequences, and they are probably related to global eustatic changes (Bond et al. 1995). The latest dating suggests that several SALMAs of the Paleogene overlap with each other, such as the Vacan and the "Sapoan" with the Riochican (Woodburne et al. 2014; Krause et al. 2017) (Table 3.1).

Table 3.1 Sequences of the Paleogene Period (Paleocene-Oligocene Epochs)
Paleogene Period (Krause et al. 2017; Woodburne et al. 2014; Clyde et al. 2014; Flynn and Swisher 1995)

Deseadan **Salma**	29–21 Ma
Tinguiririican **Salma**	36–29 Ma
Oligocene Epoch	33.9–23 Ma
Divisaderan **Salma**	42–36 Ma
Mustersan **Salma**	38–37 Ma?
Barrancan **Salma**	42.1–38.16 Ma (Krause et al. 2017)
Vacan **Salma**	46–44 Ma (Krause et al. 2017)
"Sapoan" **Salma**	48.5–47 Ma (Woodburne et al. 2014)
Riochican **Salma**	46.7–42.2 Ma (Krause et al. 2017) 51.4–42.2 Ma (Krause et al. 2017)
Itaboraian **Salma**	53.0–50.0 Ma (Woodburne et al. 2014) 56–51.4 Ma (Krause et al. 2017)
Eocene Epoch	56–33.9 Ma
Cardonia Zone	62.22–59.24 Ma
Peligran **Salma**	62.5–59.0 Ma
Tiupampan **Salma**	64.5–62.5 Ma
Paleocene Epoch	66–56 Ma

3.2 Tiupampa, Bolivia

The Tiupampan fauna represents the oldest South American Cenozoic faunal association found so far in the Early Paleogene (Early Cenozoic), despite the lack of *radioisotopic* dates and previous disagreements (Gelfo et al. 2009; Marshall et al. 1997; Goin et al. 2016). A slightly older therian *bunodont* mammalian molar (*Cocatherium*) has been found in Chubut Province, however, which is placed tentatively with the *polydolopimorphian marsupials* (Chap. 4) and is dated from the earliest *Danian* Paleocene (the Danian Stage is dated at 61.7–65.5 Ma) (Goin et al. 2006). The slightly later Tiupampan fauna was described about 95 km southeast of Cochabamba, Bolivia, and it has been dated to about 64.5–62.5 million years before the present time. The locality gives its name to the Tiupampan SALMA of South America (de Muizon 1991, 1998; de Muizon and Cifelli 2000, 2001; de Muizon and Marshall 1992; de Muizon et al. 1983). The list of mammals discovered in this fauna is given below (Table 3.2).

Table 3.2 Tiupampan local fauna (http://www.paleocene-mammals.de/pal-sa.htm) (Jehle 2006)
Location: 95 km southeast of Cochabamba, Bolivia
Age: type locality of Tiupampan (Pascual and Ortiz-Jaureguizar 1991b; Bond et al. 1995; Goin et al. 2016)

Infraclass Metatheria Order Peradectia	
Family Peradectidae	*Peradectes* cf. *austrinum*
Family Caroloameghiniidae	*Roberthoffstetteria nationalgeographica*
Order Didelphimorphia	
Family Didelphidae	
	Incadelphys antiquus
	Mizquedelphys pilpinensis
	Andinodelphys cochabambensis
	Tiulordia floresi
	Szalinia gracilis
Order Sparassodonta	*Allqokirus australis*
	Mayulestes ferox
Order indeterminate	*Pucadelphys andinus Andinodelphys cochabambensis Jaskhadelphys minutus*
Order Microbiotheria	
Family Microbiotheriidae	*Khasia cordillerensis*
Cohort Placentalia Order Proteutheria (o Cimolesta)	
Family Palaeoryctidae Family indet.	cf. *Cimolestes* sp.
	gen.et sp. indet.
Order Condylarthra	
Family Mioclaenidae	*Molinodus suarezi*
	Tiuclaenus minutus (= *Kollpania tiupampina*)
	T. cotasi
	T. robustus
	Pucanodus gagnieri (= cf. *Mimatuta* of de Muizon and Marshall (1991))
	Andinodus boliviensis
	Simoclaenus sylvaticus
Order Notoungulata	
Family Oldfieldthomasiidae	gen.et sp. indet. (may belong to Ornithorhynchidae; see Marshall et al. (1997))
Order Pantodonta	
Family Alcidedorbignyidae	*Alcidedorbignya inopinata*

References: de Muizon et al. (1983, 2018), Marshall and Muizon (1988), de Muizon (1991), de Muizon and Marshall (1991, 1992), Marshall et al. (1995, 1997), de Muizon and Cifelli (2000, 2001)

3.2.1 *Marsupials and Other Metatheres of the Tiupampan Fauna*

The early rich diversity of *marsupials* and other *metatheres* in the Tiupampan fauna seems too diverse for being so close to the great Cretaceous-Tertiary extinction event (the K-T extinction event; K= *Kreidezeit* in German), and many have speculated that South America must have had an evolving metatherian fauna in the Late Cretaceous, especially given the rich therian fauna that was then present in North America (Simpson 1980; Case and Woodburne 1986: Gelfo et al. 2009). This metatherian diversity exceeds that existing among the primitive ungulates of this association, suggesting an arrival before the ungulates. These animals have clear connections to the northern metatherians (Simpson 1980; de Muizon and Cifelli 2001), probably arriving in South America via some transient geographic connections formed by the volcanic and *diastrophic* (deformation of the earth's crust) conditions active in the proto-Caribbean (Pascual and Ortiz-Jaureguizar 1990, 1991a, b, 2007). So far, despite some earlier mistakes in dating Peruvian and Chilean finds as coming from the Cretaceous (erroneous) rather than the Paleocene, when they appear to have existed, no Late Cretaceous South American metatherians have been found. But it will not seem surprising, should they be found in the future. Probably the slightly earlier than Tiupampan metatherian petrosal and dental remains found in Chubut Province (Argentina) represent the earliest known marsupial of the South American Era. The fossil fits the size for the known Didelphimorphia *Derorhynchus* aff. *D. minutus* (Forasiepi and Rougier 2009) in the above list.

South American metatherians show strong connections to North American metatherians, especially by the presence of the marsupial family Didelphidae, which has a long North American history, so is shared by the two continents (Simpson 1980). Also, the order Peradectia and the family Peradectidae are known for the Late Cretaceous of North America and survived the K-T extinction event in both North and South America. The families Caroloameghiniidae, Jaskhadelphydae, and Microbiotheriidae are endemic to South America.

The best-known Paleocene metatherian is probably *Pucadelphys andinus*, of an indeterminate order found in the Tiupampan fauna in the form of complete skeletons in groups. The species was apparently fossorial, since several complete specimens in 2 family groups of 23 individuals and 12 individuals including males, females, and subadults and juveniles were found in fossil holes in the ground, probably dens that they had built themselves. The groups represent gregarious social behavior and the first such behavior detected in any metatherian. The skeletons give us a very good idea of how they looked. It is clear from the fossils that they had a very long, toothy snout (Ladevèze et al. 2011; Argot 2001, 2002; Marshall et al. 1995: de Muizon et al. 2018) (Fig. 3.2).

Fig. 3.2 *Pucadelphys andinus*, a metatherian belonging to an indeterminate order but apparently very common, social, and fossorial. (By Kelsey Van Horn)

3.2.2 Sparassodonta (Borhyaenoidea)

The endemic order Sparassodonta (Borhyaenoidea) is well-known for the interesting and various predators that evolved in this group, culminating in the amazing *Thylacosmilus* or marsupial tiger, which looked somewhat like a saber-toothed cat. This group were probably not true marsupials; they were more correctly metatherians, since they fall outside the crown or monophyletic group that contains the ancestor's marsupial descendants (Rougier et al. 1998; Asher et al. 2004). According to Asher et al. (2004), they are more generally regarded as a sister group or clade, and they are well-known for having evolved many forms convergent to placental carnivores, that is, they looked somewhat like some modern, placental forms. Convergence caused some confusion for the Argentinian paleontologist Florentino Ameghino who interpreted these animals as ancestors of modern carnivores like dogs and cats (Simpson 1980).

Sparassodonts varied in size from about 80 cm to the size of modern "big cats." The Sparassodonta had been strongly suspected of having been the ancestors of the Australian thylacine predators and would have arrived on that continent probably in the Late Paleocene along with the Microbiotheria. However, cladistics and morphological analyses and the latest molecular research suggest there is no connection (Marshall 1977; Horovitz and Sánchez-Villagra 2003; Asher et al. 2004; Meredith et al. 2008). Rather, the early appearance of the Microbiotheria and its rich diversity seem to be the ancestors of Australian marsupials.

Mayulestes is, along with *Allqokirus*, one of the two first borhyaenid sparassodonts discovered, though *Mayulestes* included a skull and part of a skeleton so that it is possible to visualize the animal (de Muizon 1994). Rougier et al. (1998) con-

Fig. 3.3 *Mayulestes ferox*. (By Kelsey Van Horn)

sidered *Mayulestes* to be outside the group that we call borhyaenoids and more closely related to *Pucadelphys*, but de Muizon et al. (2018) identify the two taxa as the only sparassadontids in this fauna. *Mayulestes* was almost certainly partially arboreal with a niche similar to that of weasels or martens. The animals probably had a prehensile tail, and the hip joints were extremely mobile, bounding and climbing with ease (Argot 2001, 2002). These sparassodonts were the largest metatherian mammals in the Tiupampan fauna and certainly filled the mammalian predaceous ecological niche (de Muizon et al. 2018) (Fig. 3.3).

Finally the first microbiotheriid (*Khasia*) was described from this fauna. This order became relatively diverse in South America during most of the Cenozoic, and molecular studies of the last existing species (*Dromiciops gliroides*) have shown that the group is more related to Australian marsupials than to South American marsupials and thus is classified with them as Australidelphia. This strongly suggests that the order could have entered Australia via Antarctica to establish the Australian marsupials (Springer et al. 1998). The alternative theory is that the Microbiotheria entered South America from Antarctica and Australia, since the earliest South American microbiotherians were described via a molar and so present the possibility of a misdiagnosis (Beck et al. 2008).

Six genera are known from various Paleogene and Neogene fossil sites in South America. A number of possible microbiotheres, again represented by isolated teeth, have also been recovered from the Middle Eocene La Meseta Formation of Seymour Island, Western Antarctica. Lastly, several undescribed microbiotheres have been reported from the Early Eocene Tingamarra local fauna in Northeastern Australia.

Marsupials reached Australia via Antarctica before 45 Ma when Australia had split off, suggesting a single dispersion event of just one species, related to South America's monito del monte (a microbiothere and the only New World australidelphian) (Nilsson et al. 2004, 2010; Li and Powell 2001). This progenitor may have rafted across the widening but still narrow sea gap between Australia and Antarctic which first began around 52 Ma (Woodburne and Case 1996), but molecular dating suggests 50 Ma as the date of splitting of *Dromiciops* from Australian marsupials. The split between the Microbiotheria (*Dromiciops*) and the Australian orders Peramelemorphia, Dasyuromorphia, and Notoryctemorphia was dated to 50 Ma, so the molecular dating of the divergence coincides with the early separation of Australia from Antarctica (Li and Powell 2001). Was the divergence then vicariance (fragmentation of the environment as by tectonic forces)-related (Nilsson et al. 2010)? In Australia, the marsupials are practically the only mammals present (except for a few austrosphenids like echidnas and platypuses, rodents, and bats), and they radiated into the wide variety we see today, island hopping some way through the Indonesian archipelago to almost complete a circumnavigation back to their homeland in China (Kemp 2005).

3.2.3 Cimolesta (Proteutheria)

Cimolesta **(Proteutheria)** is a paraphyletic (all the descendants of the last common ancestor of the group's members minus a small number of monophyletic groups of descendants, typically just one or two such groups) suborder of primitive eutherians (placental mammals) that comprises a large number of extinct forms, the earliest of which lived during the Late Cretaceous. *Cimolestes* is well-known from North America from the Late Cretaceous-Early Paleocene, so its discovery in South America has been a real eye-opener. The genus has been considered to be ancestor to creodonts (a mammalian order related to carnivores and possibly sharing a common origin, which could be *Cimolestes*) and carnivores so it may be paraphyletic and need to be split into several genera (Gunnell et al. 2007; Rose 2006; McKenna and Bell 1997). *Cimolestes* may be the origin of insectivores and primates as well.

The discovery of *Cimolestes* (prototheurid) in South America demonstrates early competition between placentals and marsupials in the insectivore niche; the marsupials won, since the prototheurids disappear from the Paleocene fossil record.

3.2.4 The "Ungulates"

The first appearance of "ungulates" in South America occurs in the Tiupampan fauna with five genera and seven species (de Muizon et al. 1983; de Muizon and Cifelli 2000). Condylarths appear to be present and are the ancestors of all South American ungulates. The name condylarth was formerly a formal taxonomic order, but now it is used in general to refer to the Late Cretaceous ungulates. Condylarths evolved into the ungulates of both the northern continents and of South America (Kemp 2005). Seven species seem to be present in this fauna from the family Mioclaenidae, a family also known from North America, though some believe that the Mioclaenidae of South America are actually another group, distinct from the Mioclaenidae of the north (de Muizon et al. 1983).

Many define 15–30 genera of condylarths from both North and South America. Studies clearly show that the condylarth *Kollpaniinae* are closely related to the South American endemic Didolodontidae and the Litopterna, so that some have named this clade Panameriungulata. The lack of dental characteristics with other groups of South American ungulates suggests that the origin of the other ungulates or Meridiungulata (Xenungulata, Notoungulata, Astrapotheria, Pyrotheria) is paraphyletic (Rose 2006; Cifelli 1993; McKenna and Bell 1997; Prothero et al. 1988).

Besides being the first ungulates for South America, the first purely endemic South American ungulates are also found in Tiupampa, including the first to be included in the endemic South American order Notoungulata, which is the most diverse and successful of the South American ungulate group with 14 families and 150 genera. The family Oldfieldthomasidae is considered by some to be the first notoungulate, although others place this fossil with the Ornithorhynchidae (Rose 2006).

3.2.5 Pantodont

Finally, evidence of a pantodont (Order Pantodonta …Family Alcidedorbignyidae *Alcidedorbignya*) has appeared in this fauna. These early placental mammals were very common in North America and Asia, and this fossil is the only evidence of its presence in South America. This is also the oldest and one of the most (not the most) primitive pantodont known. It was a very common animal at this site, judging by the many fossils recovered. Several upper and lower jaws of different ages have been found so that the teeth have been well-studied. Pantodonts are the most ancient herbivores known, and some forms became quite large on the northern continents, and they lasted into the Eocene there. However *Alcidedorbignya* was a small, primitive form. The group seems to have died out in South America during the Paleocene, since none have been found after *Alcidedorbignya* (de Muizon and Marshall 1992; Rose 2006; de Muizon et al. 2015) (Fig. 3.4).

This Tiupampan mammalian association is the only one found that has both northern groups (Pantodont, *Cimolestes*, condylarths) and southern groups (notoungulates, henricosborniids, and perutheriids) suggesting some biogeographic connection with North America must have existed. The ungulates and some marsupials also had teeth that were adapted to forest vegetation, suggesting a moist and semitropical climate (Pascual and Ortiz-Jaureguizar 1991b; Flynn and Wyss 1998), that at the time was lowland tropical forest.

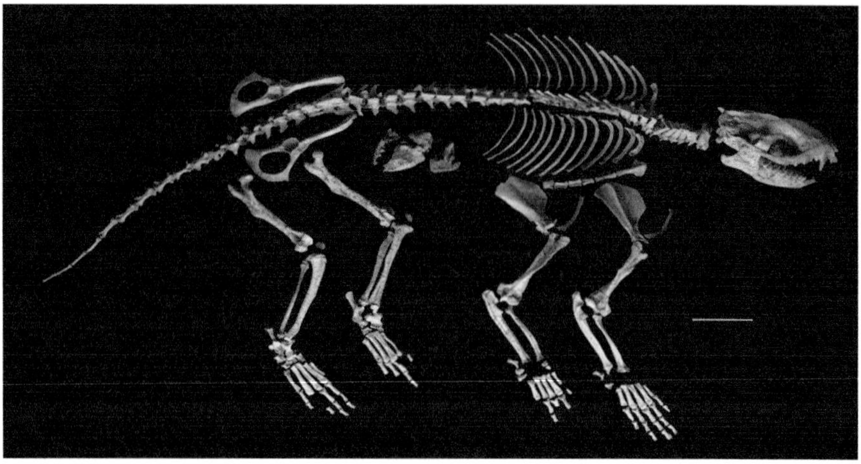

Fig. 3.4 *Alcidedorbignya* is the only pantodont that has been found in South America, and it is considered to be the most primitive known pantodont. (Permission by C. de Muizon)

3.3 Punta Peligro Local Fauna (63.2 to 63.8–59 Ma), Argentina

Punta Peligro local fauna (63.2 to 63.8–59 Ma) Chubut Province, Argentina, has been suggested to be part of the Tiupampan SALMA (Bond et al. 1995; Woodburne et al. 2014). This faunal association from southern Chubut Province, Argentina, with an age (62.5–59 Ma) (similar but slightly later than the Tiupampa) is made up of a very intriguing mix of elements suggesting a history quite different from the north. This fauna is a comingling of Mesozoic or Gondwana taxa with northern therian taxa, and this probably represents an earlier isolation of this southern region from the rest of South America (Forasiepi and Rougier 2009) (Table 3.3).

3.3.1 South American Ornithorhynchidae and Gondwanatheres

Most surprising is perhaps the astonishing evidence of a South American platypus *Monotrematum sudamericanum*, whose two molars were described recently by Pascal and associates, and is described in Chap. 2 (Pascual and Ortiz-Jaureguizar 1991a). This is, of course, clear confirmation of a previous connection with Australia. Also the gondwanatherian *Sudamerica ameghinoi* (family Sudamericidae) is now recognized as the youngest example of this generally Gondwanan Cretaceous group. Suggestions have been made of their connections to the Xenarthra and to the extinct

Table 3.3 Punta Peligro local fauna (63.2 or 63.8–59 Ma)

Clade Australosphenida	
Order Monotremata	
Family Ornithorhynchidae	*Monotrematum sudamericanum*
Subclass Theria	
Order Dryolestida	
Family Peligrotheriidae	*Peligrotherium tropicalis*
Multituberculata? Xenarthra?	
Suborder Gondwanatheria	
Family Sudamericidae	*Sudamerica ameghinoi*
Cohort Marsupialia	
Order Didelphimorphia	
Family Didelphidae	aff. *Derorhynchus*
Infraclass Placentalia	
Order Condylarthra	
Family Didolodontidae	*Escribania chubutensis* (=*Raulvaccia peligrensis*)
Cohort Meridiungulata	
Order Litopterna	
Family Notopteridae	*Requisia vidmari*

Woodburne et al. (2014)

Multituberculata, but their phylogenetic position is still unclear. *Peligrotherium tropicalis* has been identified as an ancient dryolestid, a subgroup of the eupantotheres, primitive mammals that may have been ancestors of modern mammals and that were quite common in South America during the latter part of the Cretaceous, though they had become extinct on the northern land masses (Laurasia) (see Chap. 2) (Gelfo and Pascual 2001; Bonaparte et al. 1993).

These two mammals show continued ancient and southern (Gondwanan) connections with other southern continents, since the fact that at this time South America was still connected to Antarctica and through Antarctica to Australia is reflected in this early fauna of Patagonia. Strong southern Gondwana floral connections during the Late Cretaceous and Early Paleogene are evident through many floral elements shared by southern South America and the Australasian region (Iglesias et al. 2011), distinct from floras in the northern part of South America (e.g., Crisci et al. 1991; Moreira-Muñoz 2007).

But the beginning of influences from the north is also seen in the didelphoid marsupial and two ungulates present (a condylarth and a litoptern). The northern influences strongly dominate southern Argentina a few million years later in the Cañadon Hondo SALMA and fauna of 58–54 Ma years which overlaps that of the Itaboraí SALMA.

3.4 The Itaboraí Local Fauna, Brazil

The third oldest mammalian fauna from Brazil, previously dated for the Late Paleocene, has now been more precisely dated for the Early Eocene at about 53–50 Ma, the Itaboraí local fauna (Woodburne et al. 2014). This fauna has an even more diverse marsupial fauna; it is the most diverse known from either North or South America with 22 species from five orders! This rich diversity suggests that the ecology of this fauna was probably more forest and humid than the Tiupampan community (Bergqvist et al. 2000; Woodburne et al. 2014) (Table 3.4).

Placental mammals are richly represented in the Itaboraí in the form of ungulates – they are the most abundant fossils found in the Itaboraí basin. Didolodontid and sparnotheriodontid condylarths have three species, while four orders of the Meridiungulata are present (Litopterna, Notoungulata, and the first appearances of the Astrapotheria and Xenungulata) (Bergqvist 2008).

Table 3.4 Itaboraí local fauna (Oliveira and Goin 2011) Itaboraí near Rio de Janeiro, Brazil
Age: Mixture of Itaboraian and Riochican faunas (59–57 Ma)

Infraclass Marsupialia	
Cohort "Ameridelphia"	
Family Sternbergiidae	*Carolopaulacoutoia itaboraiensis*
	Itaboraidelphys camposi
	Didelphopsis cabrerai
	Didelphopsis sp.
Family Derorhynchidae	*Derorhynchus singularis*
Family Protodidelphidae	*Carolopaulacoutoia itaboraiensis*
	Guggenheimia brasiliensis
	Guggenheimia crocheti sp. nov.
	Protodidelphis vanzolinii
	Protodidelphis mastodontoides
	Periprotodidephis bergqvistae
	Zeusdelphys complicatus
Family indeterminate	*Eobrasilia coutoi*
	Marmosopsis juradoi
	Marmosopsis sp.
	Gaylordia macrocynodonta
	Gaylordia sp.
	Minusculodelphys minimus
	Minusculodelphis sp.
Order Didelphimorphia	
Superfamily	Peradectoidea
Family Peradectidae	*Epidolops ameghinoi* (= *Epidolops gracilis*)
Family Caroloameghiniidae FF	*Procaroloameghinia pricei* cf. *Nemolestes*

(continued)

Table 3.4 (continued)

Order Paucituberculata	
Family indeterminate	*Riolestes capricornicus*
Order Sparassodonta	
Family Hathliacynidae	*Patene simpsoni* (= *Ischyrodidelphis castellanosi*)
	Palaeocladosictis mosei
Family Borhyaenidae	cf. *Nemolestes* sp.
Order Microbiotheria	
Family Microbiotheriidae	*Mirandatherium alipioi* (= *Mirandaia ribeiroi*)
Order Microbiotheria?	
Family Microbiotheriidae?	*Monodelphopsis travassosi*
Cohort Placentalia	
Order Condylarthra	
Family Didolodontidae	*Paulacoutoia protocenica* (= *Ernestokokenia protocenica*)
	Lamegoia conodonta
Order Condylarthra?	
Family Sparnotheriodontidae	*Victorlemoinea prototípica*
Order Litopterna	
Family Protolipternidae	*Protolipterna ellipsodontoides*
	Miguelsoria parayirunhor (= *Ernestokokenia parayirunhor*)
	Asmithwoodwardia scotti
Family Anisolabididae	*Paranisolambda prodromus* (= *Anisolambda prodromus*)
Order Notoungulata	
Family Henricosborniidae	*Camargomendesia prístina*
Family Oldfieldthomasiidae	*Colbertia magellanica* (= *Henricosbornia magellanica*)
	Itaboraitherium atavum
Order Astrapotheria	
Family Trigonostylopidae	*Tetragonostylops apthomasi* (= *Trigonostylops apthomasi*)
Order Xenungulata	
Family Carodniidae	*Carodnia vieirai*
Order Xenarthra	
Family Dasypodidae	*Riostegotherium yanei* (= *Prostegotherium* aff. *P. astrifer* of)

References: Simpson (1967), Cifelli (1983), Marshall et al. (1983), Marshall and Muizon (1988), Muizon and Brito (1993), McKenna and Bell (1997), Goin et al. (1998), Oliveira and Goin (2011), Bergqvist et al. (2004), Billet and de Muizon (2013)

The first Astrapotheria known, *Tetragonostylops apthomasi* (= *Trigonostylops apthomasi*), defines a bizarre group of large rhinoceros-like ungulates with large, tusklike canines and proboscises. They have been found in the Paleocene to the Miocene with 34 species and 16 genera (Kramarz and Bond 2013). Their sizes ranged from about 500 kg to 2.5 or 3.5 tons. Because they had slender legs, they are thought to have been semiaquatic, a conclusion that seems to me to be curious, since hippopotamus have thick legs. The astrapotheres have been suggested as being a sister group of the notoungulates (Billet 2010).

Fig. 3.5 *Carodnia vieirai.* (By Roman Uchytel)

The Xenungulata, *Carodnia vieirai*, was the largest mammal found in southern South American Paleocene, and it is known from various places (including Argentina) and with an almost complete skeleton. This herbivorous ungulate measured about 2.2 m long and 1 m high and weighed about 170 kg and was far larger than other Paleocene mammals. Three species have been described, although there is disagreement. Some dental characteristics appear to link this order to the Pyrotheria, but there is no agreement (Cifelli 1993) (Fig. 3.5).

Lastly in this fauna, the first xenarthran, apparently an armadillo, appears in the form of osteoderms, the part most often preserved from these ancient animals. Osteoderms are bony scales that form in the skin of many xenarthrans and were often the only evidence left of an ancient animal like glyptodonts, armadillos, and ground sloths. They were also found in some dinosaurs and are known in crocodilians and some reptiles.

Table 3.5 Cañadón Hondo **Location: Cañadón Hondo, Provincia Chubut, Argentina, http://www.paleocene-mammals.de/pal-sa.htm; Age: Itaboraian (Bond et al. 1995) ("*Kibenikhoria* faunal zone" of Simpson (1935a, 1935b); (54–58 Ma)**

Cohort Marsupialia	
Order Polydolopimorphia	
Family Polydolopidae	*Polydolops kamektsen*
Order Sparassodonta	
Family Hathliacynidae inc. sed	*Patene* sp.
Order inc. sed.	
Family Gashterniidae	*Gashternia ctalehor*
Cohort Placentalia	
Order Condylarthra	
Family Didolodontidae	*Ernestokokenia yirunhor*
Order Condylarthra?	
Family Sparnotheriodontidae	*Victorlemoinea* sp.
Order Litopterna?	
Family inc. sed.	gen. et sp. Indet.
Order Notoungulata	
Family Henricosborniidae	*Henricosbornia waitehor*
	Peripantostylops? orehor
Family Notostylopidae?	*Seudenius cteronc*
Family Isotemnidae	*Isotemnus*? sp.
Family Interatheriidae	*Transpithecus*? sp.
Family Oldfieldthomasiidae	*Kibenikhoria get*
Order Astrapotheria	
Family Trigonostylopidae	*Shecenia ctirneru*

3.5 Cañadon Hondo, Argentina

Cañadon Hondo, from the same province as Punta Peligro but from at least one million years later, has yielded at least three species of marsupials and nine species of ungulates, but no representatives of ancient Gondwanan orders, illustrating the new dominance of modern northern mammals. The new mammals became dominant throughout the continent, only sharing their domination with later caviomorph rodents and platyrrhine primates as they made their way to South America from Africa (Table 3.5).

References

Arenillas I, Arz JA, Grajales-Nishimura JM, Murillo-Muñetón G, Alvarez W, Camargo-Zanoguera A, Molina E, Rosales-Domínguez C (2006) Chicxulub impact event is Cretaceous/Paleogene boundary in age: new micropaleontological evidence. Earth Planet Sci Lett 249:241–257

Argot C (2001) Functional-adaptive anatomy of the forelimb in the Didelphidae, and the paleobiology of the Paleocene marsupials *Mayulestes ferox* and *Pucadelphys andinus*. J Morphol 247:51–79

Argot C (2002) Functional-adaptive analysis of the hind limb anatomy of extant marsupials and the paleobiology of the Paleocene marsupials *Mayulestes ferox* and *Pucadelphys andinus*. J Morphol 253:76–108

Argot C (2003a) Functional adaptations of the postcranial skeleton of two Miocene borhyaenoids (Mammalia, Metatheria), Borhyaena and Prothylacinus, from South America. Palaeontology 46(6):1213–1267

Argot C (2003b) Functional-adaptive anatomy of the axial skeleton of some extant marsupials and the Paleobiology of the Paleocene marsupials Mayulestes ferox and Pucadelphys andinus. J Morphol 255:279–300

Argot C (2003c) Postcranial functional adaptations in the South American Miocene borhyaenoids (Mammalia, Metatheria): Cladosictis, Pseudonotictis and Sipalocyon. Alcheringa 27(4):303–356. https://doi.org/10.1080/03115510308619110

Asher RJ, Horovitz I, Sánchez-Villagra MR (2004) First combined cladistics analysis of marsupial mammal interrelationships. Mol Phylogenet Evol 33:240–250

Beck RMD, Godthelp H, Weisbecker V, Archer M, Hand SJ (2008) Australia's oldest marsupial fossils and their biogeographical implications. PLoS One 3(3):e1858. https://doi.org/10.1371/journal.pone.0001858

Bergqvist LP (2008) Postcranial skeleton of the upper Paleocene (Itaboraian) "Condylarthra" (Mammalia) of Itaboraí basin, Brazil. In: Sargis EJ, Dagosto M (eds) Mammalian evolutionary morphology: a tribute to Frederick S. Szalay. Springer, New York, pp 107–133

Bergqvist LP, Mansur K, Rodriguez MA, Rodriguez-Francisco BH, Perez R, Beltrãu MdaC (2000) Itaboraí basin, state of Rio de Janeiro. Sigep.cprm.com.br 123:1–19

Bergqvist LP, Abrantes EA, Dos Santos AL (2004) The Xenarthra (Mammalia) of Sao José de Itaboraian basin (upper Paleocene, Itaboraian), Rio de Janeiro, Brazil. Geodiversitas 26(2):323–337

Billet G (2010) New observations on the skull of Pyrotherium(Pyrotheria, Mammalia) and new phylogenetic hypotheses on South American ungulates. J Mamm Evol 17(1):21–59

Billet G, de Muizon C (2013) External and internal anatomy of a petrosal from the late Paleocene of Itaboraí, Brazil, referred to Notoungulata (Placentalia). J Vertebr Paleontol 33(2):455–469

Bonaparte JF, Van Valen L, Kramarz A (1993) La Fauna Local de Punta Peligro, Paleoceno inferior, de la provincia de Chubut, Patagonia, Argentina. Evol Monogr 14:1–61

Bond M, Carlini AA, Goin FJ, Legarreta L, Ortiz-Jaureguizar E, Pascual R, Uliana MA (1995) Episodes in South American land mammal evolution and sedimentation: testing their apparent concurrence in a Palaeocene succession from Central Patagonia. VI Congreso Argentino de Paleontología y Bioestratigrafía, Actas, pp 47–58

Case JA, Woodburne MO (1986) South American marsupials: a successful crossing of the Cretaceous-Tertiary boundary. PALAIOS 1:413–416

Cifelli RL (1983) The origin and affinities of the South American Condylarthra and early Tertiary Litopterna (Mammalia). Am Mus Novit 2772:1–49

Cifelli RL (1993) The phylogeny of the naïve South American ungulates. In: Szalay FS, Novacek MJ, McKenna MC (eds) Mammal phylogeny: placentals. Springer, New York, pp 195–216

Clyde WC, Wilf P, Iglesias A, Slingerland RL et al (2014) New age constraints for the Salamanca Formation and lower Río Chico Group in the western San Jorge Basin, Patagonia, Argentina: Implications for cretaceous-Paleogene extinction recovery and land mammal age correlations. GSA Bull 126(3-4):289–306. https://doi.org/10.1130/B30915.1

Cowen R (1999) The K-T extinction. University of Berkeley, California. In: History of life. Blackwell Scientific Publications, Cambridge, MA. http://www.ucmp.berkeley.edu/education/events/cowen3b.html. Accessed 10 Nov 2012

Crisci JV, Cigliano MM, Morrone JJ (1991) A comparative review of cladistic approaches to historical biogeography of southern South America. Aust Syst Bot 4:117–126

de Muizon C (1991) La fauna de mamíferos de Tiupampa (Paleoceno inferior, Formación Santa Lucía), Bolivia. In: Suárez Soruco R (ed) Fósiles y facies de Bolivia, vol 1. Vertebrados. Revista Técnica de Yacimientos Petrolíferos Fiscales Bolivianos 12/3–4:575–624

de Muizon D (1994) A new carnivorous marsupial from the Paleocene of Bolivia and the problem of marsupial monophyly. Nature 370:208–211

de Muizon D (1998) *Mayulestes ferox*, a boryhaenoid (Metatheria, Mammalia) from the early Palaeocene of Bolivia. Phylogenetic and palaeobiologic implications. Geodiversitas 20:19–142

de Muizon C, Cifelli RL (2000) The "condylarths" (archaic Ungulata, Mammalia) from the early Palaeocene of Tiupampa (Bolivia): implications on the origin of the South American ungulates. Geodiversitas 22(1):47–150

de Muizon C, Cifelli RL (2001) A new basal "Didelphoid" (Marsupialia, Mammalia) from the early Paleocene of Tiupampa (Bolivia). J Vertebr Paleontol 21(1):87–97

de Muizon C, Marshall LG (1991) Nouveaux Condylarthres du paléocene inférieur de Tiupampa (Bolivie). Muséum national d'histoire naturelle 13(3/4):201

de Muizon C, Marshall LG (1992) *Alcidedorbignya inopinata* (Mammalia: Pantodonta) from the Early Paleocene of Bolivia: phylogenetic and paleobiogeographic implications. J Paleontol 66(3):499–520

de Muizon C, Gayet M, Lavenu A, Marshall LG, Sigé B, Villarroel C (1983) Late Cretaceous vertebrates, including mammals from Tiupampa, southcentral Bolivia. Geobios 16(6):747–753

de Muizon C, Billet G, Argot C, Ladevéze S, Goussard F (2015) *Alcidedorbignya inopinata*, a basal pantodont (Placentalia, Mammalia) from the early Palaeocene of Bolivia: anatomy, phylogeny and palaeobiology. Geodiversitas 37(4):397–634

de Muizon C, Ladevèze S, Selva C, Vignaud R, Goussard F (2018) *Allqokirus australis* (Sparassodonta, Metatheria) from the early Palaeocene of Tiupampa (Bolivia) and the rise of the metatherian carnivorous radiation in South America. Geodiversitas 40 (in press)

Flynn JJ, Swisher CC (1995) Cenozoic South American land mammal ages: correlation to global geochronologies. In: Berggren WA (ed) Geochronology time scales and global stratigraphic correlation. SEPM Special Publication No. 54, SEPM Society for Sedimentary, Tulsa, pp 317–333

Flynn JJ, Wyss AR (1998) Recent advances in South American mammalian paleontology. Trends Ecol Evol 12:449–454

Forasiepi AM, Rougier G (2009) Additional data on early Paleocene metatherians (Mammalia) from Punta Peligro (Salamanca Formation, Argentina): comments based on petrosal morphology. J Zool Syst Evol Res 47(4):391–398

Gelfo JN, Pascual R (2001) *Peligrotherium tropicalis* (Mammalia, Dryolestida) from the early Paleocene of Patagonia, a survival from a Mesozoic Gondwanan radiation. Geodiversitas 23(3):369–379

Gelfo JN, Goin FJ, Woodburne MO, de Muizon C (2009) Biochronological relationships of the earliest South American Paleogene mammalian faunas. Palaeontology 52(1):251–269

Goin FJ, Oliveira EV, Candela AM (1998) *Carolocoutoia ferigoloi* nov. gen. and sp. (Protodidelphidae), a new Paleocene "opossum-like" marsupial from Brazil. Palaeovertebrata 27(3–4):145–154

Goin FJ, Pascual R, Tejedor MF, Gelfo JN, Woodburne MO, Case JA, Reguro MA, Bond M, López GM, Cione A, Udrizar Sauthier D, Balarino L, Scasso RA, Medina FA, Ubaldón MC (2006) The earliest Tertiary therian mammal from South America. J Vertebr Paleontol 26(2):505–510

Goin FJ, Woodburne MO, Zimicz AN, Martin GM (2016) A brief history of South American metatherians: evolutionary contexts and intercontinental dispersals. Springer, New York, 246 p

Gunnell GF, Bown TM, Bloch JI, Boyer DM (2007) Proteutheria. In: Janis CM, Gunnell GF, Uhen MD (eds) Evolution of Tertiary Mammals of North America, vol 2. Cambridge University Press, Cambridge, pp 63–81

Horovitz I, Sánchez-Villagra MR (2003) A morphological analysis of marsupial mammal higher-level phylogenetic relationships. Cladistics 19:181–212

Iglesias A, Artabe AE, Morel EM (2011) The evolution of Patagonian climate and vegetation from the Mesozoic to the present. Biol J Linn Soc 103:409–422. https://doi.org/10.1111/j.1095-8312.2011.01657.xVillagra

Jehle M (2006) Paleocene mammals of the world. http://www.paleocene-mammals.de/index.htm. Accessed 2014

Keller G (2014) Deccan volcanism, the Chicxulub impact, and the end-Cretaceous mass extinction: coincidence? Cause and effect? In: Keller G, Kerr AC (eds) Volcanism, impacts, and mass

extinctions: causes and effects, GSA special papers, vol 505. Geological Society of America, Boulder, pp 29–55. https://doi.org/10.1130/SPE505
Kemp TS (2005) The origin and evolution of mammals. Oxford University Press, Oxford
Kramarz A, Bond M (2013) On the status of *Isolophodon* Roth, 1903 (Mammalia, Astrapotheria) and other little-known Paleogene astrapotheres from Central Patagonia. Geobios 46:203–211
Krause JM, Clyde WC, Ibañez-Mejía M, Schmitz MD, Barnum T, Bellosi E, Wilf P (2017) New age constraints for early Paleogene strata of Central Patagonia, Argentina: implications for the timing of South American Land Mammal Ages. Geol Soc Am Bull 129(7/8):886–903. https://doi.org/10.1130/B31561.1
Ladevèze S, de Muizon C, Beck MD, Germain D, Cespedes-Paz R (2011) Earliest evidence of mammalian social behavior in the basal Tertiary of Bolivia. Nature 474:83–86
Li ZX, Powell CMA (2001) An outline of the paleogeographic evolution of the Australasian region since the beginning the Neoproterozoic. Earth Sci Rev 53(3–4):237–277
MacLeod N, Rawson PF, Forey PL, Banner FT, Boudagher-Fadel MK, Bown PR, Burnett JA, Chambers P, Culver S, Evans SE, Jeffery C, Kaminski MA, Lord AR, Milner AC, Milner AR, Morris N, Owen E, Rosen BR, Smith AB, Taylor PD, Urquhart E, Young JR (1997) The Cretaceous-Tertiary biotic transition. J Geol Soc Lond 154(2):265–292
Marshall LG (1977) Cladistic analysis of borhyaneoid, dasyuroid, didelphoid, and thylacinid (Marsupialia: Mammalia) affinity. Syst Zool 26:410–425
Marshall LG, Muizon C (1988) The dawn of the age of mammals in South America. Natl Geogr Res 4:23–55
Marshall LG, Hoffstetter R, Pascual R (1983) Mammals and stratigraphy: geochronology of the continental mammal-bearing Tertiary of South America. Palaeovertebrata, Montpellier, Mem, Extr, pp 1–93
Marshall LG, de Muizon C, Sigogneau-Russell D (1995) *Pucadelphys andinus* (Marsupialia, Mammalia) from the early Paleocene of Bolivia. Muséum national d'Histoire naturelle, Paris, 164 p. (Mémoires du Muséum national d'Histoire naturelle, 165)
Marshall LG, Sempere T, Butler RF (1997) Chronostratigraphy of the mammal-bearing Paleocene of South America. J S Am Earth Sci 10(1):49–70
McKenna MC, Bell SK (1997) Classification of mammals above the species level. Columbia University Press, New York, p 211
Meredith RW, Westerman M, Case JA, Springer MS (2008) A phylogeny and timescale for marsupial evolution based on sequences for five nuclear genes. J Mamm Evol 15:1–36
Moreira-Muñoz A (2007) The Austral floristic realm revisited. J Biogeogr 34:1649–1660
Muizon C, Brito I (1993) Le basin calcaire de Sao Jose de Itaborai (Rio de Janeiro, Bresil): les relaciones faunique avec le site de Tiupampa (Cochabamba, Bolivie). Ann Paleontol 79:233–269
Nilsson MA, Arnason U, Spencer PBS, Janke A (2004) Marsupial relationships and a timeline for marsupial radiation in South Gondwana. Gene 340:189–196
Nilsson MA, Churakov G, Sommer M, Tran NV, Zemann A et al (2010) Tracking marsupial evolution using archaic genomic retroposon insertions. PLoS Biol 8(7):e1000436. https://doi.org/10.1371/journal.pbio.1000436
Oliveira EV, Goin FJ (2011) A reassessment of bunodont metatherians from the Paleogene of Itaboraí (Brazil): systematics and age of the Itaboraian Salma. Rev Bras Paleontolog 14(2):105–136
Pascual R, Ortiz-Jaureguizar E (1990) Evolving climates and mammal faunas in Cenozoic South America. J Hum Evol 19:23–60
Pascual R, Ortiz-Jaureguizar E (1991a) Evolutionary pattern of land mammal faunas during the late Cretaceous and Paleocene in South America: a comparison with the North American pattern. Ann Zool Fenn 28(3/4):245–252
Pascual R, Ortiz-Jaureguizar E (1991b) The Gondwanan and South American episodes: two major and unrelated moments in the history of the South American mammals. J Hum Evol 19:23–60
Pope KO, Baines KH, Ocampo AC (1999) Impact winter and the Cretaceous/Tertiary extinctions: results of a Chicxulub asteroid impact model. Earth Planet Sci Lett 128:719–725

Prothero DR, Manning EM, Fischer M (1988) The phylogeny of the ungulates. In: Benton MJ (ed) The phylogeny and classification of the tetrapods. 2: Mammals. Systematics Association Special Volume 35B. Clarendon Press, Oxford, pp 201–234

Pascual R, Ortiz-Jaureguizar E (2007) The Gondwanan and South American episodes: two major and unrelated moments in the history of the South American mammals. J Mammal Evol 14(2):75–137

Renne PR, Deino AL, Hilgen FJ, Kuiper KF, Mark DF, Mitchell WS III, Morgan LE, Mundil R, Smit J (2013) Time scales of critical events around the Cretaceous-Paleogene boundary. Science 339(6120):684–687. https://doi.org/10.1126/science.1230492

Rose KD (2006) The beginning of the age of mammals. The Johns Hopkins University Press, Baltimore

Rougier GW, Wible JR, Novacek MJ (1998) Implications of *Deltatheridium* specimens for early marsupial history. Nature 396:459–463

Schulte P, Alegret L, Arenillas I et al. (2010) The Chicxulub asteroid impact and mass extinction at the Cretaceous-Paleogene boundary. Science 327:1214–1218

Simpson GG (1967) The meaning of evolution: a study of the history of life and its significance for man. Yale University Press, New Haven

Simpson GG (1980) Splendid isolation: the curious history of South American mammals. Yale University Press, New Haven

Simpson GG (1935a) Descriptions of the oldest known South American mammals, from the Río Chico Formation. Am Mus Novitates no. 793, p 25

Simpson GG (1935b) Early and middle Tertiary geology of the Gaiman region, Chubut, Argentina. American Museum novitates no. 775, p 30

Springer MS, Westerman M, Kavanagh JR, Burk A, Woodburne MO, Kao DJ, Krajewski C (1998) The origin of the Australasian marsupial fauna and the phylogenetic affinities of the enigmatic monito del monte and the marsupial mole. Proc R Soc Lond Ser B 265:2381–2386

Taylor KWR, Hollis CJ, Pancost RD (2011) Reconstructing post Cretaceous/Paleogene boundary climate and ecology at mid-Waipara River and Branch Stream, New Zealand. Berichte Geol B-A 85 (ISSN 1017-8880) – CBEP 2011, Salzburg, 5–8 June

Woodburne MO, Case JA (1996) Dispersal, vicariance, and the late Cretaceous to early Tertiary land mammal biogeography from South America to Australia. J Mammal Evol 3(2):121–161

Woodburne MO, Goin FJ, Raigemborn MS et al (2014) Revised timing of the South American early Paleogene land mammal ages. J S Am Earth Sci 54:109–119

Chapter 4
Marsupials and Other Metatheres of South America

4.1 Introduction

For many years marsupials had been regarded as the probable ancestors of the placental mammals; this is mostly based on our ignorance of the fossil record. Recent finds make it clear that marsupials and placental animals are actually parallel lines of evolution reaching more than 125 million years into the past and perhaps as ancient as 160–190 Ma (Kumar and Hedges 1998; Penny et al. 1999; Cifelli and Davis 2003; Woodburne et al. 2003; Luo et al. 2011).

The undoubted Cretaceous diversity of fossil marsupials in North America made it seem probable that the group had originated there, so it was with great surprise that the 125 million-year-old fossil *Sinodelphys szalayi* was found in China in 2003 (Luo et al. 2003). This beautiful and complete skeleton has many marsupial characteristics (apomorphies), though its tooth formula and lack of inflected lower mandible angle cause it to fall outside the marsupial group, to be more correctly seen as a basal metatherian (marsupials and relatives) (Luo et al. 2003). Nevertheless, this and other Asiatic finds made it clear that North American marsupials arrived from Asia (Fig. 4.1).

The discovery of an undoubted 160-year-old eutherian *Juramaia sinensis* from the Jurassic of China makes it clear that the eutherian-metatherian divergence had not yet occurred (Luo et al. 2011). The age of this fossil provides a minimum divergence date in agreement with recent molecular data (Phillips et al. 2009) (Maríe López, grupo de Evolución y Ecología de Mamíferos Neotropicales, Universidad Nacional de Colombia).

T. Defler, *History of Terrestrial Mammals in South America*, Topics in Geobiology 42, https://doi.org/10.1007/978-3-319-98449-0_4

Fig. 4.1 *Sinodelphys szalayi*, a metathere belonging to the group that includes both the marsupials and their closest living relatives and includes Deltatheroidea and Asiadelphia (based on skeleton) (By M. Joelle Giraud, Grupo de Evolución y Ecología de Mamíferos Neotropicales, Universidad Nacional de Colombia)

4.2 Marsupial, Deltatheroida, Asiadelphia

The Cretaceous Deltatheroidea of Asia and North America seem to be closer to the "true" marsupials, although they are without tribosphenic teeth and probably are best considered a sister group of the marsupials. Some of these include *Atokatheridium, Deltatheridium, Deltatheroides, Oxlestes*, and *Sulestes* (Rougier et al. 1998).

So how do we recognize marsupials from other mammals, since soft tissues such as the marsupium (the marsupial pouch) are not preserved in fossils? With fossils we need to reference characters like the *epipubic* (marsupial) *bone* (a pair of bones projecting forward from the pelvic bones of modern marsupials and of some fossil mammals: multituberculates, monotremes, and even some eutherians, epipubic bone, Wikipedia), an *auditory bulla* composed mostly of the alisphenoid (the bones forming the wings of the *sphenoid*, which is a compound bone at the base of the cranium), large openings in the palate, an inflected angular process of the dentary, more upper than lower incisors and tribosphenic teeth without a hypocone but with a wide *stylar shelf* with multiple cusps, three simple premolars and four molars with the lower molars often with a unreduced paraconid and twinned hypoconulid and entoconid (Wikipedia). *Sinodelphys* did not have an epipubic bone nor an inflected mandibular process, and the dental formula is distinct from early marsupials (Luo et al. 2003, 2011). So can we truly call it a marsupial? These definitions are human inventions, and evidence shows *Sinodelphys* to be closer to marsupials than to early placental mammals, so some do call *Sinodelphys* the earliest marsupial, but more correctly, I suppose we can call it an early metatherian, including marsupials and their relatives.

The Late Cretaceous (88 million years) *Asiatherium* (order Asiadelphia) from Mongolia is probably the oldest marsupial known that is somewhat less controversial as a marsupial, although some still place it into a separate clade (group of organisms classified together as descendants of a common ancestor). The jaw of this fossil has three premolars and four molars with a wide stylar shelf, one or more stylar cusps, and no hypocone, placing in squarely into the marsupial camp, though there are some dissenters (Trofimov and Szalay 1994; Rose 2006) (Fig. 4.2).

Probably the North American *Kokopellia*, *Lugomortiferum*, and *Anchistodelphys* can be placed among the most primitive North American marsupials. *Kokopellia* from the Late Cretaceous (98 million years) had a dental formula like marsupials with three premolars and four molars, but it lacks typical marsupial *conulids* (i.e., small accessory cusps on a postcanine tooth) of the lower mandible, so the placement of the genus is not neat. The younger *Anchistodelphys* and *Lugomortiferum* of Campanian times (83.5–70.6 million years) were also very similar to other marsupials, although they lacked slightly in the development of the stylar cusps (de Muizon et al. 1997).

Although there are several classificatory schemes for South American marsupials, I use McKenna and Bell (1997) here, which is supported by Archer and Kirsch (2006). The first marsupials probably entered South America sometime in the Upper Cretaceous at around 75–65 Ma, judging by their Early Paleocene diversity in South

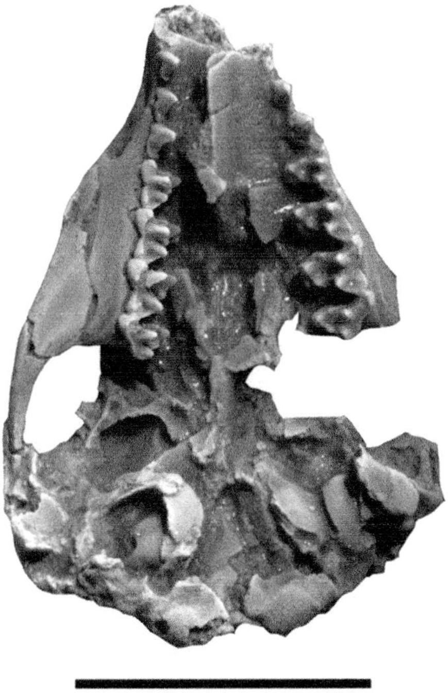

Fig. 4.2 *Asiatherium* from Mongolia represents the oldest fossil marsupial without controversy. (From Szalay and Trofimov 1996)

America (Case and Woodburne 1986; Iturralde-Vinent 2006). There probably existed a Central American volcanic island arc system connecting Central and South America (Iturralde-Vinent and MacPhee 1999) that consisted of volcanic and non-volcanic islands surrounded by shallow water and that during uplifts and sea level lowering would have provided a series of stepping stones and *sweepstake dispersal* mechanisms (Iturralde-Vinent 2006), and that explains tetrapods that invaded South America from the north at this time. Times of greater connectivity could have been geologically short or ephemeral, but they did allow time for an invasion of marsupials, metatheres, condylarths, and other fauna into South America. An abrupt end would have come to this shallow connectivity with the impact of Chicxulub 65.6 Ma ago and the ensuing destruction of ecosystems that initiated the Paleogene.

4.2.1 Order Didelphimorphia

The Didelphimorphia are generally accepted as the most basal marsupial lineage for South America. When they arrived in South America, the Paucituberculata (shrew opossums) evolved from them, probably before the end of the Cretaceous around 69 Ma, although a better estimate for arrival in South America of marsupials would be about 75 Ma, according to one mitochondrial study (Nilsson et al. 2004) or very early in the Paleocene, according to another study by the same group (Nilsson et al. 2003). According to many, the North American *Alphadon* (order Didelphimorphia) is the source of all Early Paleocene marsupials and as such may have given rise to the Peradectidae, a group shared by North American, Europe, and South America and perhaps gave rise to all other Cenozoic marsupials. Fossils labeled *Alphadon* from the Cretaceous have variously been reassigned to other genera, though the forms are very similar dentally with didelphoid-like molars. These animals were generally small, reaching 30 cm with one species, and they are known mostly from their teeth (Fig. 4.3).

Fig. 4.3 *Alphadon* of North America (by M. Joelle Giraud, grupo de Evolución y Ecología de Mamíferos Neotropicales, Universidad Nacional de Colombia)

Actually it is not settled that all South American marsupials spring from an *Alphadon*-like ancestor, though mouse-sized *Peradectes* have been found in the Early Paleocene not only in South America but also in North America and Europe. Nevertheless, it seems fairly clear that South American marsupials are descendants of North American marsupials, even though the affinities of the earliest South American marsupials from Tiupampa are closer to the later Paleocene metatherians from Itaboraí, thus illustrating considerable early endemic diversification in South America (de Muizon and Cifelli 2001). The North Americans were quite diverse in the latter part of the Cretaceous, but after the end of the Cretaceous, that diversity disappeared, and the only fossil marsupial known there after the disaster of the meteorite Chicxulub was *Peradectes*, which managed to last in North American as well as Europe until the end of the Eocene as registered by fossils in Germany (Fig. 4.4).

In South America the first and earliest mammalian community known to date was found in Bolivia and named the Tiupampan South American Land Mammal Age, dated at about 64.5–62.5 million years ago. The fauna is basically made up of metatherians and ungulates, so that around 12 species have been recognized, including the problematic borhyaenid group, which seems to be more metatherian than true marsupials, but even the most conservative estimates identify at least 5 families of true marsupials. These early marsupials have been distinguished using relatively minor dental differences. All were probably didelphoid.

The Didelphimorphia is the most diverse, although *paraphyletic* (Ladevèze and Muizon 2010), especially during the Paleocene with at least 12 genera recognized from 3 main sites and perhaps as many as 16 or 17 genera from Itaboraí (Early Eocene), Brazil, Tiupampa (Early Paleocene), and Laguna Umayo, Peru (Late Paleocene).

Fig. 4.4 *Peradectes* sp. has been found in South America, North America, and Europe (by M. Joelle Giraud, grupo de Evolución y Ecología de Mamíferos Neotropicales, Universidad Nacional de Colombia).

Fig. 4.5 *Pucadelphys andinus* based on complete fossil. (By Kelsey Van Horn)

The Tiupampan marsupial *Pucadelphys* is one of the best-known species with complete skulls, skeletons, and a long, nonprehensile tail known from the fossils. As the various skeletons where found in fossil burrows, it apparently was fossorial and terrestrial, though it could probably climb as well (Marshall et al. 1995). The specimens were found in 2 groups, one group of 23 and another of 12 with multiple males, females, and young, so they must have been social. Judging by the many frogs found among the mammal fossils, it is interpreted that these marsupials lived in burrows close to a river that disastrously flooded, sealing the animals in their burrows (Muizon 1998; Argot 2001; Muizon and Argot 2003) (Fig. 4.5).

4.2.2 Order Paucituberculata (Polydolopimorphia)

The Paucituberculata now includes the living rat opossums *Caenolestes* (Caenolestidae), previously included the Carloameghiniidae as well as other families like Argyrolagidae and Polydolopidae (Goin et al. 2009; Chornogubsky et al. 2009; Chornogubsky 2010). Sometimes the peradectids have been classified in the same order as the Caroloameghiniidae with the polydolopoids as a separate order or the order Sudameridelphia has been created to receive the polydolopoids and the borhyaenoids (Marshall 1982a; Archer and Kirsch 2006). But the relationships have not yet been worked out. The Caenolestidae *Carlopaulacoutoia* is the oldest genus known for this family from the Itaboraí Early Eocene fauna of Brazil (Goin et al. 2009). In the Late Eocene, the strange *Groeberia* appeared with long, cutting incisors and now classified in the family Groeberiidae (Patterson 1952; McKenna and Bell 1997), though some would justify a new order Groeberida for the combination of unique *autapomorphies* (a distinctive anatomical feature, known as a derived trait, that is unique to a given terminal group) that have been defined, such as powerful gnawing teeth and enlarged, curved incisors (two pairs in the upper mandible and one pair on the lower mandible) (Pascual et al. 1994; Simpson 1984).

A most spectacular find has been the Caroloameghiniidae *Chulpasia*, found both in Eocene Argentina and Australia in the Murgon fossil site of the *Tingamarra* fauna located in southeastern Queensland. This illustrates the marsupial connection between the two continents by way of the Antarctic and may even represent a pan-South American-Antarctica-Australian fauna (Sigé et al. 2009), although this identification of *Chulpasia* in Australia has been challenged (Woodburne et al. 2014; Goin et al. 2016).

Chulpasia mattaueri and *Chulpasia tingamarra* were two species of the same genus found in the Early Peruvian Caroloameghiniidae of the Paucituberculata also found in Queensland, Australia, also in the Early Eocene. The animal was a small omnivore, related to the shrew opossums (Sigé et al. 2009) (from Australia's Lost Kingdoms, http://www.admc.hct.ac.ae/tjohnson/1_ASTEROIDS/DINOSAURS/ASTEROIDS/www.lostkingdoms.com/snapshots/eocene_animals.htm, Australian Museum, 2000).

The most primitive members of the Paucituberculata are known to be the Carolameghiniidae (caroloameghinioids), and the Tiupampan *Roberthoffstetteria* is considered to be the oldest genus known for the group. These marsupials are thought to have evolved from didelphoids by some, although others find a relation more closely associated with the Australidelphia (Horovitz and Sánchez-Villagra 2003). The Paucituberculata contains four families of derived marsupials from the Paleocene and Eocene of South America, although two genera are known from Antarctica.

4.2.3 *Caroloameghiniidae*

Whether the Caroloameghiniidae are didelphimorphs or are more correctly seen as paucituberculates, their evolution is interesting inasmuch as they exhibit characters that are convergent with primates: "strong dentaries, brachydont cheek teeth, and bunodont molars with wide trigon and talonid basins," while some features are unique among metatheres: "proportionately enormous protocone and additional crests between stylar cusps in the upper molars, etc." (Goin 2006). These animals were probably frugivorous like many early primates, and their evolution tracked the extreme planet warming and expansion of tropical and subtropical rainforests of Late Paleocene–Early to Middle Eocene, South America, while their extinction corresponds with the global cooling of the Late Eocene-Oligocene and perhaps to the arrival and competition of the early platyrrhine primates (Woodburne et al. 2014). Since we now know that the primates arrived in South America at least as Early as the Late Eocene, the possibility of this evolutionary interaction and probable competition fascinates (Bond et al. 2015). The primates won.

4.2.4 Superfamily Argyrolagoidea

This superfamily was divided into three families: Argyrolagidae, Patagoniidae, and Groeberiidae, and the argyrolagids were considered perhaps the strangest of the three (von Koenigswald and Pascual 1990; Flynn and Wyss 1999; Goin and Abello 2013). But lately the Patagoniidae and the Groeberiidae have been associated with the Gondwanatheres and not with the marsupials (Chimento et al. 2015).

The Argyrolagidae seem to exhibit adaptations to deserts and are very convergent with some placental rodents such as kangaroo rats and jerboas. They have only two digits on the feet and greatly elongated legs, showing extreme convergence to placental rodents such as the Heteromyidae from North America and the Dipodidae from the Old World (Simpson 1970; Villarroel and Marshall 1988). They probably moved about via bipedal jumps, balanced by a long tail. Additionally the skull has very large orbits, suggesting night activity, and they are placed very far back on the skull, leaving almost no room for jaw muscles and leaving a long, bony tube of a nose (Simpson 1980). Four genera and eight species are known from the Late Oligocene of 27 Ma to Pleistocene times about 1.5 Ma (Villarroel and Marshall 1988). Marsupial radiations seem to have reached levels of adaptation equal to those in Australia (Simpson 1970, 1980) (Fig. 4.6).

4.2.5 Order Microbiotheria

The singular order Microbiotheria first appears in the Tiupampan fauna (64.5–62.5 Ma), singular because of the order's systematic connection with the Australidelphia or Australian marsupials, rather than with the Ameridelphia or South American marsupials (Marshall 1982b; Asher et al. 2004; Horovitz and

Fig. 4.6 *Argyrolagus* possessed a long, tubelike nose, large orbits, and a long tail. (By M. Joelle Giraud, grupo de Evolución y Ecología de Mamíferos Neotropicales, Universidad Nacional de Colombia)

Fig. 4.7 *Dromiciops australis*. (By José Luis Bartheld), the only South American representative of the Australidelphia, which includes all Australian marsupials

Sánchez-Villagra 2003). The evidence strongly suggests that the Australian marsupials evolved from the South American Microbiotheria and that divergence was about 46–43 Ma (Beck 2012; Nilsson et al. 2003, 2004). The contrary view of Microbiotheria having evolved from an Australian form has been also suggested by some (Kirsch et al. 1991).

The Microbiotheria were fairly diverse and found in most South American faunas throughout the Cenozoic. They were also found on Seymour Island, Antarctica, evidence of the South American-Antarctic link (Goin et al. 2007). The first known species is from the Early Paleocene Tiupampan fauna, *Khasia cordillerensis*. The group probably evolved from a Didelphimorphia ancestor very early in the Paleocene, since they are unknown in North America. They are now represented by only one living species in South America, *Dromiciops australis* from the Cordillera of Chile (Woodburne and Case 1996) (Fig. 4.7).

4.3 Order Sparassodonta or Borhyaenoidea

The Sparassodonta (Borhyaenidae) were a clade of carnivorously adapted metatherians that evolved (Paleocene–Pliocene) many wondrous forms, some very large, that must have preyed on the many South American ungulates. These mammals maintained the position of top predators throughout the Cenozoic, although the predaceous terror birds, the Phorusrhacidae, were their competitors. Because many species were convergently evolved to appear like some placental carnivores such as dogs, Florentino Ameghino believed that some were the ancestors of the canids, this to the detriment of his reputation (Simpson 1980, 1984; Marshall 1978). Many sparassodont characteristics distinguish the group from other mammals and establish this unique group. The appearance of endemic groups like the borhyaenoids early in the Paleogene suggests a very early, probably Cretaceous appearance in South America.

Sparassodonts (borhyaenoids) were previously considered to be marsupials but are now generally conceded to be a sister group to the marsupials, thus metatherians (Sinclair 1905). Metatherian synapomorphies (i.e., a character or trait that is shared by two or more taxonomic groups and is derived through evolution from a common ancestral form, Merriam-Webster) are defined as an absence of the tympanic process of the alisphenoid (bony process of the sphenoid bone), absence of the epipubic bones (pair of bones projecting forward from the pelvic bones of modern marsupials and of some fossil mammals, multituberculates, monotremes, and even basal eutherians, the ancestors of placental mammals), and other characteristics (de Muizon 1998, 1999). Of course, some of these "bones," such as the epipubic bones, were probably cartilaginous and so would not have appeared as fossils.

Borhyaenoids have also been compared to the Australian thylacinids and were suggested as being their ancestors (Archer 1976), but phylogenetic analyses show unequivocally that they are more closely related to South American didelphimorphs (Argot 2004a; de Muizon et al. 1997: Marshall 1977b), which were most probably their ancestors. There are about six families known: Mayulestidae, Hathliacynidae, Borhyaenidae, Proborhyaenidae, Prothylacinidae, and Thylacosmilidae (Argot 2004a; Naish 2012).

4.3.1 Mayulestidae

The most ancient member known *Mayulestes ferox* (along with the lesser known *Allqokirus*) is classified in the Mayulestidae in the Early Paleocene at Tupinamba. *Mayulestes* was a small, hedgehog-sized animal weighing less than 1 kg, whose skeleton is known (de Muizon 1994, 1998) (Fig. 4.8).

Mayulestes is the oldest borhyaenoid known and has been characterized as a partially arboreal predaceous mammal capable of jumping and fast but short runs, necessary for prey capture. *Mayulestes* was agile and perhaps had an ecological niche close to that of weasels or martens, although more arboreal than the former (Muizon and Argot 2003; Argot 2001, 2004a). Several arboreal features of *Mayulestes* are also found in *Pucadelphys*, a didelphid marsupial from the same locality, a genus also regarded as partially arboreal (de Muizon 1998). *Mayulestes* had a dental formula of 5/4-1/1-3/3-4/4 like most other primitive marsupials. Lately there has been some doubt whether *Mayulestes* actually belongs to the Borhyaenoidea (Rougier et al. 1998; Forasiepi et al. 2006), but recent phylogenetic analysis confirms it (de Muizon et al. 2018).

Allqokirus australis is another member of the Mayulestidae from Tiupampa, but we can conclude much less about this mammal due to the paucity of fossil material

Fig. 4.8 *Mayulestes ferox* from Tiupampa. (By Roman Uchytel)

(de Muizon 1994). Nevertheless, it was probably about the same size as *Mayulestes* and was carnivorous (Argot 2004a).

4.3.2 Hathliacynidae

About 21 species are known for this family, the largest family of this clade. Their known evolutionary history is Pliocene/Pleistocene (Marshall 1977c; Forasiepi et al. 2000, 2006). The many species, found mostly in Argentina, Bolivia, and Brazil, seemed to have been very similar to dogs or martens. They were probably hunters of small animals and may often have had climbing abilities (Marshall 1981). A functional analysis showed that *Cladosictis lustratus* was a semiterrestrial/semi-arboreal predator on small animals and might best be compared to the tropical weasel *Eira barbara*. It weighed about 3–9 kg (Ercoli and Prevosti 2011; Argot 2003a) (Fig. 4.9).

A species *Sipalocyon myctoderos* is known by an almost complete skeleton and a calculated body mass of about 1–5 kg. The animal was probably arboreal and had good manipulative abilities because of a pseudo-opposable thumb (Forasiepi et al. 2006; Argot 2003a). Another species, the fox-like *Acyon myctoderos*, was recently discovered in the Middle Miocene of Quebrada Honda, Bolivia, and analysis suggests that it is closely allied to *Cladosictis*. Based on dentition, it is the largest hathliacynid known with a skull about 180 mm having a long and narrow snout; the animal had a weight of about 12–13 kg (Forasiepi et al. 2006; Zimicz 2014).

Fig. 4.9 *Cladosictis lustratus* (4–6 kg body mass) hunting small mammalian prey. (By Charles Knight in W.B. Scott's 1913. *A History of Land Mammals in the Western Hemisphere*. New York: The Macmillan Company)

4.3.3 Borhyaenidae

This family seems to be in taxonomic flux; previously it contained around ten genera that have been reduced to three. The known species tended to be husky predators, often compared to dogs or thylacines, and they have been found from the Early Eocene to Early Pliocene (Marshall 1978). The best-known species *Borhyaena tuberata* has been functionally analyzed and compared to *Prothylacinus*. It was a terrestrial hypercarnivorous ambush predator weighing 25–36 kg, according to various authors (Ercoli and Prevosti 2011; Prevosti et al. 2012; Argot 2003b). *Borhyaena tuberata* was similar in size to a small hyena or wolf and was probably a terrestrial predator with some running ability (Argot 2003b). Other, much less well-known genera in this family are *Nemolestes* (from the Early Eocene and the oldest) and the youngest two genera, *Eutemnodus* and *Parahyaenodon* from the Early Pliocene (Fig. 4.10).

4.3.4 Prothylacinidae

Many of this group were of medium-sized, strongly built predators. *Prothylacinus patagonicus*, known from the Santa Cruz Formation of Patagonia (Late Early Miocene) has been analyzed as a "scansorial ambush predator with a forefoot

Fig. 4.10 *Borhyaena* weighed about 25–36 kg or the size of a wolf (which are variable in size and was probably a strong hunter). (By Roman Uchytel)

capable of supination and grasping and a flexible vertebral column" (Argot 2003b). *Prothylacinus* was probably terrestrial but able to climb in trees. They were *hypercarnivores* of around 30 kg, so they could overcome a wide variety of prey (Ercoli and Prevosti 2011; Prevosti et al. 2012) (Fig. 4.11).

Some fossils provide enough information to give some idea of the animal that left them. From the Prothylacinidae or the Borhyaenidae family was the 15 kg *Lycopsis longirostris* from the Late Miocene (12 million years) La Venta of Colombia (Ercoli and Prevosti 2011; Marshall 1977a, 1979). Another closely related species was *Lycopsis torresi* was from the Santa Cruz (Middle Miocene), Argentina (Marshall 1977a). *Lycopsis longirostris* is probably derived from *L. torresi* (Fig. 4.12).

The Colombian *L. longirostris* have been analyzed as having weights of around 20 kg. Diet was undoubtedly small animals; the remains of the rodent *Scleromys colombianus* were found in the body cavity of *L. longirostris* remains. Because of the similarity of the two species of *Lycopsis* (*L. torresi* is slightly smaller than *L. longirostris*), it is probable that *L. torresi* is ancestor to *L. longirostris* (Marshall 1977a).

The large prothylacine known as *Dukecynus magnus* was also discovered at La Venta, Colombia, and probably weighed in the neighborhood of about 68 kg; it is the largest predator so far found at La Venta. It was sympatric with *Lycopsis longirostris* (above). The animal must have looked vaguely like a wolf (Goin 1997), and it was more carnivorous than other members of the family (Fig. 4.13) (Argot 2004b).

Fig. 4.11 Life restoration of *Prothylacinus patagonicus*, an arboreal prothylacine (from Argot 2003). (By Charles R. Knight, Life restoration of *Prothylacinus patagonicus* and *Interatherium robustum* from W.B. Scott's 1913. *A History of Land Mammals in the Western Hemisphere*. New York: The Macmillan Company)

Fig. 4.12 This Colombian La Venta *Lycopsis longirostris* (weight calculated at 30 kg by Ercoli and Prevosti 2011) is currently classified as a *Lycopsis*, which were very common in Early–Middle Miocene Patagonia. However, the genus might eventually be broken up into several genera (by Ryan Somma).

Fig. 4.13 *Dukecynus magnus* was the largest predator known from La Venta, though ecological densities of all predators are low, making them more difficult to detect. (By Roman Uchytel)

4.3.5 Proborhyaenidae

The Proborhyaenidae evolved some of the largest marsupial predators known and have been referred to as marsupial bears. There are only four genera, but each one was a spectacular beast. They had large heads for their bodies. Both proborhyaenids and thylacosmilids (see below) possessed upper canines that continued to grow throughout the lifetime of the animal.

Callistoe vincei was found in the Early Eocene Patagonia and was about the size of a jaguar and had a skull 24 cm long, a size that beat the *Thylacosmilus* by 4 cm. This species is represented by the most complete skeleton collected so far of the family. This animal's skull was particularly massive with large canines, skull, and jaw but a narrow snout like a thylacine. It weighed perhaps 23–34 kg and was strictly terrestrial, probably living in humid temperate mountain forest, and it may have dug for its prey, judging by the long claws that probably graced their angular (clawed) phalanges. It was probably the largest predator of its time. A well-developed mobile thumb indicates that it was able to grab and manipulate objects, such as prey (Babot et al. 2002; Argot and Babot 2011). Digging ability was unique among the borhyaenoids (Fig. 4.14).

Arminiheringia auceta, also from the Early Miocene of Patagonia, was larger than *Callistoe*; probably it weighed around 40–80 kg. This was one of the most spe-

Fig. 4.14 *Callistoe vincei* skeleton and skull. (By Roman Uchytel)

cialized of all the borhyaenids, based on the incomplete skull that is known. It had procumbent lower canines (Bond and Pascual 1983; Babot et al. 2002) (Fig. 4.15).

An ancestral connection to *Thylacosmilus* has been suggested, although there isn't much evidence for that.

Proborhyaena gigantea is the largest sparassodont known. It has been found from Patagonia, Bolivia, and Uruguay and might have topped 600 kg (Sorkin 2008; Patterson and Marshall 1978). This huge predator had a short rostrum and saber-tooth-like canines that were ovoid rather than blade-like, thus being probably more robust than the canines of *Thylacosmilus*. Like the thylacosmilids, *Proborhyaena* had only one pair of lower incisors. This animal existed among many large herbivores and must have preyed upon them during the Deseadan SALMA at 34 Ma during the Early Oligocene (Bond and Pascual 1983). Both *Proborhyaena* and many large ungulates became extinct as the world cooled during the Early Oligocene, suggesting a predator-prey link that was influenced by climate change (Argot 2004a) (Fig. 4.16).

4.3.6 Thylacosmilidae

This family contains around seven recognized genera: *Achlysictis*, *Amphiproviverra*, *Anachlysictis*, *Hyaenodontops*, *Notosmilus*, *Patagosmilus*, and *Thylacosmilus*. *Thylacosmilus* is undoubtedly the most famous of all. The incredible *Thylacosmilus atrox* or "saber-toothed marsupial cat" seems like a product of parallel evolution,

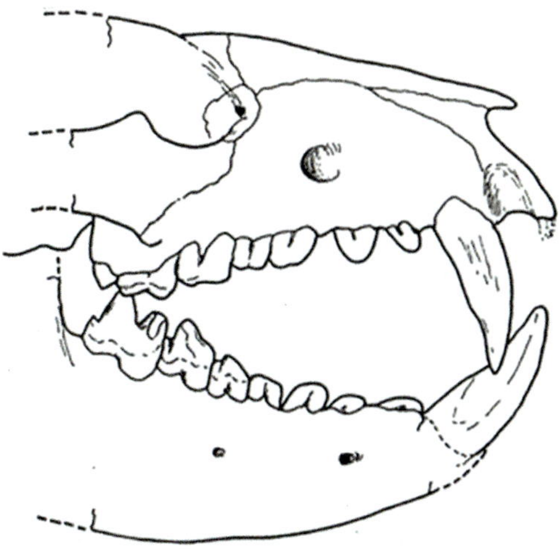

Fig. 4.15 *Arminiheringia auceta* weighed about 40–80 kg. (By Roman Uchytel)

with great canines and a cat-shaped body. It lived during the Pliocene and became extinct about two million years ago, perhaps because of competition from the saber-toothed cat *Smilodon populator*, though recent evidence seems to place the extinction of *Thylacosmilus* well before the arrival of *Smilodon*. *Thylacosmilus* weighed about the same as a large jaguar at 80–120 kg (Ercoli and Prevosti 2011). It was about 2 m long and killed its prey with the oversized canines which continued growing during the course of the animal's life. The gape of this animal's mouth could reach an incredible 90 degrees allowing it to precisely place its canines after immobilizing the prey. Neck and forelimb muscles were very strong, allowing manipulation and immobilizations of prey so that it could precisely stab it. The animal was

Fig. 4.16 *Proborhyaena gigantea* was the largest marsupial yet known. It was the size of a large bear reaching 600 kg and lived during the Oligocene in Patagonia. (By Roman Uchytel)

also able to move quickly and to run but may have lain in wait as do modern lions (Riggs 1933, 1934; Argot 2004c). An enlarged auditory bulla and hypotympanic sinus suggests acute hearing, and the small eyes and low estimated overlap of visual fields suggest the possibility that it might have hunted at night (Argot 2004c).

Another intriguing possibility is that there might have been post-lactational parental care, a behavior unknown in other metatherians. This is because the specialized killing bite of an adult *Thylacosmilus* required an important learning component, since it involved complex mobilization of the prey with the powerful forelimbs and exact and careful placement of the canines against the carotid of a struggling prey. Young *Thylacosmilus* would have had to learn these techniques in order to be efficient killers. Meanwhile they could have eaten alongside the parent, probably the mother (Argot 2004c).

It must have also been true that, because a *Thylacosmilus* could not clean up many meat scraps from its prey, a kill site must have attracted a coterie of scavengers, certainly avian phororhacoids among them, as well as a host of other small metatherian scavengers (Fig. 4.17).

4.3.7 Metatherian Carnivory

Analyses show that early carnivorous mammals comprised not only the early sparassodonts but also included members of the Peradectidae and Mayulestidae (taking this family as being apart from the sparassodonts) and that all were *mesocarnivores*

Fig. 4.17 Here a *Thylacosmilus* attacks a *Tapirus*. (By Roman Uchytel)

which supplemented their vertebrate meat diet with invertebrates and fruit (Zimicz 2014). By the late Paleocene-Early Eocene, the first medium- and large-sized sparassodont *hypercarnivores* had appeared (Forasiepi 2009; Zimicz 2012, 2014). Didelphoids evolved small terrestrial and *scansorial* carnivores from 0 to 10 kg in the range of the hathliacyanid sparassodonts, although the hathliacyanids are considered hypercarnivores and the didelphoids as meso- and hypocarnivores, permitting dietary segregation and, thus, an absence of strong competition (Zimicz 2014).

References

Archer M (1976) The dasyurid dentition and its relationships to that of didelphids, thylacinids, borhyaenids (Marsupicarnivora) and peramelids (Peramelina: Marsupialia). Australian J Zool Suppl Ser 24(39):1–34

Archer M, Kirsch J (2006) The evolution and classification of marsupials. In: Armanti PJ, Dickman CR, Hume ID (eds) Marsupials. Cambridge University Press, Cambridge, pp 1–21

Argot C (2001) Functional-adaptive anatomy of the forelimb in the Didelphidae and the paleobiology of the Paleocene marsupials *Mayulestes ferox* and *Pucadelphys andinus*. J Morphol 247:51–79

Argot C (2003) Functional-adaptive analysis of the hindlimb anatomy of extant marsupials and the paleobiology of the Paleocene marsupials *Mayulestes ferox* and *Pucadelphys andinus*. J Morphol 253:76–108

Argot C (2003a) Postcranial functional adaptations in the South American Miocene borhyaenoids (Mammalia, Metatheria): *Cladosictis, Pseudonotictis* and *Sipalocyon*. Alcheringa 27:303–356

Argot C (2003b) Functional adaptations of the postcranial skeleton of two Miocene boryhaenoids (Mammalia, Metatheria), *Borhyaena* and *Prothylacinus* from South America. Palaeontology 46:1213–1267

Argot C (2004a) Evolution of South American mammalian predators (Borhyaenoidea): anatomical and palaeobiological implications. Zool J Linnean Soc 140(4):487–521

Argot C (2004b) Functional-adaptive analysis of the postcranial skeleton of a *Laventan borhyaenoid, Lycopsis longirostris* (Marsupialia, Mammalia). J Vertebr Paleontol 24:689–708

Argot C (2004c) Functional-adaptive features and palaeobiologic implications of the postcranial skeleton of the late Miocene sabretooth boryhaenoid *Thylacosmilus atrox* (Metatheria). Alcheringa 28:229–266

Argot C, Babot J (2011) Postcranial morphology, functional adaptations and Palaeobiology of *Callistoe vincei*, a predaceous metatherian from the Eocene of Salta, North-Western Argentina. Palaeontology 54:411–480

Asher RF, Horovitz I, Sánchez-Villagra MR (2004) First combined cladistics analysis of marsupial mammal interrelationships. Mol Phylogenet Evol 33:240–250

Babot MJ, Powell JE, de Muizon C (2002) *Callistoe vincei*, a new Proborhyaenidae (Borhyaenoidea, Metatheria, Mammalia) from the Early Eocene of Argentina. Geobios 35:615–629

Beck RMD (2012) An "ameridelphian" marsupial from the early Eocene of Australia supports a complex model of Southern Hemisphere marsupial biogeography. Naturwissenschaften 99:715–729

Bond M, Pascual R (1983) Nuevos y elocuentes restos craneanos de *Proborhyaena gigantea* Ameghino, 1897 (Marsupialia, Borhyaenidae, Proborhyaeninae) de la edad Deseadense, un ejemplo de coevolución. Ameginiana 20:47–60

Bond M, Tejedor MF, Campbell KE Jr, Chornogubsky L, Novo N, Goin F (2015) Eocene primates of South America and the African origins of New World monkeys. Nature 520:538. https://doi.org/10.1038/nature14120. Supplementary information supplementary discussion

Case JA, Woodburne MO (1986) South American marsupials: a successful crossing of the Cretaceous-Tertiary boundary. PALAIOS 1:413–416

Chimento NR, Agnolin FL, Novas FE (2015) The bizarre metatherians' Groeberia and Patagonia, late surviving members of gondwanatherian mammals. Hist Biol 27(5):603–623

Chornogubsky Clerici L (2010) Sistemática de la familia Polydolopidae (Mammalia, Marsupialia, Polydolopimorphia) de América del Sur y la Antártida. PhD dissertation, Universidad de Buenos Aires

Chornogubsky L, Goin FJ, Reguero M (2009) A reassessment of Antarctic polydolopid marsupials (Middle Eocene), La Meseta Formation. Antarct Sci 21(3):285–297

Cifelli RL, Davis BM (2003) Marsupial origins. Science 302:1899–1900

de Muizon C (1994) A new carnivorous marsupial from the Palaeocene of Bolivia and the problem of marsupial monophyly. Nature 370:208–211

de Muizon C (1998) *Mayulestes ferox*, a borhyaenoid (Metatheria, Mammalia) from the early Palaeocene of Bolivia. Phylogenetic and paleobiologic implications. Geodiversitas 20(1):19–142

de Muizon C (1999) Marsupial skulls from the Deseadan (late Oligocene) of Bolivia and phylogenetic analysis of the Borhyaenoidea (Marsupialia, Mammalia). Geobios 32:483–509

de Muizon C, Cifelli RL (2001) A new basal "didelphoid" (Marsupialia, Mammalia) from the early Paleocene of Tiupampa (Bolivia). J Vertebr Paleontol 21(1):87–97

de Muizon C, Cifelli RL, Céspedes R (1997) The origin of the dog-like borhyaenoid marsupials of South America. Nature 389:486–489 de Muizon C, Ladevèze S, Selva C, Vignaud R, Goussard F (2018) *Allqokirus australis* (Sparassodonta, Metatheria) from the early Palaeocene of Tiupampa (Bolivia) and the rise of the metatherian carnivorous radiation in South America. *Geodiversitas* 40 (in print)

Ercoli MD, Prevosti FJ (2011) Estimación de masa de las especies de Sparassodonta (Mammalia, Metatherias) de edad Santacrusense (Miocene Temprano) a partir del tamaño del centroide de los elementos apendiculares: infernecias paleoecológicas. Ameghiniana 48(45):462–479

Flynn J, Wyss A (1999) New marsupials from the Eocene-Oligocene transition of the Andean main range, Chile. J Vertebr Paleontol 19:533–549

Forasiepi AM (2009) Osteology of *Arctodictis sinclairi* (Mammalia, Metatheria, Sparassodonta) and phylogeny of Cenozoic metatherian carnivores from South America. Monogr Mus Arg Cs Nat n s 6:1–174 (cited in Zimicz 2014)

Forasiepi AM, Sánchez-Villagra MR, Goin FJ, Takat M, Shigehara N, Kay RF (2000) A new species of Hathliacynidae (Metatheria, Sparassodonta) from the Middle Miocene of Quebrada Honda, Bolivia. J Vertebr Paleont 26(3):670–684

Forasiepi AM, Sánchez-Villagra MR, Goin FJ, Takai M, Shigehara N, Kay RF (2006) A new species of Hathliacynidae (Metatheria, Sparassodonta) from the middle Miocene of Quebrada Honda, Bolivia. J Vertebr Paleontol 26(3):670–684

Goin FJ (1997) New clues for understanding Neogene marsupial radiations. In: Kay RF, Madden RH, Cifelli RL, Flynn J (eds) Vertebrate paleobiology of the Miocene in Colombia. Smithsonian Institution Press, Washington, DC, pp 185–204

Goin FJ (2006) A review of the Caroloameghiniidae, Paleogene South American "primate-like" marsupials (?Didelphimorphia, Peradectoidea). In: Kalthoff DT, Martin T, Möors T (eds) Palaeontographica abteilung A 278:57–67

Goin FJ, Abello MA (2013) Los Metatheria sudamericanos de comienzos del Neógeno (Mioceno Temprano, Edad Mamífero Colhuehuapense): Microbiotheria y Polydolopimorphia. Ameghiniana 50:51–78

Goin FJ, Zimicz N, Reguero MA, Santillana SN, Marenssi SA, Moly JJ (2007) New marsupial (Mammalia) from the Eocene of Antarctica, and the origins and affinities of the Microbioteria. Rev Asoc Geol Argent 62(4):597–603

Goin FJ, Candela AM, Abello MA, Oliveira EV (2009) Earliest South American paucituberculatans and their significance in the understanding of "pseudodiprotodont" marsupial radiations. Zool J Linnean Soc 155:867–884

Goin FJ, Woodburne M, Zimicz AN, Martin GM, Chornogubsky L (2016) A brief history of South American Metatherians: evolutionary contexts and intercontinental dispersals. Springer, New York

Horovitz I, Sánchez-Villagra MR (2003) A morphological analysis of marsupial mammal higher-level phylogenetic relationships. Cladistics 19:181–212

Iturralde-Vinent MA (2006) Meso-Cenozoic Caribbean paleogeography: implications for the historical biogeography of the region. Int Geol Rev 48:791–827

Iturralde-Vinent MA, MacPhee RDE (1999) Paleogeography of the Caribbean region: implications for Cenozoic biogeography. Am Mus Nat Hist Bull 238:1–95

Kirsch JA, Dickerman AW, Reig OA, Springer MS (1991) DNA hybridization evidence for the Australasian affinity of the American marsupial *Dromiciops australis*. Proc Natl Acad Sci U S A 88:10465–10469

Kumar S, Hedges B (1998) A molecular timescale for vertebrate evolution. Nature 393:917–920

Ladeveze S, de Muizon C (2010) Evidence of early evolution of Australidelphia (Metatheria, Mammalia) in South America: phylogenetic relationships of the metatherians from the Late Palaeocene of Itaboraí (Brazil) based on teeth and petrosal bones. Zool J Linnean Soc 159:746–784

Luo Z-X, Ji Q, Wible JR, Yuan C-X (2003) An early Cretaceous tribosphenic mammal and metatherian evolution. Science 302:1934–1940

Luo Z-X, Yuan C-X, Meng Q-J, Quian J (2011) A Jurassic eutherian mammal and divergence of marsupials and placentals. Nature 476:442–445. https://doi.org/10.1038/nature10291

Marshall LG (1977a) A new species of *Lycopsis* (Borhyaenidae: Marsupialia) from the La Venta fauna (Late Miocene) of Colombia, South America. J Paleontol 51(3):633–642
Marshall LG (1977b) Cladistic analysis of borhyaenoid, dasyuroid, didelphoid, and thylacinid (Marsupialia: Mammalia) affinity. Syst Zool 26:410–425
Marshall LG (1977c) Evolution of the carnivorous adaptive zone in South America. In: Hecht MK, Goody PC, Hecht BM (eds) Major patterns in vertebrate evolution. Plenum Press, New York, pp 709–722
Marshall LG (1978) Evolution of the Borhyaenidae, extinct South American predaceous marsupials. U Calif Pub Geol Sci 117:1–93
Marshall LG (1979) Review of the Prothylacininae, an extinct subfamily of South American "dog-like" marsupials. Fieldiana Geol New Ser 7:1–120
Marshall LG (1981) Review of the Hathlyacyninae, an extinct subfamily of South American "dog-like" marsupials. Fieldiana Geol New Ser 3:1–50
Marshall LG (1982a) Systematics of the extinct South American marsupial family Polydolopidae. Fieldiana Geol New Ser 12(1339):1–109
Marshall LG (1982b) Systematics of the South American marsupial family Microbiotheriidae. Fieldiana Geol New Ser 10:1–75
Marshall LG, de Muizon C (1995) Part II. The skull. In: de Muizon C (ed) *Pucadelphys andinus* (Marsupialia, Mammalia) from the Early Paleocene of Bolivi. Mémoires du Muséum National d'Histoire Naturelle 165, Paris, pp 21–90
Marshall LG, Sigogneau-Russell D (1995) Part III. Postcranial skeleton; pp. 91–164 in C. de Muizon (ed.), *Pucadelphys andinus* (Marsupialia, Mammalia) from the Early Paleocene of Bolivia. Mémoires du Muséum National d'Histoire Naturelle 165.
Marshall LG, de Muizon C, Sigogneau-Russell D (1995) Part I. The locality of Tiupampa: age, taphonomy and mammalian fauna. In: de Muizon C (ed) *Pucadelphys andinus* (Marsupialia, Mammalia) from the Early Paleocene of Bolivia. Mémoires du Muséum National d'Histoire Naturelle 165, Paris, pp 11–20
McKenna MC, Bell SK (1997) Classification of mammals: above the species level. Columbia University Press, New York
Muizon C, Argot C (2003) Comparative anatomy of the Tiupampa didelphimorphs; an approach to locomotory habits of early marsupials. In: Jones M, Dickman C, Archer M (eds) Predators with pouches: the biology of carnivorous marsupials. CIRO Publishing, Collingwood, pp 43–62
Naish D (2012) Marsupial "dogs", "bears", "saber-tooths" and "weasels" of island South America: meet the borhyaenoids. http://blogs.scientificamerican.com/tetrapod-zoology/2012/07/12/meet-the-borhyaenoids-2012/
Nilsson MA, Gullberg A, Spotorno AE, Arnason U, Janke A (2003) Radiation of extant marsupials after the K/T boundary: evidence from complete mitochondrial genomes. J Mol Evol 57:S3–S12. https://doi.org/10.1007/s00239-003-0001-8
Nilsson MA, Arnason U, Spencer PBS, Janke A (2004) Marsupial relationships and a timeline for marsupial radiation in South Gondwana. Gene 340:198–196
Patterson B (1952) Un nuevo y extraordinario marsupial deseadiano. Revista del Museo Municipal de Ciencias Naturales de Mar del Plata 1:39–44
Patterson B, Marshall L (1978) The Deseadan, Early Oligocene, Marsupialia of South America. Fieldiana Geol 41(2):1–100
Pascual R, Goin FJ, Carlini AA (1994) New data on the Groeberiidae: unique late Eocene-Early Oligocene South American marsupials. J Vertebr Paleontol 14(2):247–259
Penny D, Hasegawa WPJ, Hendy MD (1999) Mammalian evolution: timing and implications from using the log determinant transform for proteins of differing amino acid composition. Syst Biol 48(1):76–93
Phillips MJ, Bennett TH, Lee MSY (2009) Molecules, morphology, and ecology indicate a recent amphibious ancestry for echidnas. Proc Natl Acad Sci U S A 106:17089–17904
Prevosti FJ, Forasiepi AM, Ercoli MD, Turazzini GF (2012) Paleoecology of the mammalian carnivores (Metatheria, Sparassodonta) of the Santa Cruz formation (late Early Miocene). In:

Vizcaíno SF, Kay RF, Bargo MS (eds) Early Miocene paleobiology in Patagonia: high-latitude paleocommunities of the Santa Cruz formation. Cambridge University Press, Cambridge, pp 173–193

Riggs ES (1933) Preliminary description of a new marsupial saber-tooth from the Pliocene of Argentina. Field Mus Nat Hist Geol Ser 6:61–66

Riggs ES (1934) A New Marsupial Saber-Tooth from the Pliocene of Argentina and Its Relationships to Other South American Predacious Marsupials. Trans Am Philos Soc 24(1):1–32

Rose KD (2006) The beginning of the age of mammals. The Johns Hopkins University Press, Baltimore

Rougier G, Wible FR, Novacek J (1998) Implications of *Deltatheridium* specimens for early marsupial history. Nature 396:459–463

Sigé B, Archer M, Crochet J-Y, Godthelp H, Hand S, Beck R (2009) *Chulpasia* and *Thylacotinga*, late Paleocene-earliest Eocene trans-Antarctic Gondwanan bunodont marsupials: new data from Australia. Geobios 42:813–823

Simpson GG (1970) The Argyrolagidae, extinct South American marsupials. Bull Mus Comp Zoo 139:1–86

Simpson GG (1980) Splendid isolation: the curious history of South American mammals. Yale University Press, New Haven

Simpson GG (1984) Discoverers of the lost world. Yale University Press, New Haven

Sorkin B (2008) A biomechanical constraint on body mass in terrestrial mammalian predators. Lethaia 41:333–347

Szalay FS, Trofimov BA (1996) The Mongolian late Cretaceous *Asiatherium* and the early phylogeny and paleobiogeography of Metatheria. J Vertebr Paleontol 16(3):474–509

Trofimov BA, Szalay FS (1994) New Cretaceous marsupial from Mongolia and the early radiation of Metatheria. Proc Natl Acad Sci U S A 91:12569–12573

von Koenigswald W, Pascual R (1990) The Schmelzmuster of the Paleogene South American rodentlike marsupials *Groeberia* and *Patagonia* compared to rodents and other Marsupialia. Paläontol Z 64:345–358

Villarroel AC, Marshall LG (1988) A new Argyrolagoid (Mammalia: Marsupialia) from the middle Miocene of Bolivia. J Paleontol 62(3):463–467

Woodburne MO, Case JA (1996) Dispersal, vicariance, and the late Cretaceous to early Tertiary land mammal biogeography from South America to Australia. J Mammal Evol 3(2):121–161

Woodburne MO, Rich TH, Springer MS (2003) The evolution of tribospheny and the antiquity of mammalian clades. Mol Phylogenet Evol 28:360–385

Woodburne MO, Goin FJ, Bond M, Carlini AA, Gelfo JN, López GM, Iglesias A, Zimicz AN (2014) Paleogene land mammal faunas of South America: a response to global climatic changes and indigenous floral diversity. J Mamm Evol 21:1–73

Zimicz N (2012) Ecomorfología de los marsupials paleógenos de América del Sur. PhD dissertation. Universidad Nacional de la Plata, Argentina. (citado en Zimicz 2014)

Zimicz N (2014) Avoiding competition: the ecological history of late Cenozoic metatherian carnivores in South America. J Mamm Evol. https://doi.org/10.1007/s10914-9255-8

Chapter 5
The Native Ungulates of South America (Condylarthra and Meridiungulata)

5.1 Introduction

During the Cenozoic, a great variety of native ungulates evolved in South America, unfortunately none of which now are living. These animals are thought to have evolved from a group loosely called condylarths, an assortment that is believed to have given birth to most of the ungulates in the world (Simpson 1980).

The term "ungulate" is not a technical or taxonomic term. It is used loosely to mean an animal with hooves, hooves being the horny sheath covering the toes or lower part of the foot, formerly including the artiodactyls and perissodactyls but now generally including the following groups or orders of mammals: Artiodactyla, Perissodactyla, Hyracoidea (hyrax), Proboscidea, Sirenia, Cetacea, Tubulidentata (aardvark), Embrithopoda (arsinoitheres); Meridiungulata (ungulados de SA), the Condylarthra, and perhaps the Pantodonta and Dinocerata. The present trend is to separate the Hyracoidea, Proboscidea, Sirenia, Tubulidentata, and arsinoitheres into the afrotheres, leaving the Artiodactyla (even-toed), Perissodactyla (odd-toed), and the Cetacea (whales) as the true ungulates. The ancient (and extinct) ungulates of South America are generally classified as follow in the Table 5.1.

5.2 Condylarthra

The condylarths (Fig. 5.1) were principally a group of archaic ungulates that originated in *Laurasia* (the northern, connected continents, except for India) but that arrived in South America in the Late Cretaceous or Early Paleocene. Condylarthra as an order is paraphyletic (not a natural group and does not include all of the descendants of a common ancestor); there is an absence of molecular evidence to demonstrate monophylia (descended from a common ancestor). However, most species do have "broad, low-crowned, and bunodont (molar teeth with raised pointed

T. Defler, *History of Terrestrial Mammals in South America*, Topics in Geobiology 42, https://doi.org/10.1007/978-3-319-98449-0_5

Table 5.1 Synoptic classification of South American ungulates (Rose 1996; McKenna and Bell 1997) Condylarths are almost certainly *paraphyletic* as are several of the families; Mioclaenidae, Didolodontidae, and Litopterna were united in a new order Panameriungulata by Muizon and Cifelli (2000); monophyly of Meridiungulata has not been demonstrated and is uncertain, probably placed within Typotheria

Grandorder UNGULATA of South America
Order CONDYLARTHRA
Mioclaenidae
Didolodontidae
Mirorder MERIDIUNGULATA (a taxonomic ranking just below grandorder and above order)
Order LITOPTERNA, Protolipternidae, Notonychopidae, Macraucheniidae, Adianthidae, Proterotheriidae
Order NOTOUNGULATA
Suborder NOTIOPROGONIA, Henricosborniidae, Notostylopidae
Suborder TOXODONTIA, Isotemnidae, Notohippidae, Leontiniidae, Toxodontidae, Homalodotheriidae
Suborder TYPOTHERIA, Oldfieldthomasiidae, Interatheriidae, Archaeopithecidae, Mesotheriidae, Campanorcidae
Suborder HEGETOTHERIA, Archaeohyracidae, Hegetotheriidae
Order ASTRAPOTHERIA, Eoastrapostylopidae, Trigonostylopidae, Astrapotheriidae
Order PYROTHERIA, Colombitheriidae, Pyrotheriidae
Order XENUNGULATA, Carodniidae

Fig. 5.1 An Early Eocene condylarth *Meniscotherium* (Phenacodontidae) from North America. (By Robert Bruce Horsfall from Scott's 1913). Fragmentary fossil material of condylarths in South America does not allow reconstructions

projections) molars with narrow stylar shelves (expansion of the cingulum or shelf-like ridge around the outside of an upper molar), low mesiodistally compressed trigonids (the first three cusps of a lower molar), and broad talonid (the crushing region of a lower molar tooth) basins. The conules (small cusps) tend to be prominent, and a hypocone (the distolingual cusp of an upper molar tooth) is variably developed. The incisors and premolars are usually simple though the last premolar is often molarized," (Rose 2006) all of which suggests an omnivorous diet with a large proportion of vegetation. Often the feet bore hooves, but not always. Most were terrestrial, while some few (arctocyonids) were adapted to an arboreal life (Rose 2006).

In the earliest Tiupampan Paleocene mammal community known in South America, there were seven species identified (*Molinodus suarezi*, *Tiuclaenus minutus*, *T. cotasi*, *T. robustus*, *Pucanodus gagnieri*, *Andinodus boliviensis*, *Simoclaenus sylvaticus*) all in the condylarth family Mioclaenidae, which initially would make that group the strongest candidate for the earliest South American ancestors of the ungulates or Meridiungulata, which we can call a "cohort" or "superorder" or a "mirorder" (de Muizon and Cifelli 2000). The acceptance of the Mioclaenidae as ancestors of South American ungulates has been challenged. Horovitz (2004) did a phylogenetic analysis of postcranial elements of Litopterna, Notoungulata, Astrapotheria, various North American condylarths, and the two members of the Phenacodontidae, *Phenacodus*, and *Meniscotherium* and she found that Litopterna and Notoungulata were sister groups and that they were most related to the condylarth *Meniscotherium* (Fig. 5.1) of the Phenacodontidae. *Meniscoherium* was a common condylarth of the Early Eocene of North America. So far a resolution to the problem of the ancestors of South American ungulates has not appeared, but recently a proteomic analysis from bone samples of *Toxodon* (a notoungulate) and *Macrauchenia* (a litoptern) has resulted in a consensus tree that finds that these two genera (from two distinct orders) form a monophyletic group whose sister taxon are the Perissodactyla or odd-toed ungulates of modern times (Welker et al. 2015). Future work should clarify whether all Meridiungulata are truly monophyletic and have evolved from the Perissodactyla, but progress towards understanding the evolutionary origins of South American ungulates seems palpable.

The condylarth family Didolodontidae is the only endemic South American condylarth group and appears to have evolved from the Mioclaenidae (Cifelli 1983). This group of condylarths is known principally from the Paleocene of Brazil (Itaboraí) and from the Early Eocene of Patagonia and appears to be monophyletic (Gelfo 2010). The Didolodontidae show bunodont teeth similarly to a couple of North American groups of condylarths (Phenacodontidae y Hyopsodontidae). The similarity of diet between some litopternas and the Didolodontidae suggests that the litopterns may have evolved from this group. Some have classified the mioclaenidos, didolodontidos and litopternas in a special order called the Panameriungulata, although that term is not in general use. Diversity of these archaic ungulates was highest during the warming phase of the Late Paleocene–Early Eocene during the concurrent rainforest conditions that existed. During the following cooler times in the Middle to Late Eocene, condylarth diversity decreased until they became extinct after the Early Oligocene (Gelfo 2010).

5.3 The Meridiungulata

Five orders (or suborders) are accepted as belonging to the South American Meridiungulata: (1) the Litopterna (members of which managed to survive up to about 8500 years before the present), (2) the Notoungulata (the most diverse order by far and some of which—*Toxodon*—also survived to recent times), (3) the Astrapotheria, (4) the Pyrotheria, and (5) the Xenungulata (McKenna and Bell 1997). Soria (1984, 1989) defined a sixth order, the Notopterna, but that is generally subsumed within the Litopterna (McKenna and Bell 1997) (Fig. 5.2).

5.3.1 Litopterna

The Litopterna were South American ungulates adapted to running. This adaptation appeared very Early in the Paleocene, prompting us to wonder what they were running from. The only large predators would have been sparassodont metatherians and the large avian phorusrhacid predators. Many of the known sparassodont marsupial predators seem not to have a running adaptation; they were adapted for a short run and probably hunted by hiding and waiting. In contrast, the phorusrhacids were a diverse and common avian predator that probably depended on any vertebrate prey

Fig. 5.2 The Middle Miocene *Nesodon imbricatus*, a large toxodontid, and the proterotherid liptotern *Diadiaphorus majusculus*. (By Jorge W. Moreno-Bernal)

Fig. 5.3 The "false horse" *Thoatherium* (Proterotheriidae) had digits reduction in all four feet so that they used, like horses, only one digit (By Charles Knight in 1913 A History of Land Mammals in the Western Hemisphere)

large enough to satisfy them (Fig. 5.5). These very large birds ran their prey down, dispatching them with their very large, cutting beak (Bertelli et al. 2007).

There are five families known for the Litopterna: Protolipternidae, Notonychopidae, Proterotheriidae, Macraucheniidae, and Adianthidae. Some of the earliest known litopterns (*Protolipterna*, *Miguelsoria*, and *Asmithwoodwardia* of the Late Paleocene) were often assigned to the two condylarth families until their ankle bones were studied and their cursorial adaptation was recognized as a synapomorphic (a trait that is shared by two or more taxa and inferred to have been present in their most recent common ancestor, whose own ancestor in turn is inferred not to possess the trait character) of the Litopterna and assigned to the Protolipternidae (Rose 2006). Some of the smallest litopterns perhaps jumped rather than ran and they were very common. McKenna and Bell (1997) recognize only *Protolipterna* as belonging to this family, and they doubt the family's validity.

The Notonychopidae is represented only by *Notonychops powelli* that was described for Argentina by Soria (1989) and placed in a new order, the Notopterna, and including *Amilnedwardsia* and *Indalecia*. However, McKenna and Bell (1997) do not recognize this order and place it as a synonym of the Litopterna (Rose 2006). *Notonychops powelli* is a Middle Paleocene species from Argentina (Soria 1989). *Requisia vidmari*, the most primitive litoptern known, belongs to this family, described from the Punta Peligro fauna; it shows a mixture of derived and primitive characteristics when compared to the Tiupampa condylarths (Bonaparte and Morales 1997). A third Paleocene species from the Riochico of Patagonia, *Wainka tshotshe*, represents a third genus closely related to both *Notonychops* and *Requisia* (Bonaparte and Morales 1997; Simpson 1948).

Fig. 5.4 The proterotherid, *Diadiaphorus majusculus*, a horselike litoptern browser. (By Roman Uchytel)

Proterotheriidae contain some of the smallest of the litopterns, one of which evolved convergently to the horses: *Thoatherium minusculum* or the "false horse" from the Early Miocene (23.03 to 5.33 Ma) of Patagonia. *Thoatherium* (Fig. 5.3) was only about 70 cm long, and its feet were reduced to one toe, like true horses, though true one-toed horses did not appear until the Pliocene (5.33–1.81 million years ago). In fact, *Thoatherium* was more completely one-toed than even *Equus* (Simpson 1980). *Thoatherium* teeth, however, were adapted to a softer diet than true horses; probably they fed on leaves. (Rose 2006; Cifelli 1993). The entire family has been updated by Soria (2001). The family includes 2 subfamilies and 18 genera (McKenna and Bell 1997).

Diadiaphorus (Oligocene) (Fig. 5.4) was also very horselike, but it possessed three toes instead of one, although the central toe was slender and strong, while the other two were short and small so that the central toe was probably the only one that was functional. The teeth were, like *Thoatherium*, adapted to browsing soft vegetation, unlike the horses. It was about 1.5 m long and weighted about 70 kg, like a domestic sheep. *Diadiaphorus majusculus* weighed around 80 kg and were browsing herbivores that fed on dicotyledonous plants rather than on grasses.

The family has been divided into 2 subfamilies (Anisolambdinae and Proterotheriinae), and there are about 18 recognized genera (McKenna and Bell 1997); however, Soria (2001) elevates Anisolambdinae to family level.

The Litoptern family Macraucheniidae includes perhaps as many as 18 genera grouped in 4 subfamilies of generally large mammals; the family is by far best known for the camel-like *Macrauchenia* which only became extinct in the Pleistocene. They were the last-surviving litopterns, having survived the ecological upheaval of the Great American Interchange between North and South America

Fig. 5.5 (**a**, **b**) Phorusrhacid terror birds, probable predators on *Macrauchenia* and other ungulates before the arrival of new mammalian predators from North America. The 2 m tall *Paraphysornis* (above) Early Miocene (23 Ma) of Brazil and *Kelenken* (below), the largest avian species (of any group) so far found, dated from the Middle Miocene of Patagonia (15.7 Ma) and probably reached 3 m in height. (By Roman Uchytel)

three million years ago (Saldanha Scherer et al. 2009). *Macrauchenia* was discovered by Charles Darwin in Patagonia in 1834. The recessed nasal bones of their skulls suggest they may have had a small proboscis or trunk somewhat like the saiga antelope (*Saiga tatarica*). Their hooves were similar to those of rhinoceros today with a simple ankle joint and three digits on each foot. Thus, they may have been capable of rapid directional change when running away from predators, such as the huge Phorusrhacid terror birds (Fig. 5.5a, b) and the *Smilodon* saber-toothed cats. Analysis of teeth and enamel suggests that it was a mixed browser and grazer.

Macrauchenia (Fig. 5.6) probably lived in herds because of the predators. Their bodies were about 3 m in length and an adult weight of about 1040 kg. They were hunted by Early human hunters who might have helped them along into extinction, but climate change also probably played a role in the reduction and isolation of savannahs. The last *Macrauchenia* seemed to have survived into the Early Holocene (12,240–7320 BP) (Cione et al. 2009; Borrero 2009).

Recent proteomic analysis demonstrates that "*Toxodon* and *Macrauchenia* form a monophyletic group whose sister taxon is not Afrotheria or any of its constituent clades as recently claimed but instead crown Perissodactyla (horses, tapirs, and rhinoceroses). These results are consistent with the origin of at least some South American native ungulates" (Welker et al. 2015).

Another genus, *Theosodon* (Fig. 5.7), with around six species appeared similar to a guanaco and was about 2 m in length, weighting from 40 to 140 kg, depending upon the species. They might have had a short and mobile snout somewhat like a tapir. They differed from guanacos and tapirs by having 3 toes on each foot and 44 teeth, which is very rare in herbivores.

The Adianthidae (pygmy litopterns) *Miguelsoria* and *Protolipterna* were real pygmies of the order. First known in the Early Eocene, earlier they were assigned to the condylarths (Cifelli and Soria 1983). The Notonychopidae *Notonychops* (Paleoceno tardío), *Amilnedwardsia*, *Indalecia*, and *Requisia* were placed into the Notonychopidae by Cifelli (1993), though others continue to regard them as adianthids (Rose 2006). Besides these Eocene genera, the group is unknown until

Fig. 5.6 *Macrauchenia* survived until Pleistocene times and were hunted by early humans. (By Roman Uchytel)

Fig. 5.7 *Theosodon garretorum* (left) and *Borhyaena tuberata* (right). A macrauchenid from the early Miocene (From W. B. Scott 1913 Miocene)

the Deseadan (Early Oligocene). They were small, gracile mammals with little information to describe them except for their teeth (Rose 2006). The Adianthidae formed a probably monophyletic clade of dentally advanced litopterns closely related to the Macraucheniidae and the Proterotheriidae (Cifelli and Soria 1983). The first fossil adianthid found with postcranial remains (and placed in the genus *Adianthus*) indicates that it was a selectively feeding herbivore of open habitat and was cursorily adapted (Cifelli 1991).

5.3.2 *Notoungulata*

The Notoungulata (Fig. 5.8) was the most diverse and successful order of the mirorder Meridiungulata with 14 families and more than 150 genera divided into 4 suborders (McKenna and Bell 1997). The group first appeared in the Paleocene fossil record in the Tiupampan faunal community, and they lasted until the Pleistocene, though their diversity began to decline from the Miocene onward (Rose 2006). They filled many niches similar to Perissodactyla and Artiodactyla. They seem to have evolved from the mioclaenid condylarths. The teeth were diagnostic and showed increasing hypsodonty (high-crowned teeth and enamel extending past the gum line), probably due to an increase in grassy savannas (Townsend and Croft 2008). They also had rodent-like incisors (Rose 2006). Feet and ankle bones suggest a generalized terrestrial habit and the feet were usually five-toed with later the

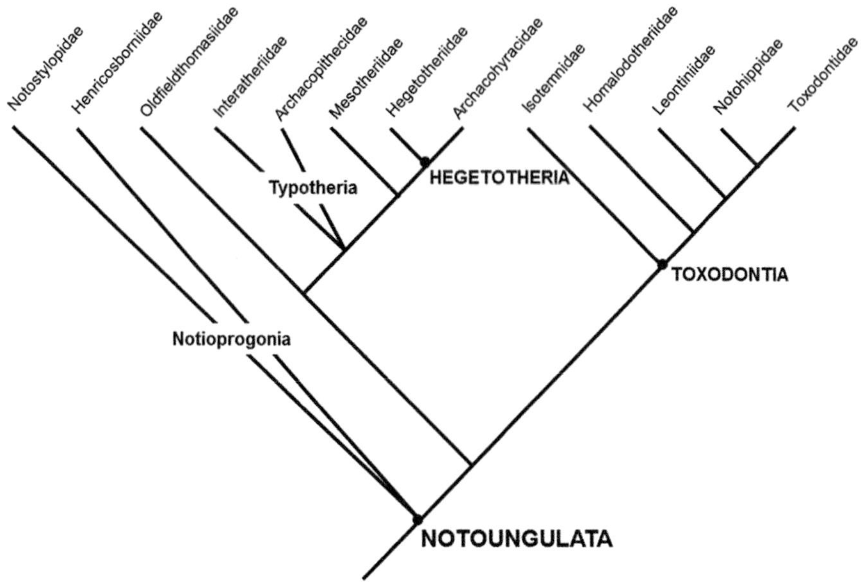

Fig. 5.8 Notoungulate relationships. (Modified from Rose 1996)

disappearance of lateral toes, evolving into three-toed forms, while some typotheres evolved only two equal-sized toes on the hind feet (Rose 2006). There was a general tendency for the notoungulates to increase in size up through the Pliocene and Pleistocene and to evolve the largest members like the *Toxodon* (Vizcaíno et al. 2012).

Although the monophyly of notoungulates is well-supported and has been divided up into four suborders: Notioprogonia, Toxodontia, Typotheria and Hegetotheria, the Toxodontia and Typotheria may belong to one clade, while Henricosborniidae, Isotemnidae and Oldfieldthomasiidae are paraphyletic (Billet 2011).

The suborder Notioprogonia is not a natural group, it is probably polyphyletic (a taxonomic group that includes members (as genera or species) from different ancestral lineages). There are two families: Henricosborniidae and Notostylopidae (Rose 2006).

The Henricosborniidae are known from the Paleocene to the Early Eocene (59–48 Ma) and include about five genera, all of which are primitive and generalized and difficult to tell apart (McKenna and Bell 1997). They may be at the base of all notoungulates and *Henricosbornia* is the sister taxon of all other notoungulates, so this family may be paraphyletic (Billet 2011).

The Notostylopidae ranged in size from 4.90 to 11.91 kg. They were more specialized than the generalized henricosbornids. Both families are noted for intraspecific variability in dental formula (Rose 2006).

Fig. 5.9 *Notostylops*, a rodent-like ungulate that was very common in the Pliocene-Pleistocene of southern South America and that may have communicated in part like modern elephants by means of infrasound. (By Roman Uchytel)

Notostylops brachycephalus (Fig. 5.9) is a well-known Pliocene-Pleistocene species of notostylopid. It had a large-bodied 75 cm long body and enormous teeth like rodents. It also had a very developed middle ear, obviously it was evolved to have very acute low frequency hearing. Based on the ratio of radii of the apical and basal turns of the cochlea studied in one skull, low frequency hearing limits ranging from 15 Hz were calculated for *Notostylops* (Macrini et al. 2013). These low frequency hearing limits could have allowed for the use of infrasound similar to that known for elephants. For Asian elephants, infrasound has a frequency of 14–24 Hz, while for African elephants, calls range from 15 to 35 Hz allowing long-distance communication for many kilometers, with a possible maximum range of around 10 km (6 mi) (Payne et al. 1986). We can speculate that infrasonic communication was used by *Notostylops* to keep in touch at long distances or in thick vegetation via appropriate vocalizations that would supply both location data and danger. Simpson (1932) considered that *Notostylops* probably was more rodent-like than ungulate-like with one set of enlarged incisors in each jaw. Various species are known, although some that have been described are probably synonyms.

so. Toxodontia

Finally, the suborder Toxodontia is known with five families (Isotemnidae, Homalodotheriidae, Leontiniidae, Notohippidae and Toxodontidae) and its monophyly is well-supported (Billet 2011). This clade includes the largest notoungulates known. Over 100 genera have been named from the five families.

Isotemnidae (53.47–351.96 kg size range) was comprised of the oldest and most primitive toxodonts and were most common during the Paleocene-Eocene. The

Fig. 5.10 *Thomashuxleya*, a primitive, sheep-sized toxodont from the Early Eocene. known from Patagonia. (By Roman Uchytel)

group may be paraphyletic or polyphyletic, but the cheek tooth patterns are sufficiently primitive to be basal to all other notoungulates except notioprogonians (Cifelli 1993; Billet 2011).

Thomashuxleya (Fig. 5.10) was one of the few notoungulates that are well-known because of complete skeletons. These were robust animals with five-toed feet that bore hoofs on each toe. They are known from the Casamayoranense of the Eocene (Naish 2012).

Homalodotheriidae (1172–1786 kg Size Range Elisamburu 2012)

The Early Miocene *Homalodotherium* (Fig. 5.11) was about 2 m long and 1150 kg and possessed long forelimbs (compared to the hindlimbs) with claws instead of hooves. It walked on the soles of its hind feet and the toes of its front feet, which would have made the animal higher at the shoulder than at the hips when it walked on all fours. The posture of this animal was unlike any other nothoungulate known. It was probably able to adopt a bipedal posture and to feed like ground sloths, pulling down tree branches with its arms while rearing up on its hind legs (Elisamburu 2010). Four species are recognized.

Notohippidae

Another primitive toxodont family was the Notohippidae (21.8–99.67 kg size range), several members of which had very high-crowned teeth, indicating a somewhat hard plant diet. Several of this group are also separated into the Leontiniidae, which did not appear until the Middle Eocene and lasted until the Late Oligocene (López et al. 2010; Elisamburu 2012).

Fig. 5.11 *Homalodotherium cunninghami* a 300 kg toxodont. (By Roman Uchytel)

Fig. 5.12 *Huilatherium pluriplicatum*, another leontinid much later than *Scarittia*, from the Middle Miocene La Venta, weighing around 800 kg. (By Rextron)

Leontiniidae (316.6–1404.4 kg Size Range)

The most successful leontinid was probably *Scarrittia* from the Early Oligocene. This was a rhinoceros-shaped mammal that lived in moist forest, near the coast, in wetlands, lakes, swamps, etc. and they ate soft vegetation, grasses, fruits and trees. Some species were omnivorous, eating also eggs and small mammals. They were not adapted for running, though their large size meant they had few enemies.

Another hervivorous leontinid, *Huilatherium* (Fig. 5.12) lived during the Colloncuran and Laventan Salmas around 16–11.8 Ma. It probably weighed up to

800 kg. It was the last-known and most specialized member of the family and probably depended on woodlands in Colombia, where it was discovered (Villarroel and Danis 1997).

Members of this family lived mostly in the Oligocene and in southern South America. Seven genera are known (Villarroel and Danis 1997).

Toxodon

Toxodons reached sizes similar to pachyderms up to 1500 kg and about 2.7 m in length, and they probably had similar habitats, many have been found in dry grassy habitat. Their upper incisors were bowed and very strong, while the inferior incisors were very broad horizontally giving the anterior part of the mandible the aspect of a spatula. The body was the form of a barrel and the legs were short but very robust, the feet plantigrade with three rather small toes. The back legs were higher than the front legs, giving an inclined appearance to the front.

The skull was very long in relation to its height and the anterior portions were broad, with probable prehensile lips for gathering grass, just like black rhinoceros. Just behind the snout the cranium narrowed as in rhinoceros and then broadened. The zygomatic arches were very large. They could eat a wide variety of course vegetation, including from trees. The teeth suggest that the toxodon had a mixed diet of herbs and grass and foliage.

The genus *Toxodon* (Fig. 5.13) was discovered by Charles Darwin in 1833 in Argentina on his trip around the world. They were very large animals, lived in herds and were very wide-spread, the closely related (*Mixotoxodon*) reaching Central America and apparently Texas during Pleistocene times (Lundelius et al. 2013). *Toxodon* survived from 2.6 Ma until about 16,500 years (Van Frank 1957; Arroyo-Cabrales et al. 2010). Four species have been described. Many fossils have arrow

Fig. 5.13 *Toxodon* (Toxodontidae). (By Roman Uchytel)

points found in them, so it is obvious they were frequently hunted by human beings (Cione et al. 2009).

Recently a toxodont tooth has been discovered in Texas that may belong to the closely related *Mixotoxodon*, since this is the toxodont most found in Central America and southern Mexico. It is the first record of toxodonts or of any other notoungulate found in the United States (Lundelius et al. 2013).

Adinotherium (Fig. 5.14) was another smaller toxodont from the Early to Middle Miocene. It weighed about 150 kg and looked similar to the larger *Toxodon*, except the forelimbs were longer in relation to the hind limbs, making the body level rather than sloping. There are several known species. The most common was *A. ovinum* that probably weighed around 100 kg and was about 1.5 m long. It ate a wide variety of plant material and may have had an ecotonic habitat, close to forests, so as to evade predators.

Nesodon imbricatus (Fig. 5.15) was very closely related to *Adinotherium*, wiegh-ing close to 500–800 kg and reaching 1.5 m. There were two or possible three spe-cies, all of them larger than the sympatric *Adinotherium* (Townsend and Croft 2008; Cassini et al. 2012). Like all the South American ungulates it was a vegetarian and probably existed on a mixture leaves and grass.

so. Typotheria

The monophyly of the suborder Typotheria is well-supported and is divided into four or five familias (Oldfieldthomasiidae, Interatheriidae, Archaeopithecidae, Campanorcidae, Mesotheriidae) (Billet 2011). Most are very much like small to medium-sized rodents with long and developed incisors with enamel on both sides instead of one side as the rodents and a tendency to live in burrows. There are 20–25

Fig. 5.14 *Adinotherium* was another smaller toxodont from the Miocene (17.5–11.6 Ma). (By Roman Uchytel)

Fig. 5.15 *Nesodon imbricatus* might have been principally a leaf-eater and perhaps even browsed on bark. (By Roman Uchytel)

known species. One genus, *Mesotherium* had four species; this genus was known as *Typotherium* from the Late 1800s to Early 1900s. *Mesotherium* was the size of a sheep at around 55 kg and lived in the Early Oligocene, surviving until the Pleistocene. The later hegetotheres (*Hegetotheria*) became rabbit-like with longer hind legs than forelegs and they can be characterized as an increasing hypsodonty and enlargement of medial incisors (Reguero and Prevosti 2010; Rose 2006) (Fig. 5.16).

The Hegetotheria (Fig. 5.17) make up two families (Hegetotheriidae and Archaeohyracidae) of mostly small notoungulates and these perhaps belong as part of the Typotheria rather than a separate suborder (Cifelli 1993; Hitz 1995; Reguero 1998; Croft 2000). These animals are best-known from the later Tertiary with a very few reported from the Paleocene or Eocene. They had large incisors with a large diastema due to loss of incisors, canine and the first premolar except for the primitive hegetotherid *Ethegotherium*, which retained primitive dentition of three incisors, a canine and four premolars. They were similar in size to interatheriids.

Hegetotheriidae (1–13.14 kg Size Range)

Hemihegetotherium (Fig. 5.17) was very common in open habitats during the middle Miocene in southern Bolivia (the most common mammal in that community). It was about the size of a house cat and represented one of the four dominant notoungulate families in the Neogene along with the Toxodontidae, Interatheriidae and the Mesotheriidae (Croft and Anaya 2006; Croft 2007). But while these other three families were quite dominant at other sites of the same age (Early–Middle Miocene), members of the Hegetotheriidae were rare. Quebrada Honda of southern Bolivia is

Fig. 5.16 (**a**) *Campanorco inauguralis* (Notoungulata, Typotheria) and (**b**) *Coquenia bondi* (Notoungulata, Toxodontia, Leontiniidae) (Late Eocene). (From Babot 2017)

Fig. 5.17 *Hemihegetotherium trilobus*, a Middle Miocene hegetothere (Typotheria) from southern Bolivia (12.5–13 Ma). (By Velizar Simeonovski)

the opposite with toxodonts, interatherids and mesotherids being represented by very limited material (one species of each), while *Hemihegetotherium* (Hegetotheriidae) (Fig. 5.17) is represented by 18 specimens. These animals may have lived in burrows and possibly in social groups.

The Archaeohyracidae include at least 15 species and Interestingly the Tinguiririca Fauna of the Andean Main Range of Chile have yielded the most diverse assemblage of archaeohyracid typotherians known with six species. This represents about 40% of the arachaeohyracid fauna known (Croft et al. 2003).

5.3.3 *Astrapotheria, (Eoastrapostylopidae, Trigonostylopidae, Astrapotheriidae)*

The astrapotheres were a group of strange ungulates found from the Late Paleocene to the Middle Miocene of Colombia (Middle Miocene) to Antarctica (Middle Eocene). They were large (some to 3 m) animals with tusk-like canines, some of which became very large and these canines grew throughout the life of the animal. Since their cheek-teeth were similar to rhinoceros, it has been proposed that they had somewhat similar habits, although they have also been hypothetically associated with moist (swampy) habitats, since because of long, slim hind pentadactylous legs they could have been amphibious (Scott 1937), a dubious conclusion. It seems to this author that this suggestion is not very logical, considering the stout strong short legs of the hippopotamus. Avilla (2005) also conjectured an aquatic habitat based on a comparison of the proportion of limbs as compared to aquatic living mammals. Astrapotheres sported a proboscis somewhat like tapirs. There are three described families.

The oldest known were the Brazilian, Itaboraian *Tetragonostylops* and the Argentinian, Riochican *Eoastrapostylops*. *Trigonostylops* found in Patagonia are distinct from other astrapotheres in their ear anatomy; they are included in the order because of otherwise similar characters. The most primitive and the smallest astrapothere known is considered to be *Eoastrapostylops riolorense*, discovered in southern Patagonia in Paleocene deposits, earlier and more primitive than *Tetragonostylops* of the Riochiquan of Itaboraí SALMA (Soria 1981). It could correspond to the Early Riochican of the Middle Paleocene (or even earlier) and is probably the ancestor of Trigonostyhlopidae and Astrapotheriidae (Soria 1981, 1987; Soria and Bond 1984). Soria (1988) suggested that the origin of the Astrapotheria as well as the Xenungulata, Notoungulata and Pyrotheria was from a condylarth group different from the Didolontidae.

Astrapotherium magnum (Figs. 5.18 and 5.19) is the best-known astrapotherium based on various fossils. It measured about 3 m long and possessed a moderately developed proboscis. It might have weighed up to 1000 kg and could have had hippopotamus-like habits (Riggs 1935).

Granastrapotherium snorki (Fig. 5.20), first discovered in the Colombian La Venta fauna, is one of the largest astrapotheres known, probably surpassed only by some *Parastrapotherium* species. It weighed between 2500 and 3500 kg and sported very developed tusks that might have been used in dominance displays and to break vegetation, somewhat like elephants. Besides being known from La Venta, Colombia the species is known from southern Peru of Santa Rita. There are other ungulates in common with the two sites, such as *Theosodon* spp. These faunistic similarities probably embody a landscape connection between Colombia and southern Peru that would have provided habitat to a particular and related fauna that lived in a slightly higher elevation habitat of well-watered open grassy savanna with gallery forests, west of the immense Amazonian wetlands that at the time were found to the east.

Fig. 5.18 *Astrapotherium magnum.* a well-kown astrapotherium from Early to Middle Miocene times and known for southern parts of South America (By Roman Uchytel)

Fig. 5.19 Head of *Astrapotherium magnum.* (From 1913 *History of Land Mammals in the Western Hemisphere* by W. B. Scott)

Hilarcotherium castanedaii was a newly discovered, medium-sized astrapotherium found in Middle Miocene deposits equivalent to La Victoria Formation in the La Venta, Colombia area, but 69 km to the southeast of La Venta. The species belongs in the Uruguaytheriinae clade along with *Granastrapotherium, Xenastrapotherium*, and *Uruguaytherium* and probably weighed around 1303–1369 kg, based on the lower molar row length, comparable in mass to the black

Fig. 5.20 *Granastrapotherium snorki* a very large astrapothere first found in La Venta in Colombia, later in southern Peru. It is the most massive mammal known from La Venta (By Roman Uchytel)

rhino (*Diceros bicornis*) and the giraffe (*Giraffa camelopardalis*) (Johnson and Madden 1997; Kramarz and Bond 2008, 2010; Vallejo-Pareja et al. 2014).

5.3.4 Pyrotheria, (Colombitheriidae, Pyrotheriidae)

The pyrotheria were medium-sized to large and robust ungulates from 1.8 to 2.5 m in length and a height of 70 cm–2.5 m, weighing 200–350 kg. The bodies were compact, like tapirs and the four feet had five toes (Shockey and Anaya 2004). They had a small proboscis that was probably used to manipulate vegetation. Their incisors were stick-like in the upper and lower jaws and they had the appearance of primitive elephants or tapirs. There are six genera known and they lasted from 30 Ma (Early Oligocene)–3 Ma (Pliocene) to about the time that North and South America became connected. They are known from their fossils from Brazil, Peru and Argentina and are divided into seven genera.

Pyrotherium romeroi (Figs. 5.21 and 5.22) known from a complete skeleton and the only skull that has been found, was the size of a rhinoceros with a very compact body weighing more than 3500 kg. They may have lived in swampy areas about 30 Ma ago. These animals had the appearance of large, digitigrade, tapir-like mammals with relatively short, slender limbs and five-toed feet with broad, flat phalanges and strong, procumbent, chisel-shaped incisors. Their fossils are restricted to Paleocene through Oligocene deposits of Brazil, Venezuela, Peru and Argentina.

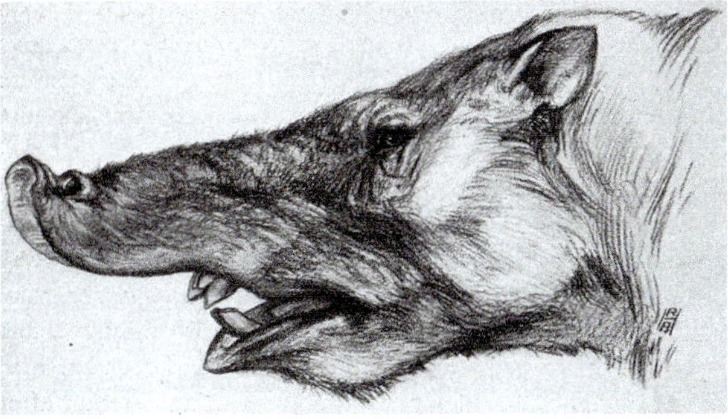

Fig. 5.21 Study of head of *Pyrotherium romeroi* from Scott (1913). (By Robert Bruce Horsfall)

Fig. 5.22 *Pyrotherium romeroi*. (By Roman Uchytel)

Two or three species of *Pyrotherium* are known. A recent analysis of skull anatomy supports the nesting of the pyrotheres within the Notoungulata via an exclusive relationship with *Notostylops* (Patterson 1977). This relationship is supported by both cranial and dental anatomy. The analysis also supports the position of Astrapotheria as the sister group of the Notoungulata (Billet 2010). This conclusion is supported via a previous analysis of *Colombitherium tolimense* by Avilla (2005). The clustering of Pyrotheria, Notoungulata, and Astrapotheria supports an isolated evolutionary history of these ungulates in South America similar to that of *afrotherian* mammals in Africa.

Some experts place the clade Xenungulata (which contains three genera, including *Carodnia*) within Pyrotheria but their dentition is distinct so that for most people the two orders remain separated. The pyrotheres have also been associated with the notoungulates (Billet 2010) and the astrapotheres were seen as a sister-group to the ungulates, thus revealing a clade of these three orders. The discovery of *Pyrotherium macfaddenis* in the Late Oligocene of Bolivia (weighed about 900 kg) showed a very distinctive dorsoventral compression of certain pedal bones, suggesting a very unusual mode of locomotion (graviportal = adapted only for moving slowly over land, due to a high body weight; and plantigrade = walking with the podials and metatarsals flat on the ground) (Shockey and Anaya 2004).

Ameghino (1895) believed that *Pyrotherium* was Cretaceous rather than Oligocene in origin, apparently due to the first fossils being packed together with dinosaur bones. He concluded that *Pyrotherium* was an ancestor to the proboscideans, though Gaudry (1909) rejected this conclusion.

Colombitherium tolimense (Fig. 5.23) is represented by a jaw fragment that has been identified as both right maxillary and left dentary and has tentatively been placed in the Pyrotheria in the family Colombitheriidae. This strange animal is probably from the Late Eocene of Tolima, Colombia, although even its age is unclear. Its morphology has nothing in common with other pyrotherians except for its bilophodont molars (=having teeth with two fused ridges, of the type found one per tooth in lophodont teeth). cheek teeth (Hoffstetter 1970; Billet et al. 2010).

Baguatherium jaureguii gen. et sp. nov. from the Early Oligocene of northern Peru is the best known pre-Deseadan pyrothere (Salas et al. 2006).

Fig. 5.23 A hypothetical reconstruction of the little-known pyrothere, *Colombitherium tolimense*. (By Zimices [Julián Bayona])

5.3.5 *Xenungulata*

The order Xenungulata has been found in several places throughout the south of the continent. They are represented by four species in two families. The earliest discovery was from the Itaboraí fauna of Brazil from the Early Eocene but it has also been found in Patagonia of Argentina (de Paula Couto 1952). Molar structure of *Carodnia* connects the group to the South American pyrotheres as well as to the uintetheres of the Dinocerata from North America and Asia, but this connection is considered by most to be parallel evolution, especially since *Etayoa* molars lack such structures. Foot bones of xenungulates are quite different from any of the other Meridiungulates.

Carodnia vieirai (Fig. 5.24) is the best-known species, a tapir-sized herbivorous animal of about 2 m in length and about 150–200 kg in weight; it represented the largest mammal found in the Itaboraí community and the largest known for the Early Eocene (changed from Late Paleocene) (Avilla and Paglarelli Bergqvist 2005). There are three genera recognized, *Carodnia*, *Notoetayoa* and *Etayoa*, both known only from the Paleocene (Gelfo et al. 2008; Villarroel 1987).

Fig. 5.24 *Carodnia vieirai* is the best-known xenungulate based on skeletal material (Avilla and Paglarelli Bergqvist 2005). (By Roman Uchytel)

References

Ameghino F (1895) Premiére contribution à connaissance de la fauna mammalogique de couches à *Pyrotherium*. Boletín Instituto Geográfico Argentino 15:603–660

Arroyo-Cabrales J, Polaco OJ, Johnson E, Ferrusquía-Villafranca I (2010) A perspective on mammal biodiversity and zoogeography in the Late Pleistocene of México. Quat Int 212:187–197

Avilla LS (2005) A revision of *Colombitherium tolimense* Hoffstetter (Pyrotheria: Mammalia) and its significance on Pyrotheria relationships. Ameghiniana 42(4R):60R

Avilla LS, Paglarelli Bergqvist (2005) Sobre o hábito locomotor de *Carodnia vieirai* Paula-Couto, 1952 (Mammalia: Xenungulata). Anuário do Instituto de Geociȩcias – UFRJ 28(1):185–186

Babot (2017) Mamíferos paleógenos del subtrópico de Argentina: síntesis de estudios estratigráfico cronológicos y taxonómicos. In: Muruaga CM, Grosse P (eds) Ciencias de la tierra y recursos naturales del NOA. Relatorio del XX Congreso Geológico Argentino, San Miguel de Tucumán, pp 730–753. isbn:978-987-42-6666-8

Bertelli S, Chiappe LM, Tambussi C (2007) A new phorusrhacid (Aves: Variamae) from the Middle Miocene of Patagonia, Argentina. J Vertebr Paleontol 27(2):409–419

Billet G (2010) New observations on the skull of *Pyrotherium* (Pyrotherium, Mammalia) and the new phylogenetic hypotheses on South American ungulates. J Mammal Evol 17:21–59

Billet G (2011) Phylogeny of the Notoungulata (Mammalia) based on cranial and dental characteristics. J System Palaenotol 9(4):481–497

Billet G, Orliac M, Antoine P-O, Jaramillo C (2010) New observations and reinterpretation on the enigmatic taxón *Colombitherium* (?Pyrotheria, Mammalia) from Colombia. Palaeontology 53(2):319–325

Bonaparte JF, Morales J (1997) Un primitive Notonychopidae (Litopterna) del Paleoceno Inferior de Punta Peligro, Chubut, Argentina. Estudios Geol 53:263–274

Borrero LA (2009) The elusive evidence: the archeological record of the South American extinct megafauna. In: Haynes G (ed) American megafaunal extinctions at the end of the Pleistocene. Springer, New York, pp 145–168

Cassini GH, Cerdeño E, Villafañe AL, Muñoz NA (2012) Paleobiology of Santacrucian native ungulates (Meridiungulata: Astrapotheria, Litopterna and Notoungulata). In: Vizcaíno SF, Kay RF, Bargo MS (eds) Early Miocene paleobiology in Patagonia: high-latitude paleocommunities of the Santa Cruz formation. Cambridge University Press, Cambridge, pp 243–286

Cifelli RL (1983) The origin and affinities of the South American Condylarthra and early Tertiary Litopterna (Mammalia). Am Mus Nat Hist Novit 2772:1–49

Cifelli RL (1991) A new adianthid litoptern (Mammalia) from the Miocene of Chile. Rev Chil Hist Nat 64:119–125

Cifelli RL (1993) The phylogeny of the native South American ungulates. In: Szalay FS, Novacek MJ, McKenna MC (eds) Mammal phylogeny: placentals. Springer, New York

Cifelli RL, Soria MF (1983) Systematics of the Adianthidae (Litopterna, Mammalia). Am Mus Novit 2771:1–25

Cione AL, Tonni EP, Soibelzon L (2009) Did humans cause the late Pleistocene-Holocene mammalian extinctions in South America in a context of shrinking open areas? In: Haynes G (ed) American megafaunal extinctions at the end of the Pleistocene. Springer, New York, pp 125–144

Croft DA (2000) Archaeohyracidae (Mammalia, Notoungulata) from the Tinguiririca Fauna, Central Chile, and the evolution and paleoecology of South American mammalian herbivores. PhD dissertatopm, University of Chicago, 311 pp

Croft DA (2007) The middle Miocene (Laventan) Quebrada Honda fauna, southern Bolivia and a description of its notoungulates. Palaeontology 50(1):277–303

Croft DA, Anaya F (2006) A new middle Miocene hegetotheriid (Notoungulata: Typotheria) and a phylogeny of the Hegetotheriidae. J Vertebr Paleontol 26(2):387–399

Croft DA, Bond M, Flynn JJ, Reguero M, Wyss AR (2003) Large archaeohyracids (Typotheria, Notoungulata) from Central Chile and Patagonia, including a revisión of *Archaeotypotherium*. Fieldiana Geol New Ser 49(1527):1–38

Elisamburu A (2010) Estudio biomecánico y morfofuncional del esqueleto apendicular de *Homalodotherium* Flower 1873 (Mammalia, Notoungulata). Ameghiniana 47(1):25–24

Elisamburu A (2012) Estimación de la masa corporal en géneros del Orden Notoungulata. Estud Geol 68(1):91–111

Gaudry A (1909) Fossiles de Patagonie: le Pyrotherium. Annales de Paléontologie 4:1–28

Gelfo JN (2010) The "condylarth" Didolodontidae from Gran Barranca: history of the bunodont South American mammals up to the Eocene-Oligocene transition. In: Madden RH, Carlini AA, Vucetich MG, Kay RF (eds) The paleontology ot Gran Barranca. Cambridge University Press, Cambridge, pp 130–142

Gelfo JN, López GM, Bond M (2008) A new Xenungulata (Mammalia) from the Pleocene of Patagonia, Argentina. J Paleontol 82(2):329–335

Hitz R (1995) Tyopothere (Notoungulata) phylogeny and proposed taxonomic revisions. J Vertebr Paleontol 15(Suppl 3):34A

Hoffstetter R (1970) *Colombitherium tolimense* pyrothérien nouveau de la Formation Gualanday (Colombia). Ann Paleontol 56:149–171

Horovitz I (2004) Eutherian mammal systematics and the origins of the South American ungulates as based on postcraneal osteology. Bull Carnegie Mus Nat Hist 36:63–79

Johnson SC, Madden RH (1997) Uruguaytheriine Astrapotheres of tropical South America. In: Kay RF, Madden RH, Cifelli RL, Flynn JJ (eds) Vertebrate paleontology in the neotropics. The Miocene fauna of La Venta, Colombia. Smithsonian Institution Press, Washington, DC, pp 355–381

Kramarz AG, Bond M (2008) Revisión of *Parastrapotherium* (Mammalia, Astrapotheria) and other Deseadan astrapotheres of Patagonia. Ameghiniana 45(3):537–551

Kramarz AG, Bond M (2010) Colhuehuapian Astrapotheriidae (Mammalia) from Gran Barranca south of Lake Colhue-Huapi. In: Madden RH, Carlini AA, Vucetich MG, Kay RF (eds) The paleontology ot Gran Barranca. Cambridge University Press, Cambridge, pp 170–181

López GM, Ribeiro AM, Bond M (2010) The Notohippidae (Mammalia, Notoungulata) from Gran Barranca: preliminary considerations. In: Madden RH, Carlini AA, Vucetich MG, Kay RF (eds) The paleontology ot Gran Barranca. Cambridge University Press, Cambridge, pp 143–151

Lundelius EL Jr, Bryant VM, Mandel R, Thies KJ, Thoms A (2013) The first occurrence of a taxodont (Mammalia, Notoungulata) in the United States. J Vertebr Paleontol 33(1):229–232

Macrini TE, Flynn JJ, Ni X, Croft DA, Wyss AR (2013) Comparative ungulate (Placentalia, Mammalia) bony labryinths and new phylogenetically informative inner ear characteristics. J Anat 223:442–461

McKenna MC, Bell SK (1997) Classification of mammals above the species level. Columbia University Press, New York

Muizon C, Cifelli RL (2000) The "condylarths" (archaic Ungulata, Mammalia) from the early Palaeocene of Tiupampa (Bolivia): implications on the origin of the South American ungulates. Geodiversitas 22(1):47–150

Naish D (2012) Obscure fossil mammals of island South America: *Thomashuxleya* and the other isotemnids. http://blogs.scientificamerican.com/tetrapod-zoology/2012/07/21/isotemnid-toxodonts-2012/

Patterson B (1977) A primitive pyrothere (Mammalia, Notoungulata) from the Early Tertiary of Northwestern Venezuela. Fieldiana Geol New Ser 33:397–421

Payne KB, Langbauer WR, Thomas EM (1986) Infrasonic calls of the Asian elephant (*Elephas maximus*). Behav Ecol Sociobiol 18(4):297–301

Paula Couto CD (1952) Fossil mammals from the beginning of the Cenozoic in Brazil. Condylarthra, Litopterna, Xenungulata and Astrapotheria. Bull Am Mus Nat Hist 99(6):359–394

Reguero MA (1998) El problema de las relaciones sistemáticas y filogenéticas de los *Typotheria* y *Hegetotheria* (Mammalia, Notoungulata): Análisis de los taxonones de Patagonia de la Edad-mamífero Deseadense (Oligoceno). PhD dissertation, Universidad de Buenos Aires, Buenos Aires

Reguero MA, Prevosti FJ (2010) Rodent-like notoungulates (Typotheria) from Gran Barranca, Chubut Province, Argentina: phylogeny and systematics. In: Madden RH, Carlini AA, Vucetich MG, Kay RF (eds) The paleontology ot Gran Barranca. Cambridge University Press, Cambridge, pp 152–169

Riggs ES (1935) A skeleton of *Astrapotherium*. Geol Ser Field Mus Nat Hist 6:167–177

Rose KD (1996) On the origin of the order Artiodactyla. Proc Natl Acad Sci USA 93:1705–1709

Rose KD (2006) The beginning of the age of mammals. The Johns Hopkins University Press, Baltimore

Salas R, Sánchez J, Chacaltana C (2006) A pre-deseadian pyrothere (Mammalia) from the northern Peru and the wear facets of molariforms teeth in Pyrotheria. J Vertebr Paleontol 26:760–769

Saldanha Scherer C, Gregis Pitana V, Ribeiro AM (2009) Proterotheriidae and Macraucheniidae (Litopterna, Mammalia) from the Pleistocene of Rio Grande do Sul State, Brazil. Rev Bras Paleontol 12(3):231–246

Shockey BJ, Anaya F (2004) *Pyrotherium macfaddeni* sp. nov. (late Oligocene, Bolivia) and the pedal morphology of pyrotheres. J Vertebr Paleontol 24(2):481–488

Scott WB (1913) A history of land mammals in the western hemisphere. The Macmillan Company, New York

Scott WB (1937) The Astrapotheria. Proc Am Philos Soc 77:309–393

Simpson GG (1932) Skulls and brains of some mammals from the 1034 Notostylops beds of Patagonia. Am Mus Novit 578:1–11

Simpson GG (1948) The beginning of the age of mammals in South America. Part 1. Introduction. Systematics: Marsupialia, Edentata, Condylarthra, Litopterna and Notioprogonia. Bull Am Mus Nat Hist 91:1–232

Simpson GG (1980) Splendid isolation: the curious history of South American mammals. Yale University Press, New Haven/London

Soria MF (1981) Un primitivo *Astrapotheria* (Mammalia) y la edad de la formación Rio Loro, Provincia de Tucumán, República Argentina. Ameghiniana 18(3–4):155–168

Soria MF (1984) Notopterna: un nuevo orden de mamíferos ungulados del Terciario Inferior de Sudamérica. Ameghiniana 25(3):245–258

Soria MF (1987) Estudios sobre los Astrapotheria (Mammalia) del Paleoceno y Eoceno. Ameghiniana 24(1–2):21–34

Soria MF (1988) Estudios sobre los Astrapotheria (Mammalia) del Paleoceno y Eoceno. Parte II: Filogenia, origen y relaciones. Ameghiniana 25(1):47–59

Soria MF (1989) Notopterna: Un nuevo orden de mamíferos ungulados neogenos de América de Sur. parte II. *Notonychops powelli* gen. et sp. nov. (Notonychopidae nov.) de la formación Río Loro (Paleoceno Medio), provincia de Tucumán, Argentina. Ameghiniana 25(3):259–272

Soria MF (2001) Los Proterotheriidae (Litopterna, Mammalia): sistemática, origen y filogenia. Buenos Aires: Museo Argentino de Ciencias Naturales "Bernardino Rivadavia" e Instituto Nacional de Investigación de la Ciencias Naturales

Soria MF, Bond M (1984) Adiciones al conocimiento de *Trigonostylops*. Ameghiniana 21(1):43–51

Townsend KEB, Croft DA (2008) Diets of notoungulates from the Santa Cruz Formation, Argentina: new evidence from enamel microwear. J Vertebr Paleontol 28(1):217–230. https://doi.org/10.1671/0272-4634(2008)28[217:DONFTS]2.0.CO;2

Vallejo-Pareja MC, Carrillo JD, Moreno-Bernal JW, Pardo-Jaramillo M, Rodriguez-Gonzalez DF, Muñoz-Duran J (2014) *Hilarcotherium castanedaii*, gen. Et sp. Nov., a new Miocene Astrapothere (Mammalia, Astrapotheriidae) from the upper Magdalena Valley, Colombia. J Vertebr Paleontol. https://doi.org/10.1080/02724634.2014.903960

Van Frank R (1957) A fossil collection from northern Venezuela 1. Toxodontidae (Mammalia Notoungulata). Am Mus Novit 1850:1–38

Villarroel C (1987) Características y afinidades de *Etayoa* n. gen., tipo de una nueva familia de Xenungulata (Mammalia) del Paleoceno medio (?) de Colombia. Comunicaciones Paleontológicas del Museo de Historia Natural del Montevideo 1(19):241–253

Villarroel C, Danis J (1997) A New leontiniid notoungulate. In: Kay RF, Madden RH, Cifelli RL, Flynn JJ (eds) Vertebrate paleontology in the Neotropics. The Miocene fauna of La Venta, Colombia. Smithsonian Institution Press, Washington, DC, pp 303–318

Vizcaíno SF, Cassini GH, Toledo N, Bargo MS (2012) On the evolution of large size in mammalian herbivores of Cenozoic faunas of southern South America. In: Patterson BC, Costa LP (eds) Bones, clones and biomes. University of Chicago Press, Chicago, pp 76–101

Welker F, Collins MJ, Thomas JA et al (2015) Ancient proteins resolve the evolutionary history of Darwin's South American ungulates. Nature. https://doi.org/10.1038/nature14249

Chapter 6
The Xenarthrans: Armadillos, Glyptodonts, Anteaters, and Sloths

6.1 Introduction

Xenarthrans are bizarre and striking mammals, representative of South America's endemic fauna. They are unknown outside of the neotropics, except for some forms that traveled the Central American highway to North America and that managed to colonize the Caribbean archipelago during the great inter-American exchange. The origin of the Xenarthra is completely unknown with only a couple of guesses. Perhaps they evolved from the small, fossorial Palaeanodonta of North America, Europe, and Asia, mammals that, like the xenarthrans, showed a reduction in dentition. Or did they evolve from specialized multituberculates that had become the strange gondwanatheres?

Previously the xenarthrans had been associated taxonomically by Cuvier (1796) with the pangolins (Pholidota) of Africa and Asia, the African aardvark (Tubulidentata), and the extinct Taeniodonta under the broad rubric Edentata, but those relationships are mostly questioned at best, while molecular evidence suggests very distant connections (Figs. 6.1 and 6.2). The discovery of the Miocene *Eurotamandua* (Fig. 6.3) from Germany excited various scientists to place this discovery with the xenarthrans, as the bauplan of this ancient mammal is strikingly like a *Tamandua*. *Eurotamandua* complicated the original idea of xenarthrans (Xenarthra proposed by Cope 1889) being South American. But being so much earlier than the appearance of *Tamandua* and lacking xenarthrous articulations and a prehensile tail make this connection seem doubtful. Convergent evolution has produced striking similarities in many unrelated animals (Rose 1999; Delsuc et al. 2001; Gardner 2007) (Fig. 6.3). The Tubulidentata were also believed to be related to the Xenarthra, but molecular biology shows that they are more related to the elephants than they are to the Xenarthra (Liu et al. 2001).

Tubulidentata Xenarthra

Xenarthra is usually classified as a magnorder with two orders, order Cingulata (armadillos and glyptodonts) and the order Pilosa, divided into two suborders, the

T. Defler, *History of Terrestrial Mammals in South America*, Topics in Geobiology 42, https://doi.org/10.1007/978-3-319-98449-0_6

Fig. 6.1 The Pholidota bauplan is very similar to modern *Tamandua*, but broad scales over the body and an absence of xenarthrous vertebrates, plus molecular results, indicate that they are very distantly related to Xenarthra and are more related to Carnivora. (By Piekfrosh)

Fig. 6.2 The aardvark, the only species of Tubulidentata that exists and thought to have been related to the xenarthrans, now known to be more closely related to the proboscideans (elephants). (By MontageMan)

Fig. 6.3 An Eocene fossil *Eurotamandua* convinced many that the xenarthrans early on had a wider distribution and that they were phylogenetically related to the pangolins. But the fact that *Eurotamandua* had no xenarthrous articulations, and other xenarthran characteristics have convinced many that it is probably an early pangolin and had no connection to the xenarthrans. (By Marie Joelle Giraud, Grupo de Evolución y Ecología de Mamíferos Neotropicales, Universidad Nacional de Colombia)

Folivora (Phyllophaga or sloths) and Vermilingua (anteaters) (Gardner 2005a, b, 2007). For the Cingulata, the extant and extinct families include the Dasypodidae (armadillos) and the Peltiphilidae, organized under the superfamily Dasypodoidea; the Glyptodontidae (glyptodonts) and Pampatheriidae (pampatheres) organized in the superfamily Glyptodontoidea; the Myrmecophagidae (anteaters) and Cyclopidae (fairy anteater) in the Vermilingua; and the Bradypodidae, Scelidotheriidae, Mylodontidae, Megalonychidae, Megatheriidae, and Nothrotheriidae, all sloths, in the Phyllophaga (Fig. 6.3) (McKenna and Bell 1997).

There are only 5 families and 13 genera alive today from a group that has had at least 200 genera (Fig. 6.4) during the Cenozoic (Simpson 1980; Wetzel et al. 2007). The majority of the fossils have been South American with some found in the Antarctic Peninsula. Several molecular studies have mostly resolved the phylogeny of the living genera (Delsuc et al. 2001, 2002, 2003, 2004). The order is characterized specifically by the presence of a special type of articulation on the thoracic vertebrae that is not found in any other group (Figs. 6.5 and 6.6). These articulations are called xenarthrous articulations, and they are the principal character of the group (Figs. 6.4 and 6.5).

The xenarthrans are recognized as the most primitive group of placental mammals known. Their origins are shrouded in mystery. Though a connection to the Pholidota of African or Asia has been suggested, molecular evidence is against this idea so that many favor a lineage independent and fairly ancient, especially because of the following: (1) their physiology, (2) their ossified sternal ribs, (3) the septomaxillary bone of the nose that is homologous with the monotremes and other Mesozoic mammals, (4) and their xenarthrous articulations. The best candidate as ancestors so far has been the gondwanatherian Ferugliotheriidae *Ferugliotherium* (Patagonia) (Chap. 2) that was found in Patagonia.

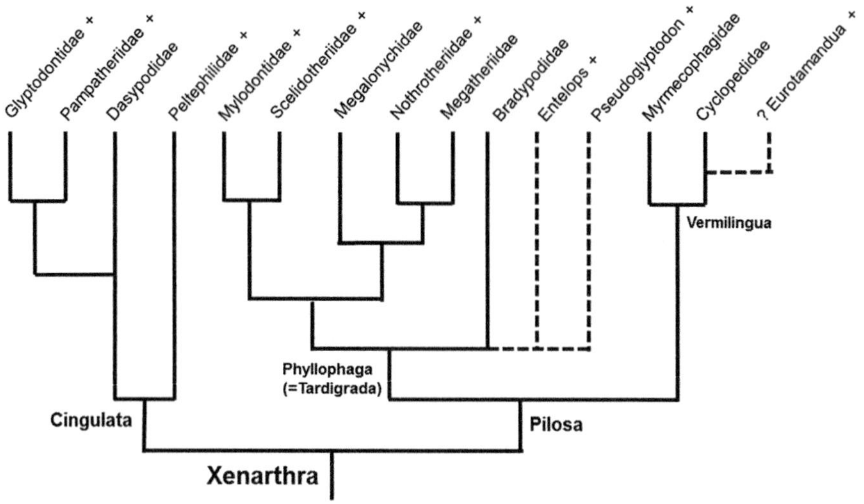

Fig. 6.4 Phylogeny of Xenarthra, including extinct taxa. Many doubt a close phylogenetic connection of *Eurotamandua* to the Xenarthra (Englemann 1985). (By Diego Casallas, Applied Biodiversity Foundation)

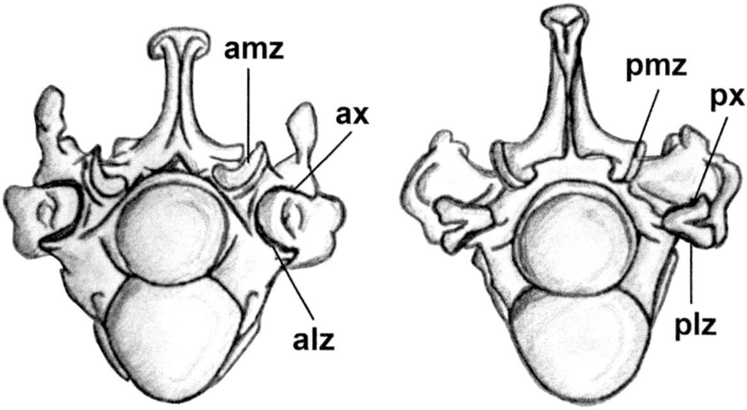

Fig. 6.5 This shows an anterior view of thoracic vertebra number 14 and a posterior view of thoracic vertebra number 13. ax = anterior xenarthrous facet; alz = anterior lateral zygapophyseal facet; pmz = posterior medial zygapophyseal facet; px = posterior xenarthrous facet; plz = posterior lateral zygapophyseal facet. (By Diego Casallas, Applied Biodiversity Foundation)

6.2 Order: Cingulata

Phylogenetic analyses suggest the emergence of the Cingulata from the Placentalia sometime around the Cretaceous/Paleogene boundary (Delsuc et al. 2001). The Cingulata or armadillos, glyptodonts, and three other less-known families are the only mammals protected by an exterior bony armor. This armor is more flexible in

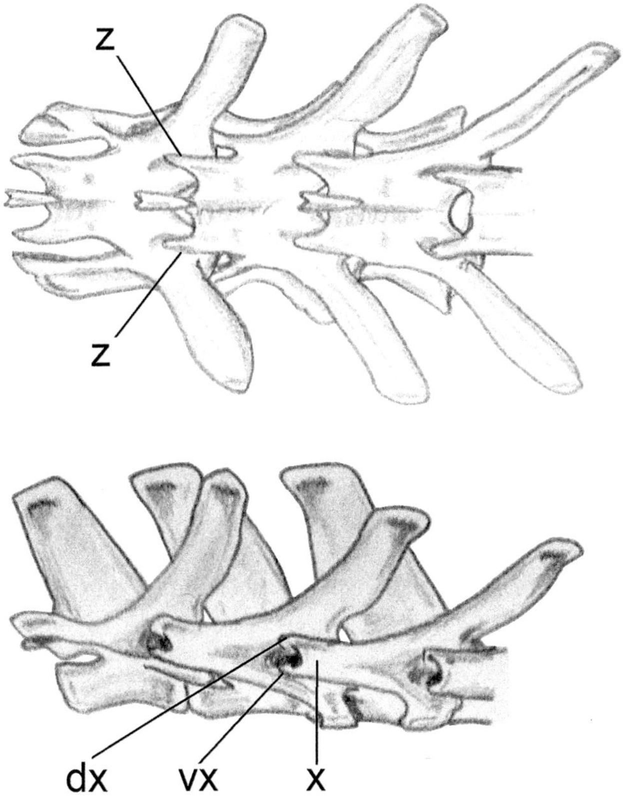

Fig. 6.6 Lumbar vertebrae of an armadillo showing normal *zygapophyses* (z) and xenarthrous joints (x. dx. vx). (By Diego Casallas, Applied Biodiversity Foundation)

the armadillos (Dasypodidae) than it was with the now extinct glyptodonts (Glyptodontidae), the two best known families in the order.

Order Cingulata
Dasypodidae – armadillos
Pampatheriidae† – pampatheres
Peltephilidae† – peltephilid armadillos
Glyptodontidae† – glyptodonts

† = Extinct

6.2.1 *Dasypodidae*

The Dasypodidae or armadillos (Fig. 6.8) are represented by about 42 genera (8 still living) of which there are 21 living species (McKenna and Bell 1997). The very earliest known xenarthran (*Riostegotherium*) was an armadillo found in the Early

Eocene of Brazil and Argentina in the form of osteoderms (Fig. 6.7) (Bergqvist et al. 2004) and some forelimb elements that might have been associated with *Riostegotherium* (this was never confirmed); this is the sole evidence of earliest armadillos. The osteoderms (Fig. 6.7) are the remains of the bony scutes or plates that form a mosaic that makes up the body shield, the basis for the external armor of the armadillos and of glyptodonts and sloths. The humeri from this same association (if they were associated) show a possible relationship to paleonodonts, an early Tertiary group of small fossorial mammals with reduced dentition (Rose 2009). Since the paleonodonts have been linked to the Pholidota as well, this would provide a link between the Xenarthra and the Pholidota. Nevertheless, molecular and some morphological data suggests a monophyletic origin of Xenarthra from a primitive eutherian clade, distinct from the Pholidota, (which seem linked to the Carnivora) (Osborn 1904; Simpson 1931; Rose and Emory 1993; Madsen et al. 2001; Delsuc et al. 2002; Fernicola et al. 2008). The basic phylogenetic position seems to indicate that Xenarthrans are a sister group to the Afrotheria (Hallström et al. 2007).

A second species of *Riostegotherium* as well as an assemblage of other primitive armadillos from the Laguna Fría and La Barda from Chubut Province, Argentina, argues for an age of Early to Middle Eocene, during the Eocene climatic optimum. Two more dasypodids are known from the Vacan fauna that probably corresponds to early Casamayoran SALMA. *Riostegotherium* is absent here, but a more derived *Stegosimpsonia* is present. Finally the Barrancan fauna, corresponding to the late Casamayoran SALMA (Middle Eocene), yields *Utaetus* (Fig. 6.9), the first skeletal material for an armadillo (Fig. 6.8).

The first preserved skeletal material of an early armadillo is Early Eocene of about 60 Ma when the dasypod *Utaetus* was found in Patagonia (Fig. 6.9). This material

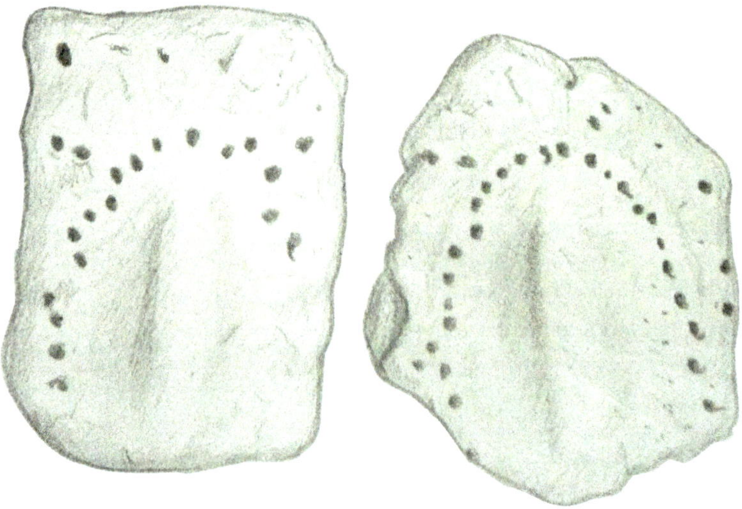

Fig. 6.7 Osteoderms of *Riostegotherium*; their form suggests that of armadillos. (By Diego Casallas, Applied Biodiversity Foundation)

Fig. 6.8 The common 12-banded armadillo (*Dasypus*), found in wide extensions of South America, has also invaded parts of the southern United States. (By http://www.birdphotos.com)

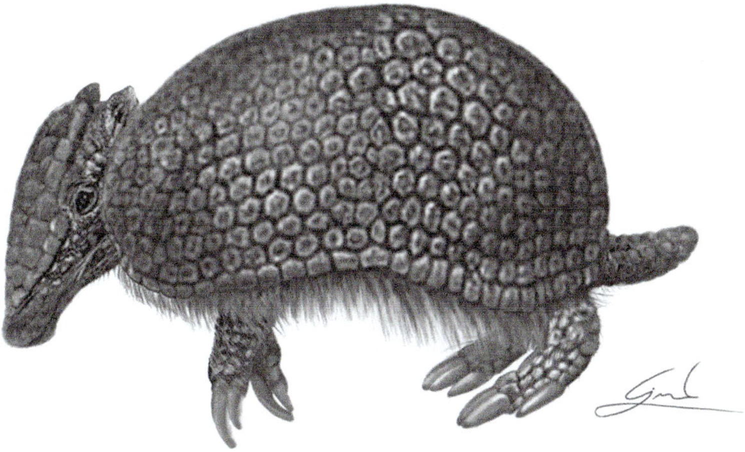

Fig. 6.9 The primitive armadillo *Utaetus* from the Early Eocene of 60 Ma. (By Marie Joelle Giraud, Grupo de Evolución y Ecología de Mamíferos Neotropicales, Universidad Nacional de Colombia)

includes a lower mandible with ten peglike and cylindrical teeth. So this armadillo still retained enameled teeth, which in later forms is lost, leaving the dentine.

The largest Dasypodidae armadillo known was the Early Eocene *Macroeuphractus outesi* with a skull nearly 28 cm long and head and body almost 2 m long. The three recognized species of this genus lived in southern Argentina and were probably carnivore specialists, as suggested by their large head and second upper caniform molars (Scillato-Yane 1975; Vizcaíno and De Iuliis 2003).

An important fauna from Santa Rosa, Peru, illustrates the only tropical faunal assemblage for the period of Late Eocene to Early Oligocene. This fauna has already supplied the earliest known rodents for South America at about 41 Ma, and the known cingulate fauna illustrates interesting diversities not concurrent with the decrease in diversity and temperatures further south in Argentina (Antoine et al. 2011; Ciancio et al. 2013; González 2010). Nor does the Santa Rita fauna correspond to any Patagonian Paleogene fauna, suggesting that the evolutionary histories were not the same during this period. This is a good example of why it is so important to study assemblages from the lower latitudes that have probably not been impacted by the strong climate and temperature changes known to have occurred further south (Ciancio et al. 2013). All other xenarthrans are descended from the dasypod xenarthrans, characterized by the top and sides of the body covered with horny scutes over a bony, flexible carapace.

The Dasypodidae had a mostly invertebrate diet but also ate carrion and some plants. Fossils from the extinct Stegotheriinae subfamily from the lower Miocene in South America had elongated anteater-like skulls with much reduced cylindrical cheek teeth; they were highly specialized for an insectivorous diet. The paleofauna of the Santa Cruz Formation (Early–Middle Miocene, 18–16 Ma) is part of the richest mammalian assemblage known for South America and has yielded the richest community of armadillos known with more than 20 described species (included is *Peltophilus*, usually classified as a separate family from the Dasypodidae) (Vizcaíno et al. 2006, 2012) and up to 8 species of glyptodonts (Vizcaíno et al. 2012; Croft 2016). This is all the more remarkable since the fossil-producing areas are between 50° and 51°S latitude. They were homogeneous in size and locomotor adaptations and with a great range of dietary adaptations, perhaps explaining such species richness located so far to the south (Vizcaíno et al. 2012). It must be mentioned here that the Santa Cruz SALMA took place during the Miocene optimum, when temperatures were much higher than they are now and the environment consisted of open vegetation in relatively dry conditions with marked seasonality with some closed forest (Tauber 1997; Vizcaíno et al. 2012).

6.2.2 *Pampatheriidae*

Another group and family of xenarthrans, often called giant armadillos (Pampatheriidae), are often associated in the mind of many with the Dasypodidae, but they were in fact an independent line of xenarthrans of large size and more closely related to the Glyptodontidae than to the Dasypodidae (Patterson et al. 1989; Mead et al. 2007). This family has its origins in the Eocene from around 45 to 48 Ma (Patterson and Pascual 1972; Englemann 1985; Gaudin and Wible 2006). They were not as diverse as the Dasypodidae with only four genera known, but they did attain weights of 225 kg or so. Their armor was more flexible than glyptodonts

Fig. 6.10 *Holmesina* was a pampathere genus that extended from southern North America to South America. Other species lived in North and South America. These were much larger than typical armadillos often reaching 225 kg. (By Roman Uchytel)

with three moveable bands of scutes, while the osteoderms were covered by one keratinized scute, as opposed to armadillos which had more than one scute per osteoderm. Studies of their jaws suggest that they were powerful and able to move side to side like a cow and were probably grazers of course vegetation as was found in the Pampas of Argentina (Iuliis et al. 2000).

One genus, *Holmesina* (Fig. 6.10), became distributed in both North and South America and possibly evolved first in North America from a more generalized pampathere (*Kraglievichia*?) originally from South America and then seemingly reinvaded South America where the youngest species were quite similar to each other, suggesting recent arrival (Scillato-Yané et al. 2005). This genus reached 225 kg and was 2 m long and was a grazer, like other pampatheres. There are seven known species of *Holmesina* (Góis et al. 2012; Aguilar and Laurito 2009). *Scirrotherium* is the only giant armadillo known for La Venta but it is known from other sites as well (Goís et al. 2013).

The genus *Pampatherium* had 2–3 species in the latter part of their reign. They may have been more adapted to dry, open areas than was *Holmesina*, which seems to have required a more humid environment (Scillato-Yané et al. 2005).

6.2.3 Peltephilidae

Another extinct relative of the armadillo is the family Peltephilidae (Early Eocene to Late Miocene in South America) represented by 5 genera and 13 known species, all extinct. They had short, broad skulls and possessed hornlike structures (Fig. 6.10)

Fig. 6.11 The horned armadillo, *Peltephilus ferox*, of the Oligocene and Miocene of Argentina. *Peltophilus ferox* was probably carnivorous and a predator on small animals. (By Roman Uchytel)

on the head armor including hornlike osteoderms on the nasals and maxillaries and a dorsal carapace. The osteoderms have a morphology that is distinct for the family, making it easy to identify fossils of this family, and they have been defined as specialized flesh eaters, probably as predators according to some (González-Ruiz et al. 2013; Krmpotic et al. 2009). However another analysis of the masticatory muscles and tooth distribution interprets them as having been feeders on soft, tough plant items from underground (roots?) (Vizcaíno and Fariña 1997). Lately this group has been proposed as a sister taxon to all other cingulates (Gaudin and Wible 2006) (Fig. 6.11).

6.2.4 *Glyptodontidae*

The extinct family Glyptodontidae is distinguished from the Dasypodidae by the possession of a rigid turtle-like carapace. Glyptodonts had an inflexible carapace with fused vertebrae that did not move. Their head was capped with dense bone, probably for protection, and their teeth were closely packed columnar trilobed molariform teeth with no incisors and canines. As with all the xenarthrans, the teeth lacked enamel. Each species had a particular type of osteoderm and pattern that allows us to recognize the species using only the osteoderm. During Pleistocene times, these animals were hunted by human beings, and their carapaces, because of their large size, were even occasionally used as shelter by early humans (Politis and Gutiérrez 1998; Gutiérrez et al. 2008).

Fig. 6.12 *Doedicurus clavicaudatus* was perhaps the largest of the glyptodonts. This animal flourished during the Pleistocene and had a height of 1.5 m (5 feet) and an overall length of around 3.6 m (12 feet). Two *Macrauchenia* are in the background. (By Roman Uchytel)

Mastication was from front to back rather than sideways (Fariña and Vizcaíno 2001). They were herbivorous and may have shared a common ancestry with the Pampatheriidae (Patterson and Pascual 1972; Englemann 1985; Gaudin and Wible 2006).

Glyptodonts appear in the fossil record from the Eocene to the Late Pleistocene in South and North America and were represented by about 68 genera (McKenna and Bell 1997; Croft et al. 2007; Fernicola 2008; Fernicola 2005). They had moved to North America by the Pleistocene and gradually increased in size from smaller forms to larger forms throughout their history, increasing to a length of 5 m (16.5 in.) with a rigid 3 m (10 feet) shell on its back. A cladistics analysis divides the family into 4–5 subfamilies (McKenna and Bell 1997).

The largest glyptodonts found seem to come from the end of their existence in the Early Holocene (8480 ± 130 years 14C cal. BPA) *Doedicurus* cf. *D. clavicaudata* (Fig. 6.12) which probably weighed 1,900–2370 kg (Fariña et al. 1998; Soibelzon et al. 2012). Species of the genera *Doedicurus*, *Panochthus*, *Glyptodon* (Fig. 6.13), and *Plaxhaplous* were among the largest of the known glyptodonts and date from the end of the Pleistocene to the beginning of the Holocene. All had species reaching up to and over 2000 kg. Perhaps this gigantism was a response to the influx of large predators from the north (including humans) (Soibelzon et al. 2012) and competition with other herbivores (Gutiérrez et al. 2010). Nevertheless, studies of limb mechanics show that even the largest animals could briefly sustain themselves bipedally, perhaps for dominance interactions or for breeding (Vizcaíno et al. 2011). During their evolution during the Miocene, they would have had to contend

Fig. 6.13 *Glyptodon* was a very large, armored glyptodont and probably would not have had trouble defending itself from jaguars, when they appeared in South America about five million years ago. (By Roman Uchytel)

with the top predator Phorusrhacids birds, and size would have been a defense against the deadly beaks of these birds. Feeding specializations are supported by the adaptations seen in the masticatory apparatus (Gutiérrez et al. 2010). The most recent glyptodont known was a *Doedicurus clavicaudata* dated at 8000–7000 years BP (Politis and Gutiérrez 1998). On the other hand, the four genera of glyptodonts known for the Early Miocene Santa Crucian Formation were slightly more than 100 kg, much smaller than the Middle Miocene to Pleistocene glyptodonts. Differences in jaw anatomy suggest habitat repartition and diet specialization (Vizcaíno et al. 2012).

A new Santacrucian glyptodontid *Parapropalaehoplophorus septentrionalis* (and new genus) was found from northern Chile with a distinct osteodermal pattern from any other Santacrucian glyptodontid and with many other mixed characters distinctive from other known glyptodonts of this SALMA (Fig. 6.14). The animal was probably around 100 kg, large for Santacrucian glyptodonts (Croft et al. 2007).

6.3 Order: Pilosa

The second large group (suborder) in the Xenarthra is called the Pilosa. The Pilosa includes the anteaters (Vermilingua) and the sloths (Folivora, *Phyllophaga* or Tardigrada). The first pilosan was found in the Middle Eocene La Meseta deposits of Seymour Island in Antarctica, but most think that the group as such was around even from the Paleocene, just as the armadillos. There is, however, no fossil

Fig. 6.14 *Parapropalaehoplophorus septentrionalis* from northern Chilean Santacrucian SALMA. (By Velizar Simeonovski)

evidence as yet. Most of the remains discovered during the Eocene are limited to osteoderms, and it is not until the Miocene that several fossilized skeletons and skulls appear.

6.3.1 Suborder: Vermilingua

Although Vermilingua or anteaters were probably separately evolving even in the Eocene and perhaps in Paleocene times, there are no fossil records until the Early Miocene of Patagonia (Carlini et al. 1992), but molecular evidence suggests that the Vermilingua separated from the rest of the xenarthrans in the Early Eocene around 58 Ma (Gaudin and Branham 1998). Also, *Cyclopes* has been separated from the rest of the anteaters for at least 38 Ma (Delsuc et al. 2001). There are few fossils to record Vermilingua evolution: *Protamandua rothii* (Early Miocene), *Palaeomyrmidon* (Late Miocene), and *Neotamandua conspicua* (Late Miocene–Pliocene (Bargo et al. 2012). *Neotamandua borealis* is directly ancestral to the giant anteater *Myrmecophaga*, and it is close to *Tamandua*, although it is probably *Protamandua* that is the common ancestor of these two modern genera (Hirschfeld 1976). *Nunezia Caroloameghino* from the Late Paleocene was also recognized as a probable immediate ancestor of *Myrmecophaga*, or it may be synonymous with *Myrmecophaga*. *Palaeomyrmidon incomptus* is a sister group of *Cyclopes* (Bargo et al. 2012; Hirschfeld 1976).

Most analyses place the enigmatic *Eurotamandua* outside the Vermilingua and Xenarthra, seeing that European Middle Eocene fossil as an independent and convergent evolution having nothing to do with xenarthran evolution (Delsuc et al. 2001).

6.3.2 Suborder: Folivora

Although these animals have been called *Phyllophaga* and Tardigrada as well, these are not legal designations, since those names were previously used to refer to a group of beetles (*Phyllophaga*) and to the tardigrades (Tardigrada) and the International Zoological Code makes them unavailable for the sloths and anteaters. Therefore, the group legally has been designated Folivora (Shockey and Anaya 2010).

Many sloths evolved terrestrial locomotion, and the group was very widely spread throughout South America, especially beginning in the Miocene and extending up to a few thousands of years ago. They were the first South American mammals to spread to North America and to the Caribbean, and this they managed to do around eight million years ago, before the actual physical connection of the two continents at about three million years ago. Recent discoveries in Peru also suggest the early entrance into South America of several mammalian lines such as gomphotheres, tayassuids, and camelids at about the same time as the spread of ground sloths northward (Campbell et al. 2000, 2001, 2009, 2010).

There have been about 89 genera of sloths described in six families, and these, along with the glyptodonts, represent the most successful radiation of xenarthrans, especially because both groups managed to disperse to North America. Nevertheless, the sloths diversified into at least five genera in North America and were able to claim niches in the far north as far as Alaska, something that the glyptodonts were unable to do, only evolving a few species in one genus in the southern parts of North America. Phylogenetic analysis suggests that the three traditional families of sloths, Mylodontidae, Megalonychidae, and Megatheriidae, are monophyletic (Gaudin 2004).

During Plio-Pleistocene times, many of the Folivora grew to great size; the largest sloth (and largest Xenarthra) was probably the gigantic *Megatherium americanum* (Fig. 6.15), the largest terrestrial mammal known in South America except for the immigrant proboscideans that became widespread in South America after the joining of the two Americas. This is also the first fossil ever to be mounted in a hypothetical life form in Spain (the feet were inadvertently inverted, front to back and back to front) (Piñero 1988). Illustrations (Fig. 1.1) of this early fossil were the basis of the scientific description of the species (Cuvier 1796). The *M. americanum* reached 4 tons and 6 m in length, while the two South American species of *Stegomastodon* (or *Notiomastodon*) probably topped 6 tons (Fariña et al. 1998). *Megatherium americanum* was, however, taller than the South American proboscideans. Its great height allowed it to browse on trees; its claws were useful tools for pulling branches into reach of its long, narrow muzzle that was perhaps equipped with a long, thin tongue. Because of its great claws, this sloth probably walked on the sides of its feet like a giant anteater. Its ancient tracks showed that it was also capable of bipedalism. This leads paleobiologists to believe that it was a very selective eater and probably ate leaves, small twigs, and perhaps fruit, which it could chose (Bargo and Vizcaíno 2008). Another proposal was that *M. americanum* and

Fig. 6.15 *Megatherium americanum*, the largest ground sloth known. (By Roman Uchytel)

other ground sloths were scavengers, while *M. americanum* could have also hunted (Fariña 1996; Fariña and Blanco 1996), a prospect difficult to believe considering its great bulk and phylogenetic history (Bargo 2001). Considering its size, *Megatherium* would have been capable of taking prey from any large predator, as nicely dramatized in the BBC video dramatization, "Walking with Prehistoric Beasts." There were various species of this genus (Pujos and Salas 2004; Saint-André and de Iuliis 2001).

Another giant, *Eremotherium laurillardi*, was able to disperse to North America from South America. Formerly known by various synonyms and particularly *E. rusconi*, this was one variable tropical species that, at its largest, was about equivalent to the size of *M. americanum* (Cartelle and De Luliis 1995, 2006) and was closely related to it. This species was widely distributed from southern United States to Ecuador (Tito 2008) and seemed to be ecologically flexible and able to exploit grazing to a great degree at around 6° latitude, changing to a mixed diet in caatinga around 9°–10° S in the Brazilian intertropical region (de Melo França et al. 2014). In more closed environments, *E. laurillardi* probably fed on leaves and fruits, and so it would have been important as a seed disperser (Dantas et al. 2012).

Size developed as the sloth group evolved, just as in the glyptodonts (Fig. 6.16). Studies of Santa Cruzean sloths found that the largest reached only about 100 kg and the only Vermilingua was smaller, perhaps 12 kg (Bargo et al. 2012). Based on dental and other evidence, these Early Miocene sloths were probably semiarboreal and depended on leaves for their diet (Pujos et al. 2012). The use of pedolateral stance evolved in Mylodontidae, Nothrotheriidae, and Megatheriidae, while the Mylodontidae were probably the only grazing sloths (Pujos et al. 2012).

Fig. 6.16 *Glossotherium robustum*. (By Roman Uchytel)

Pseudoglyptodon (Tinguiririca SALMA, Early Oligocene) has been reported as the earliest sloth or sloth relative (McKenna et al. 2006), although a tooth and humerus from Seymour Island, Antarctica (Middle Eocene) (45 Ma), might mark the very earliest evidence of the sloth lineage, even though molecular studies suggest that the sloths and the anteaters separated from each other only in the Oligocene at about 31.7 Ma (Pujos and de Iuliis 2007; Vizcaíno amd Scillato-Yané 1995).

Very large mammals (proboscideans, sloths) survived in the pampas of Argentina until about 7500 years BP, and they were mostly xenarthrans. The wide feeding habits and habitat preferences and differential consumption of vegetation seem to have supported successful competition with new grazers from the north like the gomphotheres.

Most surprising has been the discovery of a branch of sloths of the Nothrotheriidae that were at least partially aquatic, judging by the taphonomy and anatomy of the many specimens that have been found. The first species, *Thalassocnus natans* (Fig. 6.17), was found in marine Pliocene deposits from Peru (de Muizon and McDonald 1995). Other species of *Thalassocnus* have been found along the Pacific coast of South America (five species) in Late Miocene–Pliocene marine deposits and show different feeding adaptations that probably ranged from shallow water and beachside grazing of sea grasses (the oldest species) to specialized feeding in deep water (de Muizon et al. 2003, 2004a, b; Canto et al. 2008). Judging by the numerous bones of this sloth found in Miocene and Pliocene sediments, it would appear that they were adapted to graze on marine vegetation like Galapagos marine iguanas (*Amblyrhynchus cristatus*), especially since inland from the coast it was mostly desert. The earlier forms seemed to be less adapted to this curious lifestyle and may

Fig. 6.17 *Thalassocnus natans* was partially aquatic and fed on marine vegetation from the Pacific coast of South America. (By Roman Uchytel)

have picked up vegetation from the beaches, while later forms showed clear skeletal adaptations that would have facilitated deep-sea diving and underwater grazing on marine alga, including anatomical evidence of a strong upper lip similar to manatees (Muizon et al. 2003; Amson et al. 2014).

There are only two living genera of sloth, although McKenna and Bell (1997) list 7 families (Rathymotheriidae, Scelidotheriidae, Mylodontidae, Orophodontidae, Megatheriidae, Megalonychidae, and Bradypodidae) and 87 genera from the past, plus 2 living genera, so the group has been very diverse. Sloths are among the most diverse and dominant mammalian groups during the South American Cenozoic. Phylogenetic analyses show that the still-living *Bradypus* is sister taxon to all other sloths, while the living *Choloepus* is related to the extinct Megalonychidae, indicating that the split between *Bradypus* and *Choloepus* is very ancient at around 40 Ma. Thus, the many similarities between the two genera illustrate a dramatic example of convergent evolution (Gaudin 2004; Gaudin and Wible 2006).

In the Bradypodidae, the split between *Bradypus torquatus* and the proto-*Bradypus tridactylus*/*B. variegatus* was estimated at about 7.7 million years ago (MYA), while in the Myrmecophagidae, the first offshoot was *Cyclopes* at about 31.8 MYA followed by the split between *Myrmecophaga* and *Tamandua* at 12.9 MYA. We estimate the split between sloths and anteaters to have occurred at about 37 MYA (Gaudin 2004).

References

Aguilar DH, Laurito CA (2009) El armadillo gigante (Mammalia, Xenarthra, Pampatheriidae) del Río Tomayate, Blancano Tardío-Irvingtoniano temprano, El Salvador,América Central. Revista Geológico de América Central 41:25–36

Amson E, de Muizon C, Laurin M, Argot C, de Burfénil V (2014) Gradual adaptation of bone structure to aquatic lifestyle in extinct sloths from Peru. Proc R Soc B 281. https://doi.org/10.1098/fxp.2014.0192

Antoine PO, Marivaux L, Croft DA, Billet G et al (2011) Middle Eocene rodents from Peruvian Amazonia reveal the pattern and timing of caviomorph origins and biogeography. Proc R Soc B 279:1319–1326

Bargo MS (2001) The ground sloth *Megatherium americanum*: skull shape, bite forces, and diet. Acta Palaeontol Pol 46(2):173–192

Bargo MS, Vizcaíno SF (2008) Paleobiology of Pleistocene ground sloths (Xenarthra, Tardigrada): biomechanics, morphogeometry and ecomorphology applied to the masticatory apparatus. Ameghiniana (Rev Asoc Paleontol Argent) 45(1):175–196

Bargo MS, Toledo N, Vizcaíno SF (2012) Paleobiology of the Santa Crucian sloths and anteaters (Xenarthra, Pilosa). In: Vizcaino SF, Kay M, Bargo MS (eds) Early Miocene paleobiology in Patagonia: high-latitude paleocommunities of the Santa Cruz Formation. Cambridge University Press, Cambridge, pp 216–242

Bergqvist LP, Abrantes EAL, Avilla LS (2004) The Xenarthra (Mammalia) of Sao José de Itaboraí Basin (upper Palaeocene, Itaboraian) Rio de Janeiro, Brazil. Geodiversitas 26(2):323–337

Campbell KE Jr, Frailey DC, Romero-Pittman L (2000) The late Miocene gomphothere *Amahuacatherium peruvium* (Proboscidea: Gomphotheriidae) from Amazonian Peru: implications for the great American faunal interchange. Boletin-Instituto Beologico Minero y Metalúrgico. Serie D: Estudios Regionales 23:1–152

Campbell KE Jr, Frailey DC, Romero-Pittman L (2009) In defense of *Amahuacatherium* (Proboscidea: Gomphotheriidae). Neues Jahrb Geol P-A 252:113–128

Campbell KE Jr, Prothero DR, Romero-Pittman L, Hertel F, Rivera N (2010) Amazonian magnetostratigraphy: dating the first pulse of the great American faunal interchange. J S Am Earth Sci 26:619–626

Campbell KE Jr, Heizler M, Frailey CD, Romero-Pittman L, Prothero DR (2001) Upper Cenozoic chronostratigraphy of the southwestern Amazon basin. Geology 29(7):595–598

Canto J, Salas-Gismondi R, Cozzuol M, Yáñez J (2008) The aquatic sloth *Thalassocnus* (Mammalia, Xenarthra) from the late Miocene of North-Central Chile: biogeographic and ecological implications. J Vertebr Paleontol 28(3):918–922. https://doi.org/10.1671/0272-4634(2008)28[918:TASTMX]2.0.CO;2

Carlini AA, Scillato-Yané GJ, Vizcaíno SF, Dozo MT (1992) Un singular Myrmecophagidae (Xenarthra, Vermilingua) de Edad Colhuehuapense (Oligoceno Tardio-Miocene Temprano) de Patagonia, Argentina. Ameghiniana 29:176

Cartelle C, De Luliis G (1995) *Eremotherium laurillardi*: the Panamerican late Pleistocene megatheriid sloth. J Vertebr Paleontol 15(4):830–841

Cartelle C, De Luliis G (2006) *Eremotherium laurillardi* (Lund) (Xenarthra, Megatheriidae), the Panamerican giant ground sloth: taxonomic aspects of the ontogeny of skull and dentition. J Syst Palaeontol 4(2):199–209

Ciancio MR, Carlini AA, Campbell KE, Scillato-Yané GJ (2013) New Paleogene cingulates (Mammalia, Xenarthra) from Santa Rosa, Perú and their importance in the context of South American faunas. J Syst Palaeontol 11(6):727–741

Cope ED (1889) The Edentata of North America. Am Nat 23(272):657–664

Croft DA (2016) Horned armadillos and rafting monkeys. Indiana University Press, Bloomington

Croft DA, Flynn JJ, Wyss AR (2007) A new basal glyptodontid and other Xenarthra of the early Miocene Chucal Fauna, Northern Chile. J Vertebr Paleontol 27(4):781–797

Cuvier G (1796) Notice sur le squelette d'une très grande espèce de quadrupede inconnue jusqu'à présent, trouvé au Paraguay, et déposé au cabinet d'histoire naturelle de Madrid. Mag encyclop, ou J des Sci, de Lettres et del Art 1:303–310; 2:227–228

Dantas MAT, Dutra RP, Cherkinsky A, Costa Fortier D, Yoshino Kamino LH, Cozzuol MA, Ribeiro S, Silva Vieira FS (2012) Paleoecology and radiocarbon dating of the Pleistocene megafauna of the Brazilian intertropical region. Quat Res 79:61. https://doi.org/10.1016/j.yqres.2012.09.006

De Muizon C, McDonald HG (1995) An aquatic sloth from the Pliocene of Peru. Nature 375:224–227

De Muizon C, McDonald HG, Salas R, Urbina M (2003) A new early species of the aquatic sloth *Thalassocnus* (Mammalia, Xenarthra) from the late Miocene of Peru. J Vertebr Paleontol 23(4):886–894

De Muizon C, McDonald HG, Salas R, Urbina M (2004a) The youngest species of the aquatic sloth *Thalassocnus* and a reassessment of the relationships of the Nothrothere sloths (Mammalia: Xenarthra). J Vertebr Paleontol 24(2):387–397

De Muizon C, McDonald HG, Salas R, Urbina M (2004b) The evolution of feeding adaptations of the aquatic sloth *Thalassocnus*. J Vertebr Paleontol 24(2):398–410

Delsuc F, Ctzeflis FM, Stanhope MJ, Douzery EJP (2001) The evolution of armadillos, anteaters and sloths depicted by nuclear and mitochondrial phylogenies: implications for the status of the enigmatic fossil *Eurotamandua*. Proc R Soc Lond B 268(1476):1605–1615

Delsuc F, Scally M, Madsen O, Stanhope MJ, de Jong WW, Catzeflis FM, Springer MS, Douzery EJP (2002) Molecular phylogeny of living xenarthrans and the impact of character and taxon sampling on the placental tree rooting. Mol Biol Evol 19:1656–1671

Delsuc F, Stanhope MJ, Douzery EJP (2003) Molecular systematics of armadillos (Xenarthra, Dasypodidae): contribution of maximum likelihood and Bayesian analyses of mitochondrial and nuclear genes. Mol Phylogenet Evol 28(2):261–275

Delsuc F, Vizcaíno SF, Souzery EJP (2004) Influence of Tertiary paleoenvironmental changes on the diversification of South American mammals: a relaxed molecular clock study within xenarthrans. BMC Evol Biol 4:1–13

Englemann G (1985) The phylogeny of the Xenarthra. In: Montgomery GG (ed) The ecology and evolution of armadillos, sloths, and vermilinguas. Smithsonian Institution Press, Washington, DC, pp 51–64

Fariña RA (1996) Trophic relationships among Lujanian mammals. Evol Theory 11:125–134

Fariña RA, Blanco RE (1996) *Megatherium*, the stabber. Proc R Soc Lond 263:1725–1729

Fariña RA, Vizcaíno SF (2001) Carved teeth and strange jaws: how glyptodonts masticated. Acta Palaeontol Pol 46(2):219–234

Fariña RA, Vizcaíno SF, Bargo MS (1998) Body mass estimations in Lujanian (Late Pleistocene-Early Holocene of South America) mammal megafauna. Mastozool Neotrop 5(2):87–108

Fernicola JC (2005) Análisis filogenético de la familia Glyptodontidae Gray 1869. PhD dissertation, Universidad de la República, Montevideo

Fernicola JC (2008) Nuevos aportes para la sistemática de los *Glyptodontia* Ameghino 1889 (Mammalia, Xenarthra, Cingulata): Ameghiniana 45(3):553–574

Fernicola JC, Vizcaíno SF, Fariña RA (2008) The evolution of armored xenarthrans and a phylogeny of the glyptodonts. In: Vizcaíno SF, Loughry WJ (eds) The biology of the Xenarthra. University Press of Florida, Gainesville, pp 79–85

França LM, Trindade DAntas MA, Bocchiglieri A, Cherckinsky A, Souza Ribeiro A, Bocherens H (2014) Chronology and ancient feeding ecology of two upper Pleistocene megamammals from the Brazilian intertropical region. Quat Sci Rev 99:78–83

Gardner AL (2005a) Order Cingulata. In: Wilson DE, Reeder DAM (eds) Mammal species of the world: a taxonomic and geographic reference, vol I. The Johns Hopkins University Press, Baltimore, pp 94–99

Gardner AL (2005b) Order Pilosa. In: Wilson DE, Reeder DAM (eds) Mammal species of the world: a taxonomic and geographic reference, vol I. The Johns Hopkins University Press, Baltimore, pp 100–103

Gardner AL (2007) Cohort Placentalia Owen, 1837, Magnorder Xenarthra Cope, 1889. In: Gardner AL (ed) Mammals of South America. Vol. 1. marsupials, xenarthrans, shrews, and bats. University of Chicago Press, Chicago

Gaudin TJ (2004) Phylogenetic relationships among sloths (Mammalia, Xenarthra, Tardigrada): the creniodental evidence. Zool J Linn Soc Lond 140:255–305

Gaudin TJ, Branham DG (1998) The phylogeny of the Myrmecophagidae (Mammalia, Xenarthra, Vermilingua) and the relationship of *Eurotamandua* to the Vermilingua. J Mamm Evol 5(3):237–265

Gaudin TJ, Wible JR (2006) Chapter 6: The phylogeny of living and extinct armadillos (Mammalia, Xenarthra, Cingulata): a craniodental analysis. In: Carrano MT (ed) Amniote paleobiology: perspectives on the evolution of mammals, birds and reptiles. University of Chicago Press, Chicago, pp 153–198

Góis F, Scillato-Yané GJ, Carlini AA (2012) Una nueva especie de *Holmesina* (Xenarthra, Cingulata, Pampatheriidae) del Pleistoceno de Rondônia, sudoste de la Amazonia, Brasil. Rev Bras Paleontol 15(2):211–227

Goís F, Scillato-Yané GJ, Carlini AA, Guilherme E (2013) A new species of *Scirrotherium* Edmund & Theodor, 1977 (Xenarthra, Cinculata, Pampatheriidae) from the late Miocene of South America. Alcheringa 37:175–186

González LR (2010) Los Cingulata (Mammalia, Xenarthra) del Mioceno temprano y medio de Patagonia (Edades Santacrucense y "Friasense"). Revisión sistemática y concentraciones bioestratigráficas. Mastozool Neotrop 17(2):397–398

González-Ruiz LR, Góis F, Ciancio MR, Scillato-Yané GJ (2013) Los Peltephilidae (Mammalia, Xenarthra) de la formación Collón Curá (Colloncurense, Miocéno medio), Argentina. Rev Bras Paleontol 16(2):319–330

Gutiérrez MA, Martínez G (2008) Trends in the faunal human exploitation during the late Pleistocene and early Holocene in the Pampean region (Argentina). Quat Int 191:53–68

Gutiérrez MA, Martínez GA, Bargo MS, Vizcaíno SF (2010) Supervivencia diferencial de mamíferos de gran tamaño en la Región Pampeana en el Holoceno temprano y su relación con aspectos paleobiológicos. In: Gutierrez M, De Nigris M, Fernández P, Giardina M, Gil A, Izeta A, Neme G, Yacobaccio H (eds) Zooarqueología a principios del siglo XXI. Aportes teóricos, metodológicos y casos de estudio. Ediciones El Espinillo, Buenos Aires, pp 231–242

Hallström BM, Kullberg M, Nilsson MA, Janke A (2007) Phylogenomic data analyses provide evidence that Xenarthra and Afrotheria are sister groups. Mol Biol Evol 24(9):2059–2068

Hirschfeld SE (1976) A new fossil anteater (Edentata, Mammalia) from Colombia, S. A. and evolution of the Vermilingua. J Paleontol 50(3):419–432

Iuliis GD, Bargo MS, Vizcaíno SF (2000) Variation in skull morphology and mastication in the fossil giant armadillos *Pampatherium* spp. and allied genera (Mammalia: Xenarthra: Pampatheriidae), with comments on their systematics and distribution. J Vertebr Paleontol 20(4):743–754

Krmpotic CM, Ciancio MR, Barbeito C, Mario RC, Carlini AA (2009) Osteoderm morphology in recent and fossil euphractine xenarthrans. Acta Zool 90:339–351

Liu F-GR, Miyamoto MM, Freire NP, Ong PQ, Tennant MR, Young TS, Gugel KF (2001) Molecular and morphological supertrees for eutherian (placental) mammals. Science 291:1786–1789

Madsen O, Scally M, Douady CJ, Kao DF, DeBry RW, Adkins R, Amrine HM, Stanhope MJ, de Jong WW, Springer MS (2001) Parallel adaptive radiations in two major clades of placental mammals. Nature 409:610–614

McKenna MC, Bell SK (1997) Classification of mammals above the species level. Columbia University Press, New York

McKenna MC, Wyss AR, Flynn JJ (2006) Paleogene pseudoglyptodont xenarthran from Central Chile and Argentinean Patagonia. Am Mus Novit 3536:1–18

Mead JI, Swift SL, White RS, McDonald HG, Baez A (2007) Late Pleistocene (Rancholabrean) glypotodont and pampathere (Xenarthra, Cingulata) from Sonora, Mexico. Rev Mex Cienc Geol 24(3):439–449

Osborn HF (1904) An armadillo from the middle Eocene (Bridger) of North America. Bull Am Mus Nat Hist 20:163–165

Patterson B, Pascual R (1972) The fossil mammal fauna of South America. In: Keast A, Erk FC, Glass B (eds) Evolution, mammals and southern continents. State University of New York Press, Albany, pp 247–309

Patterson B, Segall W, Turnbull WD (1989) The ear region in Xenarthrans (=Edentata, Mammalia). Part 1. Cingulata. Fieldiana Geol New Ser 18:1–46

Piñero JML (1988) Juan Bautista Bru (1740–1799) and the description of the genus *Megatherium*. J Hist Biol 21(1):147–163. https://doi.org/10.1007/BF0012579

Politis GG, Gutiérrez MA (1998) Gliptodontes y cazadores-recolectores de la región pampeana (Argentina). Lat Am Antiq 9(2):111–134

Pujos F, de Iuliis G (2007) Late Oligocene Megatheriidae fauna (Mammalia: Xenathra) from Salla-Luribay (Bolivia): New data on basal sloth radiation and cingulata-Tardigrada Split. J Vertebr Paleontol 27(1):132–144

Pujos F, Gaudin TJ, De Iuliis G, Cartelle C (2012) Recent advances on variability, morpho-functional adaptations, dental terminology, and evolution of sloths. J Mammal Evol 19:159–169

Rose KD (1999) *Eurotamandua* and *Palaeanodonta*: convergent or related? Palaeontol Z 73(3–4):395–401

Rose KD (2009) The beginning of the age of mammals. The Johns Hopkins University Press, Baltimore

Rose KD, Emory RJ (1993) Relationships of Xenarthra, Pholidota, and fossil "edentates": the morphological evidence. In: Szalay FS, Novacek MJ, McKenna MC (eds) Mammal phylogeny of placentals. Springer, New York, pp 81–102

Saint-André PA, de Iuliis G (2001) The smallest and most ancient representative of the genus *Megatherium* Cuvier, 1796 (Xenarthra, Tardigrada, Megatheriidae), from the Pliocene of the Bolivian Altiplano. Geodiversitas 23(4):625–645

Scillato-Yane GJ (1975) Presencia de *Macroeuphractus retusus* (Xenarthra, Dasypodidae) en el Plioceno del área mesopotámia (Argentina): Su importancia bioestratigrafica y paleobiogeografica. *Ameghiniana* 12(4):322–328

Scillato-Yané GJ, Carlini AA, Tonni EP, Noriega JI (2005) Paleobiogeography of the late Pleistocene pampatheres of South America. J S Am Earth Sci 20:131–138

Shockey BJ, Anaya F (2010) Grazing in a new late Oligocene mylodontid sloth and a mylodontid radiation as a component of the Eocene-Oligocene faunal turnover and the early spread of grasslands/savannas on South America. J Mammal Evol 18:101. https://doi.org/10.1007/s10914-010-9147-5

Simpson GG (1931) *Metacheiromys* and the Edentata. Bull Am Mus Nat Hist 59:295–381

Simpson GG (1980) Splendid isolation: the curious history of south American mammals. Yale University Press, New Haven

Soibelzon LH, Aamorano M, Scillato-Yané GJ, Piazza D, Rodríguez S, Soibelzon E, Tonni EP, Cristóbal JS, Beilinson E (2012) Un Glyptodontidae de gran tamaño en el Holoceno temprano de la región pampeana, Argentina. Rev Bras Paleontol 15(1):105–112

Tauber AA (1997) Paleoecología de la formación Santa Cruz (Mioceno inferior) en el extremo sudeste de la Patagonia. Ameghiniana 34(4):517–529

Tito G (2008) New remains of *Eremotherium laurillardi* (Lund, 1842) (Megatheriidae, Xenarthra) from the coastal región of Ecuador. J S Am Earth Sci 26:424–434

Vizcaíno SF, De Iuliis G (2003) Evidence for advanced carnivory in fossil armadillos (Mammalia: Xenarthra: Dasypodidae). Paleobiology 29(1):123–138

Vizcaíno SF, Fariña RA (1997) Diet and locomotion of the armadillo Peltephilus: a new view. Lethaia 30(79–86):79–86

Vizcaíno SF, Scillato-Yané GJ (1995) Short note: an Eocene tardigrade (Mammalia, Xenarthra) from Seymour island, West Antarctica. Antarct Sci 7(4):407–408

Vizcaíno SF, Bargo MS, Kay RF, Milne N (2006) The armadillos (Mammalia, Xenarthra, Dasypodidae) of the Santa Cruz Formation (early-middle Miocene): an approach to their paleobiology. Palaeogeogr Palaeoclimatol Palaeoecol 237:255–269

Vizcaíno SF, Blanco RE, Bender B, Milne N (2011) Proportions and function of the limbs of glyptodonts. Lethaia 44:93–101

Vizcaíno SF, Fernicola JC, Bargo MS (2012) Paleoecology of the Santacrucian glypdodonts and armadillos (Xenarthra, Cingulata). In: Vizcaino SF, Kay M, Bargo MS (eds) Early Miocene paleobiology in Patagonia: high-latitude paleocommunities of the Santa Cruz formation. Cambridge University Press, Cambridge, pp 194–215

Wetzel RM, Gardner AL, Redford KH, Eisenberg JF (2007) Order Cingulata Illiger, 1811. In: Gardner AL (ed) Mammals of South America. Marsupials, xenarthrans, shrews, and bats. University of Chicago Press, Chicago, pp 128–157

Chapter 7
The Caviomorphs: First South American Rodents

7.1 Introduction

During the first part of the Cenozoic Era, there were no rodents in South America, just as there were no primates. But sometime, apparently during the Eocene, a group of rodents suddenly appeared that seem to be allied with a group of Hystricognathi in Africa, the phiomorphs. The suborder Hystricognathi are thought to have originated in Asia where they somehow were able to extend themselves to Africa (perhaps during the earliest part of the Eocene when world temperatures reached their highest point of the past 65 million years) (Huchon and Douzery 2001; Martin 2004, 2005; Pierre-Olivier et al. 2011).

The caviomorphs are distinctive in being both Hystricomorpha (suborder) and Hystricognathi (infraorder). Caviomorphs are characterized by the structure of the cranium (Hystricognathi, the masseter anchors in an angular process at the back of the lower mandible) and especially because the masseter muscle (the muscle used for mastication) passes in part through the large foramen (orifice) infraorbital and connects to the bone on the opposite side in front of the eye socket and cheekbone (hystricomorph) (Fig. 7.1) (Eisenberg 1981). The Hystricognathi contain 18 families divided into two suborders, the Phiomorpha (principally Africa) and the Caviomorpha (principally South America but also Central and North America). The Phiomorpha are considered to be the most primitive of the Hystricognathi.

The first appearance of the Phiomyidae as fossils in Africa is dated at about 37 Ma from the El Fayum fauna of Egypt (Late Eocene–Early Oligocene). These fossil animals had many characteristics of modern African phiomorph rodents and also of caviomorph rodents, even though the latest data from South America has shown the existence of a caviomorph community from about 41 Ma in the mid-Eocene. Thus, South American caviomorph rodents must have arrived in South America soon after they became established in Africa as primitive phiomorph

T. Defler, *History of Terrestrial Mammals in South America*, Topics in Geobiology 42, https://doi.org/10.1007/978-3-319-98449-0_7

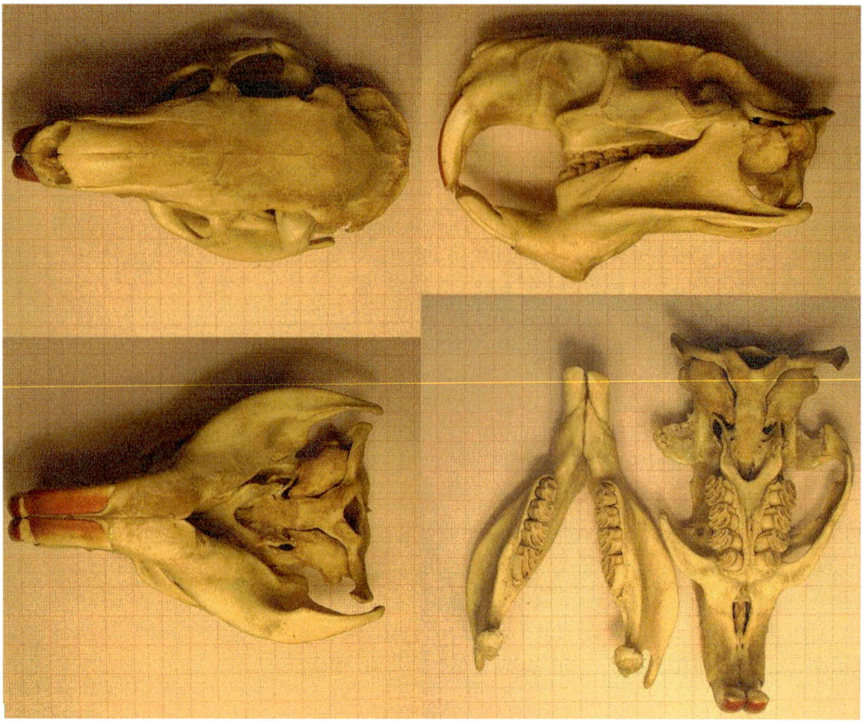

Fig. 7.1 Skull and lower mandible of *Myocastor coypus* hystricognathous lower jaw and hystricomorphous zygomasseteric system. The relative size of the infraorbital foramen through which part of the masseter medialis passes, connecting to the bone on the opposite side of the skull defines the Hystricognathi. (By Siebe)

rodents (Rose 2006; Sallam et al. 2009). Caviomorph dental patterns seem to connect the caviids to the African phiomorphs, since *Eoincamys* and *Incamys* from the Santa Rita local fauna are similar to *Gaudeamus* (Thryonomyidae) from the Late Eocene El Fayum fauna in Africa (Frailey and Campbell 2004).

7.1.1 Arrival of Rodents in South America

The arrival of these animals from Africa presents biogeographical problems, since Eocene distances between South America and Africa were extensive at about 1800 km. Both rafting and island hopping have been proposed as modes of travel from Africa to South America (Bandoni de Oliveira et al. 2009). This is hard to accept for many, who cannot imagine rodents (and primates) rafting such a distance over a great ocean to successfully land in South America (Morton 2013). It is incontrovertible that vegetation from Africa has reached South American shores and paleo-winds are thought to have been the same as modern winds, east to west. We have evidence as well of a series of very large islands that existed in the South Atlantic during the mid to late Eocene that might have made such a trip much shorter (Bandoni de Oliveira et al. 2009). However, for me the thought of the millions of years available to get the conditions just right makes it quite understandable that small African animals could make the trip. Ideally, with the right winds, it could have been 2 weeks or less. Rafts of vegetation would carry food for very small mammals (early South American rodents and primates were very small), either fruits, small invertebrates, or other parts of vegetation. A rainy period over the Atlantic could easily supply enough fresh water for these travelers. Licking wet leaves for their moisture is commonly observed in many mammals.

A synchronous arrival of both rodents and primates has been proposed in various publications with no evidence why this should have been (Lavocat 1980; Voloch et al. 2013). Perhaps the rafting hypothesis has appeared so unlikely to many that the possibility of two rafting events was difficult to swallow. However, it seems to this author that a synchronous arrival of rodents and primates on the same raft seems rather illogical and, although the dates now accepted for the arrival of primates have narrowed the gap between those accepted for rodents and the latest dates for primates, there seems to be no logical reason why the two groups need to be connected. More recently molecular data have flirted with the possibility that the two groups might have arrived simultaneously (Lavocat 1980; Perelman et al. 2011; Loss-Oliveira et al. 2012; Poux et al. 2006). But a study of both lineages using the same molecular data set dates for caviomorph rodents at between 45.4 +/− 4.1 (or 43.7 +/− 4.8) Ma, including the IRBP nuclear gene, and 36.7 +/− 3.7 (or 35.8 +/− 4.3) Ma without the IRBP. The platyrrhine primates' arrival has been calculated at 37.0 +/− 3.0 Ma with the IRBP, and without the IRBP nuclear gene, they were calculated at 38.9 +/− 4.0 Ma (Poux et al. 2006). My own view is that, given the millions of years for these events to occur, they did finally occur, probably after many, many failures. Thus, sweepstake events can finally be recognized, a posteriori for both groups, but most likely separately.

7.1.2 Cachiyacu River Rodents (41 Ma)

Very recently an extremely ancient rodent fossil community was discovered in the Cachiyacu River, Loreto, Perú. The community has been confidently dated with ashes to 41 Ma or around the Mid-Eocene Climatic Optimum – ECO (Pierre-Olivier et al. 2011). This agrees perfectly with the molecular calculation of 45.4–36.7 Ma made using three nuclear genes and IRBP (Poux et al. 2006). More recently the date 41 Ma has been questioned and discussed with the suggestion that the Contamana local fauna could actually be of the same age as the Santa Rita local fauna (whose age is unconfirmed) or even younger, based on various arguments which include the lack of detailed knowledge of the associated paleofauna and the usefulness of the $^{40}AR/^{39}Ar$ date, which was taken from silt 47 m above the actual fossiliferous level (Bond et al. 2015).

Fig. 7.2 A hypothetical illustration of the tiny 40 g *Canaanimys* rodent, among the first rodents to reach South America. (By M. J. Orliac)

Three small rodents from the Contamana local fauna have been described, *Cachiyacuy contamanensis*, *C. kummeli* (30–40 g), and *Canaanimys maquiensis* (40 g) (Fig. 7.2) as well as the teeth of *Eobranisamys* sp. (Dasyproctidae) and *Eospina* sp., earlier discovered and described from the Santa Rosa local fauna, Peru, and dated to about 34 Ma. The animals were clearly from a tropical forest. This changes our view of the earliest rodent communities that previously had been seen as having evolved in association with drier types of vegetation. Additionally the size of these rodents is much more similar to many phiomorph rodents of Africa, suggesting that these South American rodents were very close to the progenitors of caviomorph rodents. The teeth of these middle Eocene rodents clearly connect them to the African phiomorph rodents as well (Rose 2006). These early neotropical rodents probably ate soft seeds and plant parts. At the time the specimens were deposited, the area was a lush tropical forest, probably rich in animal life. Near the rodent fossils, the researchers discovered remains of crawfish, reptiles and crocodiles, other mammals, opossum-like marsupials, and even an armadillo as well as other fauna related to tropical forests (Pierre-Olivier et al. 2011, 2016).

Over time, various projects have attempted to discover the earliest age of the phiomorph/caviomorph split using various *molecular clocks*. A recent study established 45–35.7 Ma as the time of the split (Poux et al. 2006), and this was considerably more ancient times than previous attempts. Another calculated date for the appearance of the caviomorphs was 33.4–39 Ma (Sallam et al. 2009).

Tooth microstructure of enamel patterns from the Conamana rodents suggests that the caviomorphs developed from various invasions from Africa rather than one alone. Because enamel is laid down in distinct patterns that can be recognized (*schmelzmuster*), some patterns seem to be related to different African groups of phiomorphs, suggesting that they could have arrived in South America at varying times. It does not seem unreasonable to imagine that such (initially) small mammals could have crossed the Atlantic on rafts of vegetation during various times in their history (Martin 2004). Nevertheless, various molecular data produce phylogenies for the Caviomorpha showing them to be monophyletic (Nedbal et al. 1994). A type of *schmelzmuster* from 20 isolated rodent incisors found in the Santa Rita local fauna seems to be shared with the African Thryonomyoidea (Martin 2005).

7.2 Other Ancient Caviomorph Communities

Other important ancient caviomorph communities have been described, and, for a time, each community was considered to be the earliest known caviomorph community. They are (1) a community discovered at Santa Rita, Peru, from late Eocene or perhaps Early Oligocene of about 43–34 Ma (Frailey and Campbell 2004; Campbell 2004); (2) a community discovered in Chile that defined a new SALMA, the Tinguiririca, dated from about 31 Ma (Wyss et al. 1993; Flynn et al. 2003); and (3) a Patagonian community discovered from the Deseadan SALMA dated at about 28 Ma (Late Oligocene) (Wyss et al. 1993, 1994).

7.2.1 Santa Rita Rodents (43–34 Ma)

The age of the Santa Rita, Peru, rodent community is unresolved but considered to be Middle to Late Eocene? (43–34) but pre-Tinguiririca (36–29 Ma) and probably post-Cachiyacu (41 Ma). The Santa Rita rodents were until recently the oldest recognized South American rodent community. Constrained in this way, the Santa Rita local fauna could have an age of about (40–37 Ma), though this has not been suggested by its discoverers. Even at that age the families, Agoutidae, Erithizontidae, Echimyidae could be recognized. Obviously considerable evolution had taken place, arguing for a more ancient arrival in South America. It is evident that caviomorph lineages, in contrast to platyrrhine lineages, diversified very early during Eocene times and that these early lineages are still identifiable (Campbell 2004; Frailey and Campbell 2004). All Santa Rosa rodents share a common morphological pattern of the teeth, suggestive of a basal place in the caviomorph radiation (Frailey and Campbell 2004). Even recently the idea of a vicariant origin from the ancestral (Hystricognathi) (Frailey and Campbell 2004) has been considered, although currently the tide may have finally turned for a rafting origin from Africa (Bandoni de Oliveira et al. 2009). The cusps of these rodents were low or *brachydont*, unlike that of the later Tinguiririca rodent and ungulate fauna which possess a high degree of *hypsodonty* (high crowns).

7.2.2 *Tinguiririca Rodents (36–29 Ma)*

The discovery of the fossils of a rodent in a central Chilean Andean Early Oligocene mammal community was put forward for awhile as the earliest rodents in South America at about 36 Ma. This mammal community was dated from before the Deseadan SALMA and is the source material for naming the Tinguiririca SALMA, named from the Tinguiririca River of central Chile. Dating has suggested an age of between 37.5 (Late Eocene) and 31.5 Ma (Early Oligocene). Dental characteristics from this fauna also argue for a relationship to African phiomorphs (Wyss et al. 1993; Flynn et al. 2003). In contrast to the Santa Rita local fauna, the Tinguiririca rodents (and herbivore ungulates) have the world's oldest hypsodont (high-crowned) cheek teeth, and related evidence suggest an open, relatively dry habitat in accordance with the deteriorating world climate at the beginning of the Oligocene (Flynn et al. 2003; Croft et al. 2008). This is the first occurrence of open grassland known in the world and contrasts strongly with the brachydont cheek teeth of rodents and ungulates in the closed forest Santa Rita local fauna (Flynn et al. 2003; Croft et al. 2008). The Tinguiririca rodents belonged to the Dasyproctidae and Chinchillidae families.

7.2.3 *Platypittamys brachyodon*

Platypittamys brachyodon (Fig. 7.3) from Patagonia was previously known as the oldest caviomorph (from the family Octodontidae as an Acaremyinae) (Wood 1949); however, even more primitive octodontids have been found in the remarkable

Fig. 7.3 *Platypittamys brachyodon* was previously thought to be the oldest caviomorph rodent from the Octodontidae. (By Valeria Cadena)

Middle Eocene rodent fauna of the Peruvian Amazon (Pierre-Olivier et al. 2011). This early rodent is known from a complete skeleton in association with the lower Oligocene Deseadan ungulate *Scarrittia canquelensis* from an age between 33 and 24 Ma (Wood 1949; Ribeiro et al. 2010). The cheek teeth of this animal are low-crowned and may actually correspond to a specialization rather than to primitive characters (Frailey and Campbell 2004).

7.3 Molecular Phylogenies

Molecular evidence for the infraorder Hystricognathi (Woods and Kilpatrick 2005) supports evidence for originally two caviomorph lineages (Chichilloidea + Octodontoidea) and (Cavioidea + Erethizontoidea) at around 37.9 Ma which then separated out into four clades (Huchon and Douzery 2001; Voloch et al. 2013). The resulting four molecular clades are very well-defined and represent the ancient lineages that, according to some, diverged from the phiomorphs at around 43 Ma (Voloch et al. 2013). These are the Cavioidea (represented by *Cavia*) and Erethizontidae (represented by *Erethizon*) which separated at about 33.9 Ma; Chinchilloidea (represented by *Chinchilla* and *Dinomys*) and Octodontoidea (represented by *Echimys*) may then have separated at 35.0 Ma (Voloch et al. 2013).

New World caviomorph ages for families that separated from their superfamilies have been inferred as follow: Ctenomyidae and Octodontidae divergence at 23.4 Ma, Capromyidae divergence from Echimiyidae at 17.2 Ma, and Dinomyidae and Chinchilloidea at 21.3 Ma, while Cuniculidae and Caviidae separated at around 22.6 Ma (Voloch et al. 2013; Woods 1982; Huchon and Douzery 2001; Rowe and Honeycutt 2002; Rowe et al. 2010; Upham and Patterson 2012; Honeycutt et al. 2003; Opazo 2005; Patterson and Wood 1982; Vucetich et al. 1999; Woods 1982).

7.3.1 Cavioidea: Cavies and Maras

This group of caviomorphs is found throughout the neotropical region in many different habitats (prairies, steppes, forests, flooded areas, highlands). The group includes the families †Cephalomyidae and Dasyproctidae - agoutis and acouchis; Cuniculidae, pacas; †Eocardiidae and Dinomyidae, pacarana; and Caviidae, cavies, capybaras, and guinea pigs. The middle Miocene turnover of caviomorphs involved the extinction of the Eocardiidae and differentiation of the crown group into Hydrochoerinae, Cardiomyinae, Dolichotinae, and Caviinae (Vucetich and Pérez 2011). Using molecular data, the radiation of the Caviodea was estimated at 18.5 +/−2.5 Ma (Opazo 2005) despite the first fossil occurrence of Agoutidae (Dasyproctinae) with eight species in the Santa Rita local fauna (Frailey and Campbell 2004) (Fig. 7.4).

Fig. 7.4 A cavioid rodent, *Cuniculus paca*, Agoutidae family. (By Hans Hillewaert)

Eocardia (Fig. 7.5) (Eocardiidae) were cavioids whose family were probably the ancestors of the Caviidae (the cavies) of today and the Hydrochoeridae (the capybaras) (Kramarz 2006; Vucetich et al. 2013a, b). *Eocardia* is possibly the best known of the eight genera described in the Eocardiidae, but they are not found in the fossil record after Early-Middle Miocene (McKenna and Bell 1997; Simpson 1980; Candela et al. 2012). The eocardids were first recorded in Patagonian Argentina in the Deseadan SALMA (29–21 Ma), but they have not been found for the same time in Bolivia or Peru.

Early Deseadan Oligocene rodents from Argentina contained Echymidae, Dasyproctidae, and Dynomidae.

7.3.2 Erethizontoidea: Porcupines

The ancient monophyly of the group often classified as Erethizontidae (sometimes included as part of the Erethitzontoidea but probably is Cavioidea) is the only family defined for this superfamily. Only 15 species are known and, although it is controversial, the genera probably number only three. We count 13 species of *Coendou*, and one species for each of the two genera *Chaetomys* and *Erethizon* (Voss et al. 2013). The widespread genus *Coendou* is a sister genus to the North American *Erethizon*, which must have evolved from the largely tropical *Coendou* as it became a northern temperate genus (Voss et al. 2013). The group was first identified in the Santa Rita local fauna (late Eocene—early Oligocene). *Eopululo* seems to be the earliest genus established for the group and one of the largest species collected from

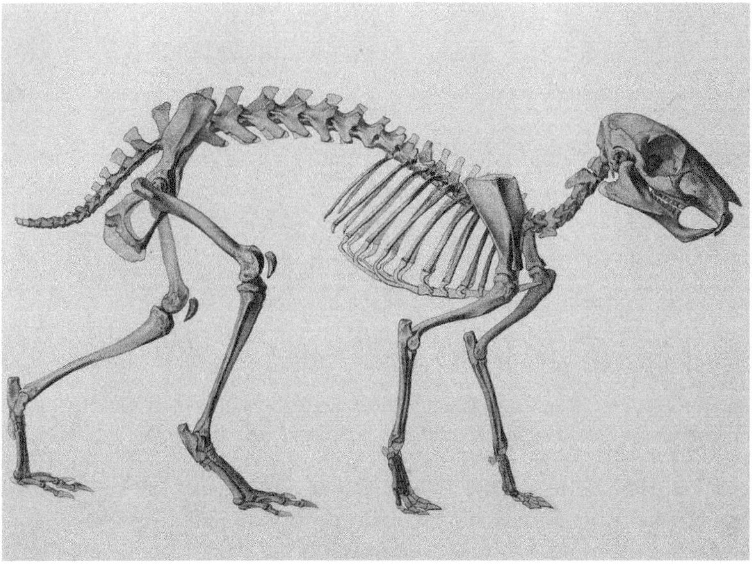

Fig. 7.5 *Eocardia* of the Early to Middle Miocene were probably ancestors of the cavies, capybaras, and maras of today (By R. B. Horsefall, 1903, from *Reports of the Princeton University Expedition to Patagonia, 1896–1899, 1901–1932*, vo. 1)

the Santa Rica local fauna. Analysis of this fossil suggests that the Erethizontinae lineage is closest to the ancestral caviid condition and predates the origins of *Playpitamys* (Frailey and Campbell 2004). The species *Coendou prehensilis* throughout its very large range from 27° of latitude and 25° of longitude has been found to be genetically very homogeneous, probably the result of a rapid trophic niche shift and range expansion (Voss et al. 2013).

The monospecific *Erethizon dorsatum* (Fig. 7.6) has extended itself from northern Mexico to northern Alaska and Canada and represents the most successful invasion of South American fauna into North America, since it, unlike the widely distributed (to Alaska) terrestrial sloth *Megalonyx jeffersonii*, is still extant (Woods 1973). The species appears to have evolved from a *Coendou* ancestor, and its shift to a temperate climate must have been key to its success.

7.3.3 Chinchilloidea: Chinchillas and Viscachas

This group is made up of three families, †Neoepiblemidae (extinct); Chinchillidae, chinchillas and viscachas; and Abrocomidae, chinchilla rats. Studies of the family Chinchillidae show a close relationship to Dinomyidae (Spotorno et al. 2004; Huchon and Douzery 2001) (Fig. 7.7).

Fig. 7.6 A porcupine, *Erethizon dorsatum* Erethezontoidea (Erethezontidae). (By Mattnad, via Wikimedia Commons). This North American species is the only erethezontid to have left South America to be widely distributed in North America

Fig. 7.7 *Chinchilla*, Chinchilloidea (Chinchillidae). (By Thirteen squared). Chinchillas live in the southern Andes Mountains at around 4000 m formerly in Chile, Peru, Argentina, and Bolivia in “herds” of about 14–100 individuals. They are now reduced mostly to living in Chile because of their constant pursuit for their fine pelts

7.3.4 *Octodontoidea: Degus, Rock Rats, Tuco-Tucos, and Nutrias*

This superfamily contains about three-quarters of the living caviomorphs and is comprised of six families (Octodontidae, degus and relatives; Ctenomyidae, tuco-tucos; Echimyidae, spiny rats; Myocastoridae, nutria; Capromyidae, hutias; †Heptaxodontidae, giant hutias) a total of about 193 species for the superfamily. New species continue to be described (Upham and Patterson 2012). The group is extremely diverse and found in most major habitats of the neotropics, including the Antilles, and they are most closely related to the Chinchilloidea). One study calculates the split of Chinchilloidea and Octodontoidea (Fig. 7.8) for the Early Oligocene, though fossils from the Santa Rita local fauna identifying Echimyidae seem to place it earlier, even from the Late Eocene (Opazo 2005; Frailey and Campbell 2004). The clade seems to be monophyletic (Nedbal et al. 1994). Late Eocene global cooling retracted southern high-latitude forests that gave origin to open biomes in high-latitude South America, allowing the evolution of the octodontids as open habitat rodents (Fig. 7.9), while the echimyids continued their adaptations to tropical forest habitats. This has caused a southerly distribution for the octodontids and the development of hypsodont species in the south (Verzi et al. 2015; Patterson et al. 2012).

The first Echimyidae (Octodontoidea) (Fig. 7.8) have been recognized from the Santa Rita local fauna of Peru from about 36–34 Ma (Eocene?), but they did not begin to diversify until the Middle Miocene at around 20.5–18.8 Ma when the three recognized main branches—Octodontidae, Echimyidae, and Abrocomidae—occurred (Fabre et al. 2013). The great diversity of the family in the tropical lowlands and the now earliest fossils suggests that they first evolved in this part of the continent (Upham and Patterson 2012). Their antiquity makes them candidates as stem octodontoids (Verzi et al. 2015). Monophyly of the entire group has been dem-

Fig. 7.8 *Diplomys caniceps* Echimyidae. (By R. Mintern). The arboreal soft-furred spiny rat is found in Northwest Colombia in tropical and subtropical lowland moist forest (Emmons et al. 2015)

Fig. 7.9 *Tympanoctomys cordubensis*, extinct Octodontidae from Argentina. (By F. Ameghino (1889))

onstrated and probably dates from the explosive mid-Miocene diversification, now termed the star-phylogeny hypothesis for echimyid diversification (Lara et al. 1996), although some have not agreed (Leite and Patton 2002). Radiation occurred with minimal cranial changes and homoplasy but with maximal habitat specializations that included ground-dwelling, fossorial, semiaquatic, and arboreal habitats, resulting in the most taxonomically, ecologically, and morphologically diverse group of all the hystricomorphs (Leite and Patton 2002; Fabre et al. 2013). The small changes in morphology have made it difficult to appreciate the extreme evolutionary radiation and high species diversity until recent molecular studies that more clearly pointed out phylogenetic relationships.

A current assessment of echimyid diversity reveals about 90 extant species and 19 genera with descriptions of more species being prepared (Woods and Kilpatrick 2005). It is widely recognized that the Caribbean Capromyidae evolved from the Echimyidae at 17.8–19.8 Ma (Fabre et al. 2013). The Echimyidae have been amply shown to belong to the Omomyoidea (Echimyidae are most closely related to the Octodontidae and the Ctenomyidae (Honeycutt et al. 2003; Huchon et al. 2007; Upham and Patterson 2012), which is a monophyletic group containing five other families: Octodontidae, Ctenomyidae (Fig. 7.10), Myocastoridae, Capromyidae, and †Heptaxodontidae (exinct family) (Lara et al. 1996; Nedbal et al. 1994; Huchon and Douzery 2001; Leite and Patton 2002; Woods and Kilpatrick 2005). Molecular divergence times suggest the middle Miocene is the origin of many modern genera (Galewski et al. 2005). Since the majority of echimyids live in lowland tropical forest areas, it seems that a more profound understanding of their evolution will have to wait the discovery of the scarce fossils from that region.

Fig. 7.10 *Octodon degu* (Octodontidae). (By ZeWrestler)

7.4 Gigantism in the Dynomyidae and Other Rodents

The family Dinomyidae has produced the largest rodents so far known, and they were truly impressive. *Phoberomys pattersoni* became famous as the "largest rodent known" until *Josephoartigasia monesi* (below) (Fig. 7.11) was described, even though the authors of this species stated that it was probably second in size to another *Phoberomys* (Rinderknecht and Blanco 2015). This new *Phoberomys* was discovered in northwestern Venezuela in upper Miocene to early Pliocene deposits of around 7–4 Ma. The animal might have weighed around 700 kg judging from the almost complete skeleton that was found. These were aquatic rodents living along the ancient Orinoco Delta (under the influence of the sea) that probably fed on abrasive, aquatic, or swampy vegetation. Their body was about 3 m long and was about 1.5 m at the shoulder (Horovitz et al. 2006). The evolution of this species must have been influenced by the huge system of lakes that were part of the western lowlands. For many millions of years, concordant with the rise of the Andes Mountains, much water did not drain into the Atlantic Ocean, but rather drained into the Caribbean Sea, supporting a huge wetlands just east of the Andes during most of the Miocene.

Although *Phoberomys pattersoni* is related to the Dinomydae (pacarana), it is sometimes placed in the extinct family Neoepiblemidae. When the fossil was discovered in the Venezuelan Urumaco Formation, it was the largest known rodent at about 700 kg (Sánchez-Villagra et al. 2003; Horovitz et al. 2011). However, another analysis suggested that the species was much smaller at around 220–340 kg, depending on the method used (Millien 2008; Millien and Bovy 2010). Since these large rodents were somewhat anomalous to living rodents, the estimation of their mass is not straight forward.

Fig. 7.11 *Ctenomys brasiliensis* Ctenomyidae (in Alcide Dessalines d'Orbigny (1847) *Voyage dans l'Amérique méridionale*)

Several other species in the genus have been discovered, all of which have become extinct. Another *Phoberomys* found in Argentina, *Phoberomys insolita*, could possibly have weighed even more than *Phoberomys pattersoni*; however the only material available of *Phoberomys insolita* is teeth, which may give erroneous calculations of body mass, while the fossil of *Phoberomys pattersoni* includes almost all of the skeleton, including femurs that probably are more informative on the question of body mass than are teeth. Many *Phoberomys* were very large rodents and may have been associated with the great upper Amazonian late Miocene wetland (Horovitz et al. 2010) (Figs. 7.12).

A huge dinomyid rodent *Josephoartigasia monesi* was discovered recently in Uruguay that is undoubtedly the largest ever discovered to date. The 53 cm skull was discovered on the coast at the mouth of the Plate River (Río de La Plata) and is dated to the Plio-Pleistocene of about 4–2 Ma (Rinderknecht and Blanco 2008). The first weight calculation, based on the 53 cm holotypic skull, was around 1000 kg. There is some disagreement with this calculation because of the technique, and the range of reported values was 468–2586 kg (Blanco 2008). Surprisingly, a very slender zygomatic arch and small molars suggest a chewing complex adapted to very soft (perhaps aquatic) vegetation as opposed to anything abrasive (like grass). They probably were found in estuaries or swamps that were surrounded in the uplands by forest.

Bite force in *Josephoartigasia* has been calculated as being extremely high, and this is also related to the extreme procumbency of the incisors. It is likely that the bite force exerted by this large rodent was a defense mechanism against predatory phorusrhacid birds and carnivorous marsupials, as a bite from one of these caviomorphs coupled with the very long rostrum would have been effective in fending off a predatory attack (Blanco et al. 2012; Cox et al. 2015).

Arazamys castiglionii, also discovered in Uruguay, appears to have reached a size between that of *Phoberomys* and that of *Josephoartigasia monesi*, although a careful estimation has not been attempted. The fossil material consists of the poste-

Fig. 7.12 *Josephoartigasia monesi,* the largest rodent so far discovered, might have weighed upward of 1000 kg and was found on the lower Río Plata (By Roman Uchytel)

rior region of a skull, the braincase, and complete dentition except for the left P4 and part of the right incisor (Rinderkenecht et al. 2011). The age is earlier than *Josephoartigasia*; it lived during the Miocene Huayquerian SALMA (9.0–6.8 Ma).

Another very large rodent and probable dinomyid described for Peru based on a lower molar (M_2) was named *Perumys gyulavarii*. The animal's mass fell just below *Phoberomys*, based on an estimated molar length of up to 126 mm for *Phoberomys* as compared to the molar length of *Perumys* of about 92–95 mm. The fossils were found in Pliocene lacustrine deposits. Molar characteristics suggest a soft diet that would be found in such a habitat (Kretzoi and Voros 1989).

The dinomyids were quite diverse during the Miocene and especially during the late Miocene, and many species were very large. They have mostly been known from Argentinean deposits, but recently several have come to light in Venezuela (Horovitz et al. 2010). Another gigantic dinomyid found in Uruguay was *Josephoartigasia magna* (Mones 1988). Since only its upper dentition is known, it seems precarious to calculate a weight.

Telicomys gigantissimus might have been almost as large as a rhinoceros (although these comparisons never specify which species of rhinoceros, which vary in weight from about 500 to 3200 kg among the five species). *Telicomys* was cited as the largest known rodent (Nowak 1999; Savage and Long 1986) at its time of 3.6–2.588 Ma; but a comparison of the lower cheek teeth of the largest specimen known (79 mm) is only about 71% of the size of *P. pattersoni*. This comparison, I calculate, would make the large *Telicomys*, only about 200–500 kg, certainly only a very small rhinoceros, indeed.

Amblyrhiza inundata was a large rodent found in late Quaternary deposits of Anguilla and St. Martin in the northern Lesser Antilles and placed in the *paraphyletic* Heptaxodontidae which subsumes other large and very large caviomorphs from the Neogene of South America. The animal's weight has been calculated between 50 and 200 kg and was characterized by a very large head, heavy body, and hind limbs with gracile forelimbs (Biknevicius et al. 1993). Populations of this large rodent could not have been large and must have varied between 125 kyr BP when sea level was 100 m lower and several islands were conjoined and comprised a single piece of land of about 2500 km^2 and could perhaps have supported 6000–15,000 individuals, according to the actual size of the animals. In contrast, a population of 187–484 animals could be supported by the 78 km^2 of the recent (present) sea level, fragmented on several islands. Since there is an absence of evidence for exploitation by humans, the extreme population reduction from rising sea levels seems the best explanation of the extinction of this large rodent (Biknevicius et al. 1993).

Others have attempted body mass calculations based on this fossil and have come up with much smaller weights. The use of one or two characteristics based on living rodents to calculate an extinct animal's weight has limitations (Millien 2008; Millien and Bovy 2010).

The family Neoepiblemidae included two genera of very large rodents which were collected in the Urumaco Formation of Venezuela but are also known from Argentina and Brazil (Bondesio et al. 1975). The family originally included *Dabbenea*, which now is considered synonymous with *Phoberomys* (Horovitz et al. 2006), and considered by some to be a Dinomyidae.

7.5 Hydrochoeridae (Cavioidea) or Hydrochoerinae (Caviidae)?

Phylogenetic analysis based on two nuclear sequences, an intron and a mitochondrial gene, strongly support the inclusion of capybaras into the Caviidae family rather than as a stand-alone sister clade to the Caviidae (Rowe and Honeycutt 2002). These results have not generally been absorbed into a literature that continues to treat them as a sister group to the caviids (Vucetich et al. 2013a, b; Pérez and Vucetich 2011; Deschamps et al. 2009). This is despite the inclusion of capybaras into the Caviidae in the influential *Mammal Species of the World* (Woods and Kilpatrick 2005).

The evolution of the caviid Hydrochoerinae is beautifully illustrated in a new book on capybaras (Vucetich et al. 2013a, b). This family includes the largest living rodents, the capybaras (*Hydrochoerus hydrochaeris* and *H. isthmius*).

The hydrochoerins most likely evolved from eocardid rodents, thought to be the stem group for this family as well as for other caviids, but the first recognizable fossil of the group *Cardiatherium Chasicoan* was found in the Late Miocene (Casacoan, 9–7 Ma) from central Argentina. By the Late Miocene, capybaras could be found

Fig. 7.13 Female capybara with young (By Fidel León Darder).

throughout South America. The first appearance of the genus *Hydrochoerus* seems to be from the late Pliocene of the Antillas (MacPhee et al. 2000). At the joining of the two continents, the capybaras *Neochoerus* and *Hydrochoerus* could be found from Argentina to southern North America (Vucetich et al. 2013a, b).

The hydrochoerins always tended toward gigantism, and the largest members of the group appeared in the Plio-Pleistocene (5.3 Ma – 10,000 years ago) with *Chapalmatherium* (200 kg) and *Neochoerus* (200 kg) (Prado et al. 1998; Vizcaíno et al. 2012; Ghizzoni 2014), but certainly not reaching the size of the largest dinomyids (Fig. 7.13).

References

Ameghino F (1889) Contribución al conocimiento de los mamíferos fósiles de la República Argentina. Actas de la Academia Nacional de Ciencias (Córdoba) 6:1–1027

Bandoni de Oliveira F, Cassola Molina E, Marroig G (2009) Paleogeography of the South Atlantic: a route for primates and rodents into the New World. In: Garber PA, Estrada A, Bicca-Marques JC, Heymann EW, Strier KB (eds) South American primates: comparative perspectives in the study of behavior, ecology, and conservation. Springer, New York, pp 55–68

Biknevicius AR, McFarlane DA, MacPhee RDE (1993) Body size in *Amblyrhiza inundata* (Rodentia, Caviomorpha), an extinct megafaunal rodent from the Anguilla Bank, West Indies: estimates and implications. Am Mus Novit 3079:1–25

Blanco RE (2008) The uncertainties of the largest fossil rodent. Proc R Soc B Biol Sci 275: 1957–1958

Blanco RE, Rinderknecht A, Lecuona G (2012) The bite force of the largest fossil rodent (Hystricognathi, Caviomorpha, Dinomyidae). Lethaia 45:157–163

Bond M, Tejedor MF, Campbell KE Jr, Chornogubsky L, Novo N, Goin F (2015) Eocene primates of South America and the African origins of New World monkeys. Nature 520:538–541

Bondesio P, Pascual R, Vucetich MG (1975) Los Neoepiblemidae (Rodentia, Caviomorpha): su caracterización y sus relaciones filogenéticas con los Chinchillidae. Actas Primer Cong Argent Paleontol Bioestratig 2:431–447

Campbell KE Jr (2004) The Paleogene mammalian fauna of Santa Rosa, amazonian Peru, Science series, vol 40. Natural History Museum of Los Angeles County, Los Angeles, p 163

Candela AM, Rasia LL, Pérez ME (2012) Paleobiology of Santacrucian caviomorph rodents: a morphofunctional approach. In: Vizcaíno SF, Kay RF, Susana M, Bargo S (eds) Early Miocene paleobiology in Patagonia: high—latitude paleocommunities of the Santa Cruz formation. Cambridge University Press, Cambridge

Cox PG, Rinderknecht A, Blanco RE (2015) Predicting bite force and cranial biomechanics in the largest fossil rodent using finite element analysis. J Anat 226(3):215–223. https://doi.org/10.1111/joa.12282

Croft DA, Flynn JJ, Wyss AR (2008) The Tinguiririca fauna of Chile and the early stages of "modernization" of South American mammal faunas. Arq Mus Nacion, Rio de Janeiro 66(1):191–211

Deschamps CM, Vieytes EC, Olivares AI, Vucetich MG (2009) Primer registro de *Cardiatherium chasicoense* (Rodentia, Hydrochoeridae) fuera del área pampeana (Argentina) y su valor bioestratigráfico. Ameghiniana 46(2):295–305

Eisenberg JF (1981) The mammalian radiations: an analysis of trends in evolution, adaptation, and behavior. The University of Chicago Press, Chicago

Emmons LH, Leite YLR, Patton JL (2015) Subfamily Echimyinae Gray, 1825. In: Patton JL, Pardiñas UFJ, D'Elía G (eds) Mammals of South America volume 2: rodents. University of Chicago Press, Chicago/London

Fabre P-H, Galewski T, Tilak M-K, Douzery EJP (2013) Diversification of South American spiny rats (Echimyidae): a multigene phylogenetic approach. Zool Scr 42:117–134

Flynn JJ, Wyss AR, Croft DA, Charrier R (2003) The Tinguiririca fauna, Chile: biochronology, paleoecology, biogeography, and a new earliest Oligocene South American land mammal age. Palaeogeogr Palaeoclimatol Palaeoecol 195:229–259

Frailey CD, Campbell KE Jr (2004) Paleogene rodents from amazonian Peru: the Santa Rosa local fauna. In: Campbell KE (ed) The Paleogene mammalian fauna of Santa Rosa, amazonian Peru, Science series, vol 40. Natural History Museum of Los Angeles County, Los Angeles, pp 71–130

Galewski T, Mauffrey J-F, Leite YLR, Patton JL, Douzery EJP (2005) Ecomorphological diversification among South American spiny rats (Rodentia: Echimyidae): a phylogenetic and chronological approach. Mol Phylogenet Evol 34:601–615

Ghizzoni M (2014) Estimación de la masa corporal de un ejemplar cuaternario del carpincho extinto *Neochoerus* a través de medidas cráneo-dentales. Rev Bras Paleontolog 17(1):83–90

Honeycutt RL, Rowe DL, Gallardo MH (2003) Molecular systematics of the South American caviomorph rodents: relationships among species and genera in the family Octodontidae. Mol Phylogenet Evol 26:476–489

Horovitz I, Sánchez-Villagra MR, Aguilera TM, Aguileara OA (2006) The fossil record of *Phoberomys pattersoni* Mones 1980 (Mammalia, Rodentia) from Urumaco (Late Miocene, Venezuela), with an analysis of its phylogenetic relationships. J Syst Palaeontol 4(3):293–306

Horovitz I, Sánchez-Villagra MR, Vucetich MG, Aguilera OA (2010) Fossil rodents from the late Miocene Urumaco and middle Miocene Cumaca formations, Venezuela. In: Sánchez-Villagra MR, Aguilera OA, Carlini AA (eds) Urumaco and Venezuelan paleontology. Indiana University Press, Bloomington

Horovitz I, Sánchez-Villagra MR, Aguilara-S OA (2011) *Phoberomys*, un gigante entre roedores. Invest y Cienc (febrero), pp 58–63

Huchon D, Douzery JP (2001) From the old world to the new world: a molecular chronicle of the phylogeny and biography of Hystricognath rodents. Mol Phylogenet Evol 20(2):238–251

Huchon D, Chevret P, Jordan U, et al (May 1, 2007) Multiple molecular evidences for a living mammalian fossil. Proc Natl Acad Sci 104 (18) 7495-7499 https://doi.org/10.1073/pnas.0701289104

Josephoartigasia monesi https://es.wikipedia.org/wiki/Josephoartigasia_monesi

Kramarz AG (2006) Eocardiids (Rodentia, Hystricognathi) from the Pinturas formation, late early Miocene of Patagonia, Argentina. J Vertebr Paleontol 26(3):770–778
Kretzoi M, Voros I (1989) On a new caviomorph rodent from Peru. Fragm Minerol Palaeont 14:111–116
Lara MC, Patton JL, Da Silva MNF (1996) The simultaneous diversification of South American echimyid rodents (Hystricognathi) base con complete cytochrome b sequences. Mol Phylogenet Evol 5:403–413
Lavocat R (1980) The implications of rodent paleontology and biogeography to the geographical sources and origin of platyrrhine primates. In: Ciochon RL, Chiarelli BA (eds) Evolutionary biology of new world monkeys and continental drift. Plenum Press, New York, pp 93–103
Leite YLR, Patton JL (2002) Evolution of South American spiny rats (Rodentia, Echimyidae): the star-phylogeny hypothesis revisited. Mol Phylogenet Evol 25(3):455–464
Loss-Oliveira L, Aguiar BO, Schrago CG (2012) Testing synchrony in historical biogeography: the case of new world primates and Hystricognathi rodents. Evol Bioinforma 2012(8):127–137. https://doi.org/10.4137/EBO.S9008
MacPhee RDE, Singer R, Diamond M (2000) Late Cenozoic land mammals from Grenada, Lesser Antilles island-arc. Am Mus Novit 3302:1–20
McKenna MC, Bell SK (1997) Classification of mammals above the species level. Columbia University Press, New York
Martin T (2004) Incisor enamel microstructure of South America's earliest rodents: implications for caviomorph origin and diversification. In: Campbell KE (ed) The Paleogene mammalian fauna of Santa Rosa, amazonian Peru, Science series, vol 40. Natural History Museum of Los Angeles County, Los Angeles, pp 131–140
Martin T (2005) Incisor Schmelzmuster diversity in South America's oldest rodent fauna and early caviomorph history. J Mamm Evol 12(3/4):405–119
Millien V (2008) The largest among the smallest: the body mass of the giant rodent Josephoartigasia monesi. Proc Biol Sci 275(1646):1953–1955
Millien V, Bovy H (2010) When teeth and bones disagree: body mass estimation of a giant extinct rodent. J Mammal 91(1):11–18
Mones A (1988) Notas paleontológicas uruguayas. IV. Nuevos registros de mamíferos fósiles de la Formacióm San José (Plioceno-Pleistoceno inferior?) (Mammalia: Xenarthra; Artiodactyla; Rodentia). Comun Paleontol Mus Hist Nat Montevideo 20:255–277
Morton MC (2013) Setting sail on unknown seas: the past, present and future of species rafting. http://www.earthmagazine.org/article/setting-sail-unknown-seas-past-present-and-future-species-rafting
Nedbal MA, Allard MW, Honeycutt RL (1994) Molecular systematics of hystricognath rodents: evidence from the mitochondrial 12S rRNA gene. Mol Phylogenet Evol 3(3):206–220
Nowak RM (1999) Walker's mammals of the world, vol II, 6th edn. The Johns Hopkins University Press, Baltimore
Opazo JC (2005) A molecular timescale for caviomorph rodents (Mammalia, Hystricognathi). Mol Phylogenet Evol 37:932–937
Patterson BD, Wood AL (1982) Rodents from the Oligocene of Bolivia and the relationships of the caviomorph. Bull Mus Comp Zool 149(7):371–543
Patterson BD, Solari S, Velazco PM (2012) The role of the Andes in the diversification and biogeography of Neotropical mammals. In: Patterrson BD, Costa LP (eds) Bones, clones, and biomes: the history and geography of recent neotropical mammals. Chicago University Press, Chicago
Pérez ME, Vucetich MG (2011) A new extinct genus of Cavioidea (Rodentia, Hystricognathi) from the Miocene of Patagonia (Argentina) and the evolution of cavioid mandibular morphology. J Mamm Evol 18:163–183
Perelman P, Johnson WE, Roos C, Seuánez HN et al (2011) A molecular phylogeny of living primates. PLoS Genet 7(3):e1001342. https://doi.org/10.1371/journal.pgen.1001342
Phoberomys pattersoni https://en.wikipedia.org/wiki/Phoberomys_pattersoni

Pierre-Olivier A, Marivaux L, Croft DA, Billet G et al (2011) Middle Eocene rodents from Peruvian Amazonia reveal the pattern and timing of caviomorph origins and biogeography. Proc R Soc B Biol Sci 279:1319–1326. https://doi.org/10.1098/rspb.2011.1732 First published online 12 October 2011

Pierre-Olivier A, Abello MA, Adnet S, Altamirano Sierra AJ, Baby P, Billet G, Boivin M, Calderón Y, Candela A, Chabain J, Corfu F, Croft DA, Ganerød JC, Klaus S, Marivaux L, Navarrete RE, Orliac M, Parra F, Pérez ME, Pujos F, Rage J-C, Ravel A, Robinet C, Roddaz M, Tejada-Lara JV, Vélez-Juarbe J, Wesselingh FP, Salas-Gismondi R (2016) A 60-million-year Cenozoic history of western Amazonian ecosystems in Contamana, eastern Peru. Gondwana Res 31:30–59

Poux P, Chevret P, Huchon D, De Jong WW, Douzery EJP (2006) Arrival and diversification of platyrrhine primates in South America. Syst Biol 55(2):228–244

Prado JL, Cerdeño E, Roig-Juñent S (1998) The giant rodent *Chapalmatherium* from the Pliocene of Argentina: new remains and taxonomic remarks on the family Hydrochoeridae. J Vertebr Paleontol 18(4):788–798

Ribeiro AM, López GM, Bond M (2010) The Leontiniidae (Mammalia, Notoungulata) from the Sarmiento Formation at Gran Barranca, Chubut Province, Argentina. In: Madden RH, Carlini AA, Vucetich MG, Kay RF (eds) The paleontology of Gran Barranca. Cambridge University Press, Cambridge, pp 170–181

Rinderknecht A, Blanco RE (2008) The largest fossil rodent. Proc R Soc B Biol Sci 275:923–928

Rinderknecht A, Blanco RE (2015) History, taxonomy and paleobiology of giant fossil rodents (Hystricognathi, Dinomyidae). In: Cox PG, Hautier L (eds) Evolution of the rodents: advances in phylogeny, functional morphology and development. Cambridge University Press, Cambridge, p 164

Rinderknecht A, Bostelmann-T E, Ubilla M (2011) New genus of giant Dinomyidae (Rodentia: Hystricognathi: Caviomarpha) from the late Miocene of Uruguay. J Mammal 92(1):169–178

Rose K (2006) The beginning of the age of mammals. The Johns Hopkins University Press, Baltimore

Rowe DL, Honeycutt RL (2002) Phylogenetic relationships, ecological correlates, and molecular evolution within the cavioides (Mammalia, Rodentia). Mol Biol Evol 19:263–277

Rowe DL, Dunn KA, Adkins RM, Honeycutt RL (2010) Molecular clocks keep dispersal hypotheses afloat: evidence for trans-Atlantic rafting by rodents. J Biogeogr 37(2):305

Sallam HM, Seiffert ER, Steiper ME, Simons EL (2009) Fossil and molecular evidence constrain scenarios for the early evolutionary and biogeographic history of hystricognathous rodents. Proc Natl Acad Sci U S A 106(39):16722–16727

Sánchez-Villagra MR, Aguilera O, Horovitz I (2003) The anatomy of the world's largest extinct rodent. Science 301:1708–1710

Savage RJG, Long MR (1986) Mammal evolution: an illustrated guide. British Museum (Natural History), London

Simpson GG (1980) Splendid isolation: the curious history of South American mammals. Yale University Press, New Haven/London

Spotorno AE, Valladares JP, Marin JC, Palma RE, Zulueta C (2004) Molecular divergence and phylogenetic relationships of chinchillids (Rodentia: Chinchillidae). J Mammal 85:384–388

Upham NS, Patterson BD (2012) Diversification and biogeography of the neotropical caviomorh lineage Octodontoidea (Rodentia: Hystricognathi). Mol Phylogenet Evol 63(2):417–429

Verzi DH, Morgan CC, Olivares AI (2015) The history of South American octodontoid rodents and its contribution to evolutionary generalisations. In: Cox PG, Hautier L (eds) Evolution of the rodents: advances in phylogeny, functional morphology and development. Cambridge University Press, Cambridge

Vizcaíno SF, Cassini GH, Toledo N, Bargo MS (2012) On the evolution of large size in mammalian herbivores of Cenozoic faunas of Southern South America. In: Patterson BD, Costa LP (eds) Bones, clones, and biomes. The history and geography of recent neotropical mammals. The University of Chicago Press, Chicago, pp 76–101

Voloch CM, Vilela JF, Los-Oliveira L, Schrago CG (2013) Phylogeny and chronology of the major lineages of New World Hystricognath rodents: insights on the biogeography of the Eocene/Oligocene arrival of mammals in South America. BMC Res Notes 6:160

Voss RS, Hubbard C, Jansa SA (2013) Phylogenetic relationships of New World porcupines (Rodentia, Erethizontidae); implications for taxonomy, morphological evolution and biogeography. Am Mus Novit 3769:1–36

Vucetich MG, Pérez ME (2011) The putative cardiomyines (Rodentia, Cavioidea) of the middle Miocene of Patagonia (Argentina) and the differentiation of the family Hydrochoeridae. J Vertebr Paleontol 31(6):1382–1386

Vucetich MG, Verzi DH, Hartenberger J-L (1999) Review and analysis of the radiation of the South American Hystricognathi (Mammalia, Rodentia). Palaeontology 329:763–769

Vucetich MG, Deschamps CM, Pérez ME (2013a) Paleontology, evolution and systematics of capybara. In: Moreira JR, Ferraz KMPMB, Herrera EA, Macdonald DW (eds) Capybara: biology, use and conservation of an exceptional neotropical species. Springer, New York, pp 39–59

Vucetich MG, Deschamps CM, Pérez ME, Montalvo CI (2013b) The taxonomic status of the Pliocene capybaras (Rodentia) *Phugatherium Ameghino* and *Chapalmatherium Ameghino*. Ameghiniana 51(3):173–183

Wood AE (1949) A new Oligocene rodent genus from Patagonia. Am Mus Novit 1435:1–54

Woods CA (1973) *Erethizon dorsatum*. Mamm Species 29:1–6

Woods CA (1982) The history and classification of South American hystricognath rodents: reflections on the far away and long ago. In: Mares M, Genoways H (eds) Mammalian biology in South America. University of Pittsburgh Press, Pittsburgh

Woods CA, Kilpatrick CW (2005) Infraorder Hystricognathi. In: Wilson DE, Reeder DAM (eds) Mammal species of the world: a taxonomic and geographic reference. The Johns Hopkins University Press, Baltimore, pp 1538–1600

Wyss AR, Flynn JJ, Norell MA, Swisher CC III, Charrier R, Novacek MJ, McKenna MC (1993) South America's earliest rodent and recognition of a new interval of mammalian evolution. Nature 365:434–437

Wyss AR, Flynn JJ, Norell MA, Swisher CC III, Novacek MJ, McKenna MC, Charrier R (1994) Paleogene mammals from the Andes of Central Chile: a preliminary taxonomic, biostratigraphic and geochronologic assessment. Am Mus Novit 3098:1–31

Chapter 8
Platyrrhine Monkeys: The Fossil Evidence

8.1 Introduction

Many primitive Strepsirrhini primates are well-known and diverse from the Eocene of North America and Europe where they finally disappeared at the beginning of the Oligocene. But absolutely none of these more primitive strepsirrhines are known from South America. Nor has a clear phylogenetic connection been made between these more primitive prosimians and the more modern anthropoids (Gebo 2002; Fleagle 2013).

A controversy raged for some time about the origins of the first South American primates. One school of thought suggested that they had to have arrived from North America and that somehow it would have been easier for them to arrive from the north (rather than from Africa) and that possible ancestors had flourished in North America. Invasion of South America would have been facilitated by an arc of volcanic islands that probably existed in the proto-Caribbean sea, although probable paleocurrents were from the south to the north, against any migrations via the sea (Fleagle 2013). Additional arguments for the northern route are that marsupials and condylarths had arrived in South America from the north during the latter part of the Cretaceous. However, no anthropoid ancestors have ever been identified in North America nor have fossil prosimians been found in South America that could have entered from the north (Ciochon and Chiarelli 1980).

The discovery of a rich fauna of early anthropoid primates of El Fayum, Egypt, early in the twentieth century established a strong candidate group as the ancestors of platyrrhine monkeys; these were the African parapithecid monkeys (Parapithecidae) or the proteopithecids (Proteopithecidae), and they are variously dated at around 34 Ma during the period at the end of the Eocene and beginning of the Oligocene. These early, closely related anthropoid groups are represented by eight species and four genera for the Parapithecidae and just two species for the Proteopithecidae. The two closely related families have many characteristics of platyrrhine New World primates including the same dental formula (except for

T. Defler, *History of Terrestrial Mammals in South America*, Topics in Geobiology 42, https://doi.org/10.1007/978-3-319-98449-0_8

Parapithecus fraasi, which lacks lower incisors), the same arrangement of cranial plates (different from the cranial plates of the catarrhine Old World primates) (Fig. 8.1), and similar arrangements for the ear morphology. One parapithecid species *Apidium phiomense* even shares about 40% fusion of the tibia and fibula with the New World *Saimiri*.

The primitive proteopithecid *Proteopithecus* is known in part by postcranial elements that include a humerus, femur, tibia, and talus, all of which are similar to a small platyrrhine.

Because platyrrhines have many features in common that are not found among the parapithecids, it seems distinctly possible that the parapithecids could have given rise to both platyrrhines and catarrhines and that the characteristics shared between parapithecids, proteopithecids, and platyrrhines could be primitive characteristics that have not been lost in the platyrrhine monkeys. Several have questioned whether the slightly more specialized dentition of the parapithecids could have given rise to the *Platyrrhini*, and several have noted the substantial similarities of dental characteristics of *Proteopithecus* with *Branisella* (see below), making this a strong candidate for platyrrhine origins (Miller and Simons 1997; Simons 1997; Kay and Williams 1995; Takai et al. 2000). More fossil material could undoubtedly clarify this in the future (Fleagle 2013), and the discovery of *Perupithecus* with molars similar to *Talahpithecus*, a possible Oligopithecidae, adds more fuel to the mystery of platyrrhine origins (Bond et al. 2015).

Arriving in South America from Africa presents huge problems because of the great distances between the two continents, even though they were several hundreds of kilometers closer in the Late Eocene than they are now. Advantages for this sweepstakes route rather than the North American route were paleocurrents that were from Africa toward South America. Also a great deal of paleogeographic activ-

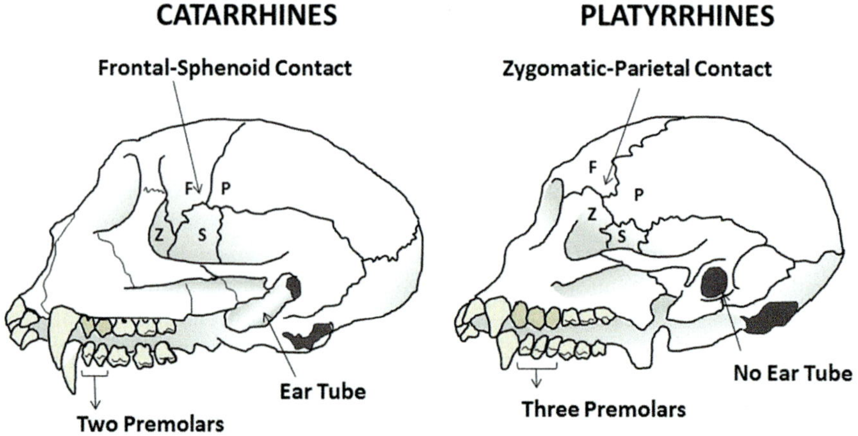

Fig. 8.1 Some distinguishing characters between the catarrhines and platyrrhines (Fulwood et al. 2016) (by Diego Casallas-Pabón, Applied Biodiversity Foundation)

ity in the South Atlantic produced some rather large islands during the Eocene and Miocene, and these could have facilitated the trip by cutting the distance in half, though the distance between the two continents was much greater than further north. The splitting of the two continents was accompanied by a pivoting movement of the continental crusts, which begun in the south to the north (Bandoni de Oliveira et al. 2009) causing more space to open up between the two continents at mid-continent as compared to the southern parts. Plate tectonics suggest the distance widens at a rate of about 4.5 cm/year = 45.55555 km/million years and 455.55 km/10 million years = **1594.425 km/35 million years** (distance from Africa to South America). Thirty-five million years ago, the distance would have been about 1006 km between South Africa and South America (Lavocat 1980).

Many have wondered how a small group of primates could have survived a water journey across an ocean (Caperton Morton 2013). The answer is, of course, they could survive if the conditions were just right. Large rafts of vegetation have been sighted in the mid-Atlantic and elsewhere that have broken off from African riverbanks, and these islands could have carried early primates (and early rodents) from one continent to another (De Queiroz 2005, 2014). A group of small mammals could survive for weeks on an inadequate diet, though some fruits, insects, and eatable leaves could have been sheltered on this vegetation. Fresh water is the most critical need, but frequent rains would have been enough to supply very small mammals that would have licked the vegetation for moisture, as do many small vertebrates today. A large island of vegetation driven by the winds of a heavy wet season could have supplied enough moisture and increased the speed of the floating island. One calculation suggests that, given the right conditions, the trip could have been made in 7–11 days given the distances (Houle 1999; De Queiroz 2014).

This same study considers the effects of freshwater scarcity. Obviously the authors of this article have never experienced a tropical downpour, such as could have occurred each day for weeks at a time and that would easily have supplied enough water to be lapped from leaves of the raft. Given the millions of years available for just the right conditions for such transportation to successfully conclude, it does not seem unreasonable to suppose that a group of small African monkeys could have made it to the shores of South America and survived, building up a population and evolving into the forms that we have discovered. Given enough deep time, many things are possible (Hofstetter 1980; Fleagle 2013; De Queiroz 2014). This scenario only seems difficult to imagine because humans are not accustomed to thinking in terms of deep time over millions of years. It is the same problem, I suppose, that many people have difficulty imagining that the oceans of the world will rise and will flood coastal cities like New York City, Miami, and London, yet logically we know that this will happen in the near future and not even over millions of years, since the physics of a warming world are well understood (Irvine 2014).

We are now pretty certain that the caviomorph rodents also arrived in South America from Africa, and it now seems clear that the caviomorphs evolved from the phiomorph rodents that continue in Africa even today (Lavocat 1980). Recent molecular and fossil evidence dates the arrival of caviomorph rodents at around 41 Ma years or shortly before (Antoine et al. 2012). The fact that rodents appeared to have made the ocean trip from Africa to South America is excellent supporting evidence suggesting that primates have done the same, however improbable the odds.

Various molecular estimates ($n = 15$) have been made using different methods to calculate divergence dates between *Catarrhini* and *Platyrrhini*, and I list some below that show a difference in dates of about 17 Ma (Perez et al. 2013).

45.3 +/− 3.1 Ma (Matsui and Hasegawa 2012)
43 Ma (Chatterjee et al. 2009)
37 +/− 3.0 Ma (and 38.9 +/− 4.0 Ma) (Poux et al. 2006)
43.5–55.7 Ma (Yoder and Yang 2004)
53.3–61.1 Ma (Yang and Yoder 2003)
35 Ma (Schrago and Russo 2003)
33 Ma (Glazko and Nei 2003; Nei and Glazko 2002)
52.5 Ma for *Cebus*; 46 Ma for ateline clade (Bauer and Schreiber 1997)

One recent analysis is particularly interesting, given that using the same techniques and genes, the authors have estimated both the diversification or arrival of caviomorph (or their ancestors) rodents and platyrrhine ancestral primates in South America (Poux et al. 2006). It is particularly encouraging that very recent fossil evidence with good dating places a caviomorph community in tropical lowland rainforest at 41 Ma (Antoine et al. 2012). Characteristics (small size) of the rodents in this community suggest that these could be very close to the earliest South American rodents. The fossil date concurs with the molecular data (45.4 +/− 4.1 (or 43.7 +/− 4.8) Ma for the arrival of the caviomorph rodents. This provides strong confidence that the date for the arrival of the platyrrhine primates or their ancestors was 37.0 +/− 3.0 (or 38.9 +/− 4.0) Ma. There are several other estimates suggesting that the divergence between the platyrrhines and catarrhines was around 35–40 Ma (Schrago and Russo 2003; Schrago et al. 2013; Glazko and Nei 2003).

The appreciation of a genetic clock using sequences from 3 nuclear genes from 60 species dates the split between Old World and New World primates at about 35–37 Ma at the end of the Eocene, also concurs with the latest and most ancient primate fossils discovered in Peru (Bond et al. 2015). Some researchers have even suggested that they find genetic evidence for two invasions into South America (Bauer and Schreiber 1997; Schrago and Russo 2003). These would be the ancestral cebids and ancestral atelids at a later time. Time and more data will surely clarify these hypotheses.

8.2 The First New World Primates

8.2.1 Santa Rosa Local Fauna

The oldest New World fossil primates have been identified from the Early Eocene, which would be close to the molecular dates calculated by Poux et al. (2006). The teeth of two distinct species, *Perupithecus ucayaliensis* and an undescribed species, have been found from the Santa Rosa local fauna on the Turúa river, Peruvian Amazon (Bond et al. 2015). The age calculation of around 39 Ma is based on other, associated faunas belonging to the Yahuarango Formation (Goin and Candela 2004; Frailey and Campbell 2004) and on the close similarities to some Eocene primates from Africa such as *Talahpithecus* and *Proteopithecus* (Bond et al. 2015). *Talahpithecus* (possibly Oligopithecidae) has been dated to around 38–39 Ma or Late and Middle Eocene (Jaeger et al. 2010), while *Proteopithecus* (Proteopithecidae) is calculated to have lived around the Late Eocene. *Perupithecus ucayaliensis* has been estimated as having been a small primate of the size of *Callimico* or some *Saguinus* (Bond et al. 2015). The associated second primate has not been named, due to inadequate material. These primates are the first early primates to be found in tropical moist forest, and they clearly demonstrate the importance of the tropics for the evolution of New World primates. While these primates cannot be easily placed within the *Platyrrhini*, it is clear that the origins of the New World primates were tropical and that many more details of their evolution will eventually be found there, rather than from southern parts of South America.

I find it useful to arrange the South American primate fossils into six groups, since each of these clusters falls into a clearly defined geographical cohort. I identify (1) the earliest known primate fossils from Peru, *Perithecia ucayaliensis* and two or three unnamed species, (2) *Branisella* and *Szalatavus* as the earliest and only Oligocene primates known for the neotropics, (3) the Miocene primates of the Southern Cone, (4) the Middle Miocene primates of La Venta, Colombia, (5) the Plio-Pleistocene primates of Amazonia and Atlantic coast, and (6) the ancient to recent primates of the Caribbean archipelago (Tejedor and Muñoz 2012). (7) We can now top the list off with the first North American primate (*Panamacebus transitus*) recently discovered in Panama.

8.2.2 *Branisella and Szalatavus*

After the Peruvian Eocene primates, the next New World primates to appear in the fossil record are from the Late Oligocene of Bolivia dated at about 26 Ma (MacFadden et al. 1985). *Branisella boliviana* (Hofstetter 1969) was roughly the size of a modern-day *Saimiri*, judging by jaw fragments and by additionally discovered material (Wolff 1984; Rosenberger 1981). Comparisons of *Branisella* with the Egyptian Late Eocene fossil *Pliopithecus* have suggested that the Pliopitheciidae might be the ancestor of the *Platyrrhini* (Miller and Simons 1997; Simons 1997; Kay and Williams 1995; Takai et al. 2000), and Rosenberg et al. (1991) suggested that *Branisella* could be the first of the callitrichine ancestry (Takai and Anaya 1996), although others demur that *Branisella* was a direct ancestor of *Platyrrhini* (Kay et al. 2002).

An analysis of the adaptations of *Branisella boliviana* based mostly on dental features suggests that this species was a (1) small (721–759 g), semiterrestrial primate (based on high molar shearing crests and strong wearing of molars) (2) that lived in open woodland with low seasonal rainfall; (3) the diet was most likely fibrous soft fruits and considerable fibrous foods such as leaves or insects. The combination of dental characteristics suggests that *Branisella* falls outside the crown group of platyrrhines, though is closely related to them (Kay et al. 2002). *Branisella* was a sister or a stem group to the platyrrhines.

Szalatavus attricuspis was defined from about the same age as *Branisella*, and some have suggested the species is a synonym for *Branisella* (Takai and Anaya 1996). Because of very high-crowned cheek teeth, the suggestion has been made that it, too, was semiterrestrial and included abrasive materials (e.g., grass) in its diet. So far, no other Oligocene primate fossils have been found, but during the approximately 10 or 12 million years of separation of *Perupithecus* from *Branisella*, there must have been other early primates, and they will undoubtedly be discovered. Evidence suggests that *Branisella* and *Szalatavus* did not live in tropical moist forest but rather in a type of xeromorphic, deciduous forest (MacFadden 1990). The evidence is twofold: (1) many associated fossil mammals have high-crowned teeth similar to notoungulates and rodents that probably fed on grass, and (2) the presence of associated calcrete ("caliche") nodules found in association with the fossils are known to form in a semiarid or arid environment (MacFadden et al. 1985; Kay et al. 2002 have discussed the formation of calcrete under moist conditions). If *Szalatavus* is synonymous with *Branisella* then it, too, is a sister group to the platyrrhines (Kay et al. 2002).

8.2.3 Primates of the Southern Cone

Tremacebus harringtoni (Rusconi 1935) (first named *Homunculus harringtoni* and later changed to *Tremacebus* by Hershkovitz (1974), was found in Patagonian deposits of the same age as *Dolichocebus* (Colhuehuapian, Early Miocene, 21–18.5 Ma), and is represented by a skull. CT imaging of the width of the holotype skull's olfactory fossa suggest that this species was probably not nocturnal as had previously been believed (Kay et al. 2004). The animal was probably 1–2 kg. The teeth are most similar to *Callicebus* and *Aotus*. Rosenberger (1979) suggested it was an ancestor to *Saimiri*, but analysis of 268 characters of the skull and dentition of 16 living extant platyrrhine genera indicate that *Dolichocebus* is a stem platyrrhine along with *Branisella, Tremacebus, Soriacebus*, and *Carlocebus* (Kay et al. 2008) (Fig. 8.2).

A complete primate skull from Andean volcanoclastic rocks (rocks and ash emitted by volcanic eruptions) in Chile. Volcanic ash, obsidian, and pumice are examples of pyroclastic or volcanoclastic materials found in central Chile that were precisely dated to 20.09 +/− 0.27 Ma. This achievement pointed out future possibilities of precisely dated fossil fauna in these types of deposits (Flynn et al. 1995). This primate, named *Chilecebus carrascoensis*, seems to belong in the cebine line and shows a dental formula 2/2, 1/1, 3/3, and 3/3 in line with its probable phylogenetic origins. It preserves as well several parapithecid and pliopithecid primitive

Fig. 8.2 The rather odd image of *Tremacebus harringtoni* was drawn by Rusconi C. 1935 based on the then known skull

characteristics. The animal's weight initially was calculated at 1000–1200 g (also, like a *Saimiri*), but later considerations of the problems of estimating weight from ancient primates that scale differently from modern primates resulted in an analysis of weight for this species as 583 g (Sears et al. 2008)! Based on the estimated weight of 583 g, *Chilecebus* exhibits an encephalization quotient that is lower than that of all living primates (Sears et al. 2008). This is the northernmost fossil so far found for the Southern Cone of South America.

Between the northernmost *Chilecebus* and the southernmost *Killikaike*, the fossils of at least five other species of primates have been found, all of these from the first half of the Miocene. *Dolichocebus gaimanensis* was a skull from the Early Miocene (SALMA Colhuehuapian, <20.5 Ma). The actual animal was probably 2.5–3 kg. The skull shows similarities to *Saimiri, Callicebus, Cebus*, and *Aotus* with a brain size similar to modern platyrrhines. An extensive analysis of cranial and dental characteristics compared to the same in modern platyrrhines and other fossils lends strong evidence that *Dolichocebus* along with *Branisella*, *Tremacebus, Soriacebus*, and *Carlocebus* are stem platyrrhines, that is, the group is closely related to all living platyrrhines, although not in the direct line of evolution, while *Dolichocebus* is sister taxon to *Carlocebus* (Kay et al. 2008; Rosenberger and Tejedor 2013). Some dental characteristics link *Dolichocebus* with African anthropoids as well (Fleagle and Kay 1989).

The discovery of a new species from the Early Miocene (20 Ma) of Gran Barranca in Chubut Province, Argentina, was a small primate around 1600 g with specialized incisors and canines and many similarities to the pithecines, so that this primate *Mazzonicebus almendrae* is thought to have also been a seed predator (Kay and Fleagle 2010). *Mazzonicebus* may be closely related to *Soriacebus* from three million years later and represents the initiation of the pithecine radiation in South America (Tejedor 2005a, b). However, an analysis using a character-taxon matrix comparing this primate and *Soriacebus* with pithecine examples strongly suggested that the two groups were evolutionarily parallel and not closely related (Kay 2010). The species represents, along with *Tremacebus* and *Dolichocebus,* the oldest Miocene finds of primates in South America.

The pithecine radiation logically seems based on a diet of hard fruits and seeds, which demand dentary adaptations leading to a strong apparatus for mastication. This would include the ability to open the strongly protected fruits (strong projecting canines), the ability to finely manipulate the seeds (procumbent incisors like tweezers), and the ability to be able to chew (crush) the seeds (large premolars) as found in both *Soriacebus* and in *Mazzonicebus.* The seeming first appearance of pithecine or related primates (sister groups) in the Southern Cone of South America might be the result of seasonally dry forests and low rainfall that led to strongly protected seeds (Hodgson et al. 2009; Kay 1990).

Soriacebus ameghinorum and *Soriacebus adrianae* were found in the Miocene Pinturas Formation and have been dated to about 17.5–16.5 Ma (Brown and Larriestra 1990; Fleagle et al. 1987, 1995; Fleagle 1990; Fleagle and Tejedor 2002; Tejedor 2005). *Soriacebus ameghinorum* is calculated to have weighed about 1800 g, while *S. andreanae* weighed in at about 800–900 g. Dental characteristics and the jaws suggest a relationship to the pithecines (Rosenberger 2011), although, as mentioned, another analysis suggested that these primates did not group with pithecines and that their dental and jaw characteristics could be a case of parallel evolution, placing them in a subfamily Soriacebinae of the family Homunculidae (Kay et al. 2008, 2010).

Carlocebus carmenensis and *Carlocebus intermedius* have been found in the Patagonian Miocene in the Santacrucian SALMA at about 17.5–16.5 Ma (dated to about 16.6 Ma by Fleagle et al. (1995). *Carlocebus carmensis* were medium-sized primates weighing about 2500 g. The fossil is a lower mandible fragment with complete teeth on the right side with small canines. It is probably related to *Callicebus* (Fleagle et al. 1988; Tejedor 1987).

Brown and Larriestra (1990) date the Pinturas Formation primates in a temporal sequence beginning with *Carlocebus carmensis,* dated at the base of the Pinturas at about 16.6 Ma then proceeding (from oldest to youngest) with *Carlocebus intermedius, Soriacebus ameghinorum, and Soriacebus adrianae* (Fleagle and Brown 1983). An upper Pinturas Formation date of 13.3 Ma has been obtained, and the other primates are suggested to be from that time (Brown and Larriestra 1990).

The southernmost primate fossil known is *Killikaike blakei* (Fig. 8.3); a complete face and dentition with the anterior part of the skull's braincase (see below) was found in southernmost Argentina on the Río Gallegos at about 51° 40'S (Tejedor et al. 2006). This is primate has been very accurately dated at 16.45 +/– 0.14 Ma, since the fossil was found in volcanic tuft. The live animal probably weighed about 1000 g when cranial characteristics are compared to selected living species. Most significant is the highly vaulted frontal bone of the fossil, reflecting a relatively enlarged forebrain as compared to non-cebine platyrrhines and suggesting evidence for the early evolution of the cebine clade. This primate's canine resembles *Saimiri,*

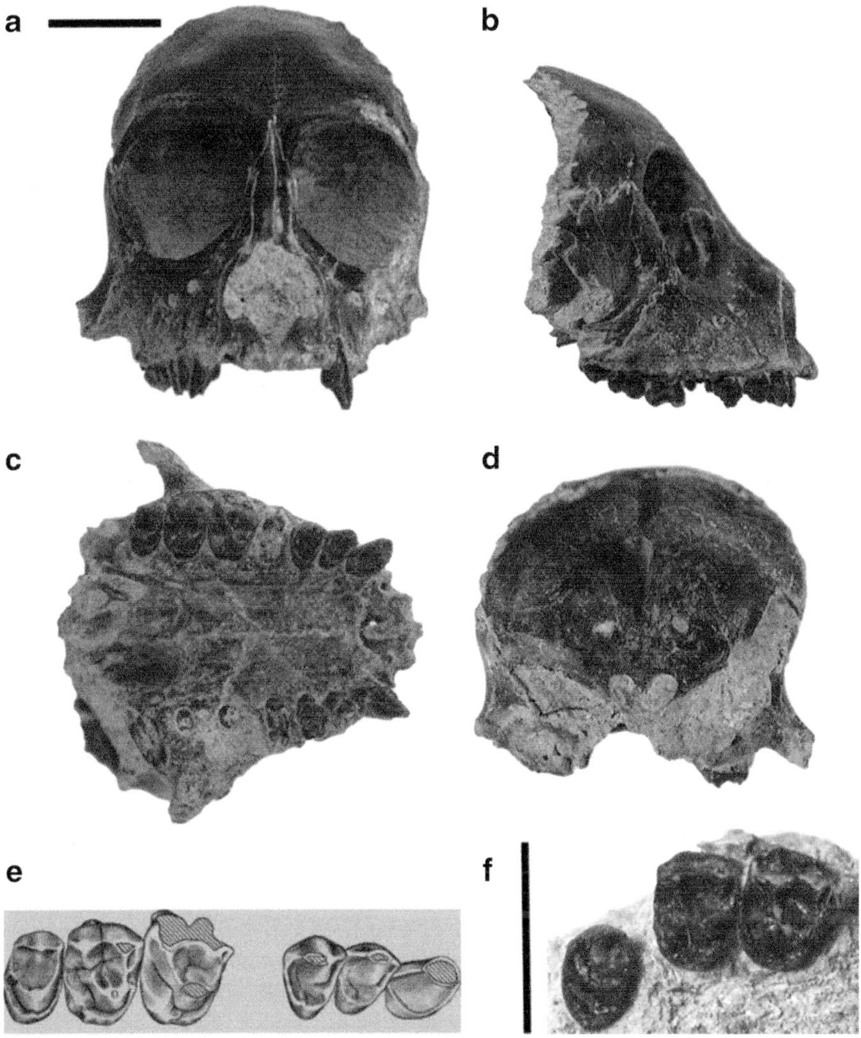

Fig. 8.3 Holotype cranium and upper molars of *K. blakei*, a new genus of platyrrhine primate from the Miocene of Argentina. Frontal (**a**), right lateral (**b**), palatial (**c**), and posterior views of MPM-PV 5000 and an illustration of the cheek teeth (**e**), (**f**) MPM-PV 1607 in occlusal view. (Scale bare, 1 cm) (Tejedor et al. 2006)

the face is elongated with a reduction of the nose rather than flattened as in *Saimiri* and *Cebus* (Tejedor et al. 2006).

Others have interpreted *Killikaike* as a synonym of *Homunculus patagonicus*, based on a comparison of cranial and maxillary measurements and other interpretations of the "vaulted frontal bone" from material identified previously as *Homunculus* (Perry et al. 2014). This would make "*Killikaike blakei*" a junior synonym of *Homunculus patagonicus*. This creates a conflict since *Homunculus* has been identified as an early pitheciine, while *Killikaike* has been identified as an early cebine, related to *Saimiri* (Tejedor and Rosenberger 2008; Tejedor et al. 2006). These conflicting views of *Killikaike* and its relationship to *Homunculus* will have to be resolved in the future.

The other most southern primate fossil from Early Miocene (Santacrucian SALMA, dated approximately 16 Ma) was collected by Carlos Ameghino and described by his brother Florentino as *Homunculus patagonicus*, in 1891 (Ameghino 1891a, b). It was the second fossil primate described for South America, since Lund (1840) had described the large Pleistocene *Protopithecus brasiliensis* from Brazil (see below). Florentino damaged his reputation by claiming that *Homunculus* was an early ancestor of the hominids from the Eocene (Ameghino 1907). The first specimen was collected around the Río Gallegos, perhaps not far north (maximum 200 km) of the site for *Killikaike*, although the precise locality was never published. In later years more material was collected (a left partial cranium and a complete mandible, a right femur, partial ulna, and left radius and right distal humerus) that was associated with the mandible and all described by Ameghino (1891a, b, 1898, 1907). Recently Tejedor and Rosenberger (2008) published a neotype for the species designating the best-preserved mandible as the neotype, since the original holotype had been lost for years. Based on the considerable material now available for the species, they analyzed craniodental and long bones concluding that *Homunculus* was a primitive pitheciine with adaptations like the quadrupedal, diurnal titi *Callicebus* (Tejedor and Rosenberger 2008). Additional material has come to light, permitting an^{40}Ar/^{39}AR date of about 16.4 Ma, the same age as *Killikaike. Homunculus* possessed large canines and because of the limb proportions of its long bones is similar to *Callicebus* (Tejedor and Rosenberger 2008). This primate lived in an environment that was less humid with more seasonal and drier conditions than typical tropical forests and was found with other fauna typical of open areas (Tejedor and Rosenberger 2008).

Proteropithecia neuquenensis was described from several isolated teeth, mandibular fragments, and a talus, which connect the species unambiguously to the pithecines, especially because of its procumbent incisors. It has been dated as being slightly younger than 15.71 +/– 0.07 Ma (SALMA Colloncuran) and thus the youngest occurrence of platyrrhine primates in Argentina and the oldest taxon that can confidently be placed into the pithecines. It weighed around 1600 g when living, and judging from the teeth, it probably ate fruits and seeds (Kay et al. 1998). The skull is unusually small when compared with the very large molars and reduced palate, a combination that is not found in other New World monkeys (Tejedor 2013).

Primates from the Southern Cone of South America and their fossil ages.

Chilecebus—20.09 Ma	(Colhuehapaian)
Dolichocebus—20.5 Ma	(Colhuehapaian)
Tremacebus—20.5 Ma	(Colhuehapaian)
Carlocebus—17.5–16.5 Ma	(Santacrucian)
Soriacebus—17.5–16.5 Ma	(Santacrucian)
Homunculus—16 Ma	(Santacrucian)
Killikaike—16 Ma	(Santacrucian)
Proteropithecia—15.7 Ma	(Frisianan)

8.2.4 La Venta: The Colombian Primates

The next group of primates to be considered come from Colombia. They represent essentially one community, since the group has been discovered and described from the La Venta deposits that capture a very short (geologically) and defined time span of around 12 Ma ago (10–14 Ma according to Flynn et al. 1997). They also represent the numerically richest ancient primate community known for South America (Kay et al. 1997). These primates show very clear connections to modern primates and can even be placed into modern families and subfamilies and, though the fossil deposits are found in the upper Magdalena River, which is now a deep river valley between two branches of the northern Andes, at the time of the deposits of the fossils, the area was actually peripheral to the proto-Amazon basin with no mountain barrier between the animals and the immense flatlands of the east. The Eastern Cordillera of the Andes had not yet begun to appear, and the local land had not been uplifted yet from its more or less 500 m elevation. Fossils from La Venta are placed in the Middle Miocene, and they have been found in two geological formations: the Victoria formation (13.5–12.9 Ma) and the Villavieja Formation (12.9–11.5 Ma) (Guerrero 1997).

This assemblage has recently been analyzed as "Andean fauna." This seems to me to be a stretch of the imagination (Flynn et al. 2012). The La Venta assemblage was a lowland tropical ecosystem peripheral to the proto-Amazonian wetlands and cannot be considered Andean for the time that it existed. These fossils represent at least 10 species of primates, typical of a lowland tropical forest with an annual precipitation of about 1500 mm, eight of which can be placed into a modern family (Kay and Madden 1997).

Two species of the ateline *Stirtonia* have been discovered (*S. tatacoensis* Stirton 1951; *S. victoriae* Kay et al. 1987), which are clearly related to *Alouatta* and are probably direct ancestors of this modern genus, dated about 12.6–13.7 Ma.

The very famous and almost complete skeleton of *Cebupithecia sarmientoi* (Stirton 1951) is a pithecine (Pitheciini) primate very similar to *Pithecia*. This animal would have weighed 2–3 kg and had a mixed locomotor repertoire of arboreal quadrupedalism, leaping, and some suspensory postures that were similar to the modern *Pithecia*, though they were structurally adapted in different ways (Ford 1990a) (Fig. 8.4).

Nuciruptor rubricae (Meldrum and Kay 1997a) also clearly belongs to the pithecids but is midway in size between the modern genera *Pithecia* and *Callicebus*.

Neosaimiri fieldsi (Stirton 1951) was a small cebid resembling *Saimiri* in many aspects. It has even been suggested that it belongs in the modern genus *Saimiri* (Delson and Rosenberger 1984). The species is now known from fragments of jaw, teeth, distal humerus, proximal ulna, proximal femur, distal tibia, calcaneus, and four tali. It shares some traits with *Aotus* as well. (Hartwig and Meldrum 2002). *Laventiana annectens* (Rosenberger et al. 1991) is a *Saimiri*-sized mandible closely related to *Saimiri* and *Neosaimiri* but with some differences. Some would include this species within *Neosaimiri* (Meldrum and Kay 1997b; Nakatsukasa et al. 1997; Takai 1994).

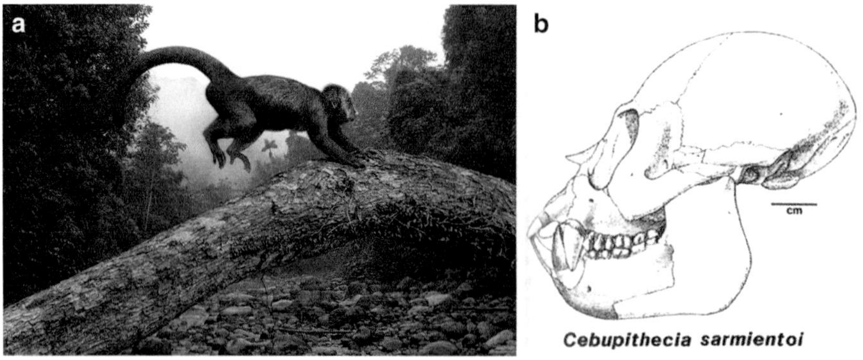

Fig. 8.4 *Cebupithecia sarmientoi* an obvious pithecine from La Venta (**a**) hypothetical drawing of a live animal, (**b**) a skull of Cebuspithecia 12 Ma (By Roman Uchytel)

Aotus dindensis (12.1–12.5 Ma) has been clearly placed within the modern genus *Aotus*, which suggests the antiquity of this genus and may explain the very similar phenotype of the many species known today, since this is a highly evolved, mostly nocturnal primate genus, well-adapted to a nocturnal life (Setoguchi and Rosenberger 1987).

Mohanamico hershkovitzi (Luchterhand et al. 1986) is a complete lower jaw, of a 1 kg primate with similarities to the callitrichine *Callimico*. The species is generally recognized as a callitrichine (Rosenberger et al. 1990).

Unfortunately the primate fossils *Patasola magdalenae*, *Lagonimico conclutatus*, and *Micodon kiotensis* are not easily placed into a known group of modern primates (Kay and Meldrum 1997; Kay 1994, Setoguchi and Rosenberger 1985a, b). But Setoguchi et al. 1985 felt that the small size of *Micodon* and a "gross resemblance" to certain *Saguinus* argued for the likelihood of the genus being allied with the callitrichines. *Patasola* and *Lagonimico* also show some similarities to the callitrichines, and *Patasola* has a large third molar like *Callimico*. But *Lagonimico* (represented by a badly crushed skull and mandible) is somewhat larger than modern callitrichines (Kay 1994). However, modern callitrichines are possibly dwarfed cebids (or callitrichids), and *Lagonimico* may come from a larger ancestor than present-day callitrichines.

Micodon kiotensis is placed with the callitrichines, mostly based on its very small size and based on an isolated upper molar and upper incisor and lower premolar (Rosenberger et al. 1990). *Kondous laventicus* was named from molars 1 + 2 that proved to be closely related to *Ateles* (Setoguchi 1985). Finally *Miocallicebus villaviejai* (Takai et al. 2001) was described from a right maxillary fragment with three molars found from the Middle Miocene of La Venta and it has been associated with the *Callicebus* monkeys, though it is much larger, about the size of a *Cebus*.

Miocallicebus was named for a fossil maxillary found by Takai et al. (2001) in La Venta. It is closely related to *Callicebus*, although the fossil animal was much larger than *Callicebus*.

8.2.5 Late Miocene Amazon and the Pleistocene Coast

The fifth group of fossil primates are those found east of the Cordillera of Los Andes from the central Amazon to the eastern coast. On the upper Acre River, the teeth of three Late Miocene primates (dated from around 9–6 Ma) have recently been found. They are a representative of *Stirtonia*, the putative ancestor of *Alouatta*. A possible ancestor of *Cebus* named *Acrecebus fraileyi* or a Sister species diagnosed from one large upper molar three or four times larger than a living *Cebus*, and *Solimoea acrensis*, a maxillary fragment with two molars, diagnosed as an early atelid weighing about 5.4–6.1 kg, the size of modern-day *Ateles* and *Lagothrix* (Kay and Frailey 1993; Kay and Cozzuol 2006; Kay 2015).

In this same group, the first primate fossil ever found in South America *Protopithecus brasiliensis*, a femur and humerus found by Lund in NE Brazil, close to Lagoa Santo (Lund 1840) but ignored for many years. However, new material thought to be of this species is now recognized as a new species *Cartelles coimbrafilhoi*. This animal is the largest platyrrhine known and would have reached about 20–25 kg (Halenar 2011; Halenar and Rosenberger 2013). Postcranial bones and the type of posited vegetation of the area indicate that it was probably semiterrestrial and somewhat similar to *Brachyteles* and definitely an ateline primate. The cranium is modified for an enlarged vocal sac and resembles *Alouatta*, yet the postcranium is modified for brachiating. The teeth have no dental adaptations to folivory. The postcranium is very robust, distinctly when compared to *Ateles* or *Lagothrix*. In recent years a skull and mandible and a complete skeleton have been found of *Cartelles coimbrafilhoi* (Hartwig and Cartelle 1996) (Fig. 8.5).

Another semiterrestrial Brazilian fossil is an almost complete skeleton *Caipora bambuiorum* found in NE Brazil a few meters away from the fossil of *Cartelles coimbrafilhoi*. The animal probably weighed about 20 kg and shows similarities to *Ateles* (Cartelle and Hartwig 1996). Both of these primate species became extinct only in the last few thousand years, so that it is difficult not to suspect some influence of the newly arrived human hunters on the scene.

Another new species *Alouatta mauroi* was discovered in Bahía State, Brazil from the Late Pleistocene in the Gruta dos Brejões (Brejões Cave) (Tejedor et al. 2008). This *Alouatta* species was about 1/3 larger than the biggest *Alouatta* alive today as calculated from the P2 (second premolar) (Cartelle and Hartwig 1996).

8.2.6 Caribbean Primates

The last group of primates that I shall discuss belong to the amazing Caribbean primates, various fossil species having been found from Hispaniola, Jamaica, and Cuba. These have been a wonderful surprise to those of us interested in the evolution of neotropical fauna. Fortunately new material continues to be found, while phylogeographic research has suggested some surprising hypotheses about how these animals might have arrived so far from South America (Hedges 2006).

Fig. 8.5 The head of the large ateline primate *Cartelles coimbrafilhoi* had a head very similar to that of *Alouatta*. (By Roman Uchytel)

In Cuba two species have been described, *Paralouatta varonai* and P. marianae. *Paralouatta varonai* has been dated to 17.5–18.5 Ma, establishing great antiquity for Caribbean primates, similar in age to Early Patagonian primates. These were originally thought to be somehow related to *Alouatta* because of the shape of the skull of *P. varonai* (Rivero and Arredondo 1991; Horovitz and MacPhee 1999). *Paralouatta varonai* included various limb bones as well as an almost complete skull, a mandible, and some teeth, so it is possible to calculate a body mass of about 9 kg. Characteristics of the limb bones suggested that it was partially terrestrial (Fleagle 2013). The species *P. marianae* is based solely on a talus that compares well with *P. varonai*.

A previously described Cuban *Ateles anthromorpha* has been carbon-dated at less than 2000 years (MacPhee and Rivero de la Calle 1996). This pretty much proves that the monkey (probably *Ateles geoffroyi*) was brought over from the Central American mainland. Animals were commonly used as trade items among the indigenous population (Lathrap 2010; Redford 1992; Stahl 2003), though these bones could have been a hoax (MacPhee and Meldrum 2006).

Jamaica has produced evidence in the form of femurs and one skull to suggest that two or three different species lived there. *Xenothrix mcgregori* and two unnamed species, based on a femur found in (MacPhee and Fleagle 1991; MacPhee and Meldrum 2006). A femur of *Xenothrix* was actually found in a Jamaican cave by Harold Anthony in 1920, but the fossils were unrecognized as early primates. They were finally described and named by Williams and Koopman in 1952, but there was no phylogenetic analysis.

A partial skull of *Xenothrix mcgregori* was found and described by Horovitz and MacPhee (1999), while other remains of a partial maxilla, fragments of two mandibles, and some long bones are being described. *Xenothrix* had a dental formula like a *Callithrix* or *Saguinus* 2.1.3.2 and probably reached about 2 kg in size. Although the species has been suggested as being related to the callitrichines (Ford 1980), it is now generally regarded as being phylogenetically linked to the pitheciine (*Callicebus*) (Williams and Koopman 1952; Rosenberger 1977; Horovitz and MacPhee 1999). The molars were very distinct from modern callitrichines, while the known postcranial bones suggest a very slow quadrupedal locomotion like nothing known. This fossil primate lived as recently as 1700 before present, according to carbon dating, and was probably a prey species of humans, who could have easily wiped them out because of their slow quadrupedal locomotion (MacPhee and Rivero de la Calle 1996; MacPhee and Horovitz 2002; MacPhee and Hoffmann 2008). A femur found on Jamaica is probably another, unnamed primate species (MacPhee and Horovitz 2002). *Xenothrix* was slightly larger than *Pithecia* and has been dated to just before the Pleistocene (Ford and Morgan 1986).

On Hispaniola, *Saimiri bernensis* (later changed to *Antillothrix bernensis*) was described, and it has become clear that this did not belong in the genus *Saimiri* (Rímoli 1977). It was later placed in a new genus *Antillothrix* (MacPhee and Woods 1982; MacPhee et al. 1995). The holotype, a partial right maxilla was found in the Dominican Republic and has been ^{14}C dated to 3850 years, while a further bone ascribed to this species and found in Haiti was dated 9550 year ago (MacPhee and Horovitz 2002). This primate reached 2–3 kg and fed on hard fruits and seeds and undoubtedly was hunted and probably driven extinct by human beings.

Evidence suggests that these Antillean primate species died out because of human intervention. Humans arrived on an originally forested Hispaniola about 6000–7000 years ago and most certainly began exploiting the native mammals, including the primates (Steadman and Takaro 2013). One of the most complete skull and skeletal parts of *Antillothrix* were recently found in an underwater cave, providing support for a multilineage hypothesis for the Caribbean primates, since the cranium resembles living cebine monkeys, although the fossil is more primitive than modern cebines (Rosenberger et al. 2011, 2013). To evaluate the phylogenetic relationships of *Antillothrix*, 32 craniodental characters were evaluated along with *Paralouatta varonai* of Cuba and 14 New World primate genera. Surprisingly, *Antillothrix* and *Paralouatta* were a sister clade which was closest to *Callicebus* (MacPhee et al. 1995).

Insulacebus toussaintiana, clearly related to *Xenothrix*, has also been described from Hispaniola with two carbon dates of 3450 and 9550 years before present. This animal would have reached 4.6–4.7 kg in life. Together with *Xenothrix* It has been associated with the Aotini (Cooke et al. 2011). However, other data suggest that *Callicebus* could be the origin of all Caribbean primates (MacPhee et al. 1995; MacPhee and Horovitz 2004; Horovitz 1999; Horovitz and MacPhee 1999; Cooke et al. 2011). The complete dentition of this fossil has a very unusual array of characteristics, demonstrating primitive features resembling Middle Miocene Patagonian primates and so interpreted as a radiation only distantly related to mainland platyrrhines (Cooke et al. 2011). In conclusion, it would appear that the origin of the Caribbean primates is not yet clear, although as a group they are not closely related to the South American platyrrhines.

Recently paleo-diet was studied emphasizing the Caribbean primates compared to present-day platyrrhines and using a three-dimensional geometric morphometric analysis of the dental relief on the second lower molars. Most of these primates were predicted to have been frugivores, while *Xenothrix* was predicted to have been a frugivore or frugivore/omnivore (Cooke 2011).

As presented above, there are two hypotheses of the phylogenetic affinities of these Caribbean primates. One theory suggests that a single introduction was the origin of all other monkeys and that the ancestor was a *Callicebus*-related animal of the Early Miocene or earlier (Horovitz and MacPhee 1999; MacPhee and Horovitz 2002). Surely the early and Mid-Miocene drainage of the paleo-Amazon, thought to have been located in roughly the same area of the present-day Orinoco River mouth, could have been the force that would have impelled plants and animals from the mainland toward the paleo-Antilles and in the course of which might have arrived the original primates via a sweepstakes dispersion. It should be remembered that both the original Antillean sloths and the hutias have connections with South America and perhaps arrived in the same manner as the primates.

Another multilineage hypothesis argues for origins from different mainland lineages (Iturralde-Vinent and MacPhee 1999; Rivero and Arredondo 1991; MacPhee and Woods 1982; Rosenberger 2002; Ford 1990b; Rosenberger et al. 2009). The discovery of a complete skull and skeleton of *Antillothrix bernensis* argues for the closest connection of this Hispaniolan species as a cebine, based on general form, although in the absence of apomorphic characteristics found in *Cebus, Killikaike*, and *Saimiri* (Rímoli 1977; Rosenberger et al. 2011). First placed by Rímoli (1977) into the genus *Saimiri,* others question this allocation and placed it into a new genus *Antillothrix* (MacPhee et al. 1995). Recovery of the lower mandible of this fossil specimen placed the *Antillothrix* within the *Callicebus-Aotus* lineage and allowed the authors to conclude that this primate was probably a pithecid, identifying most strongly, along with *Xenothrix* and *Insulacebus* with *Callicebus* (MacPhee and Fleagle 1991). This conclusion is now in agreement with previous work (Rosenberger et al. 2009; Horovitz and MacPhee 1999; MacPhee and Horovitz 2004; MacPhee et al. 1995) although an earlier analysis suggests that *Antillothrix* does not group with any modern clade (Kay et al. 2011). The odd and primitive nature of many characteristics of all of the five known taxa of Caribbean primates argues for a very early arrival of primates in the Greater Antilles probably during the Early Miocene (MacPhee 2005), judging from their relation to Early Miocene Patagonian primates and the dated Early Miocene *Paralouatta marianae* from Cuba. A more recent discovery of an entire skull and other skeletal elements of *Antilothrix* in the Dominican Republic has allowed more analysis, but without resolution as to the phylogenetic position of this species. On one hand, the possibility that this species is part of an ancient radiation from a *Callicebus*-like pithecine continues to be possible (Kay et al. 2011). Another opinion is that this species belongs to the cebid clade but that it may antedate modern radiations known from southern Argentina and may be the evolutionary product separate from the species known from Jamaica and Cuba (Rosenberger et al. 2010).

A third theory suggests that there were multiple introductions that originated *Paralouatta* on one hand and the three species of primates from Hispaniola and Jamaica on the other hand (Rosenberger et al. 2011; Rosenberger 2002). *Paralouatta* could have had an *Alouatta* ancestor, while *Antilothrix* could be connected to primitive cebids, while other taxa had a *Callicebus*-related ancestor (Rosenberger et al. 2011; Rosenberger 2002). Besides the original suggestion of a *Paralouatta- Alouatta* connection by Rivero and Arredondo (1991), the alouattine connection has also been argued by others (Rosenberger 2002; Rosenberger et al. 2008, 2015a).

Arrival of the earlier Caribbean primates has been posited to have occurred via sweepstakes dispersion (Hedges 1996, 2001; Rosenberger et al. 2009) or via a brief connecting peninsula during the Eocene-Oligocene transition (Iturralde-Vinent and MacPhee 1999; MacPhee and Iturralde-Vinent 1994, 1995). The postulated peninsula has been called the GAARlandia landspan (see Chap. 12). Extinction in the Caribbean primates was probably mostly due to the arrival of human colonists (MacPhee 2009).

8.3 First North American Platyrrhine

Quite recently seven new fossil teeth have been discovered and described from Early Miocene deposits along the Panama Canal, Panama (Bloch et al. 2016). The fossil teeth seem to be closest to *Cebus*. This cebid monkey, *Panamacebus transitus*, has been accurately dated at 20.9 Ma, and it was associated with North American fauna, rather than South American fauna, so it stands as the first South American element (Platyrrhini) to begin the Great American Biotic Interchange. Whether these monkeys rafted or were able to take advantage of a land connection northward is not known, but *Panamacebus* is considered to be a **crown** neotropical primate (Bloch et al. 2016), and most evidence so far argues that a land connection wasn't established until about three million years ago so that fauna previous to that date that crossed the gap (sloths, frogs, flightless terror birds, capybara, tayasuids, camelidos, gomphotheres) had to pass a water barrier, though such a barrier might not have been particularly wide (De Queiroz 2016).

References

Ameghino F (1891a) Los monos fósiles del Eoceno de la Republica Argentina. Arg De Hist Nat (Buenos Aires) 1:383–397

Ameghino F (1891b) Nuevos restos de mamíferos fósiles descubiertos por Carlos Ameghino en el Eoceno inferior de la Patagonia austral. – Especies nuevas, adiciones y correcciones [New remains of fossil mammals discovered by Carlos Ameghino in the lower Eocene of southern

Patagonia. – New species, additions, and corrections]. Revista Argentina de Historia Natural 1:289–328

Ameghino F (1898) Sinopsis geológico-paleontológica (de la Argentina). Segundo Censo de la República Argentina). 4 serie 1:112–255

Ameghino F (1907) El orígen del hombre. La Biblioteca de La Reforma, La Plata, 41 pp

Antoine PO, Marivaux L, Croft DA, Billet G, Ganerod M, Jaramillo C, Martin T, Orliac MJ, Tejada J, Altamirano AJ, Duranthon F, Fanat G, Rousse S, Gismondi RS (2012) Middle Eocene rodents from Peruvian Amazonia reveal the pattern and timing of caviomorph origins and biogeography. Proc R Soc B:279. https://doi.org/10.1098/rspb.2011.1732, first published online 12 Oct 2012

Bandoni de Oliveira F, Cassola Molina E, Marroig G (2009) Chapter 3: Paleogeography of the South Atlantic: a route for primates and rodents into the New World. In: Garber PA, Estrada A, Bicca-Marques JC, Heymann EW, Strier KB (eds) South American primates: comparative perspectives in the study of behavior, ecology, and conservation. Springer, New York, pp 55–68

Bauer K, Schreiber A (1997) Double invasion of Tertiary island South America by ancestral New World monkeys? Biol J Linn Soc 60:1–20

Bloch JI, Woodruff ED, Wood AR, Rincon AF et al (2016) First North American fossil monkey and early Miocene tropical biotic interchange. Nature 533:213–228

Bond M, Tejedor MF, Campbell KE Jr, Chornogubsky L, Novo N, Goin F (2015) Eocene primates of South America and the African origins of New World monkeys. Supplementary information supplementary discussion. Nature 520:538. https://doi.org/10.1038/nature14120

Brown TM, Larriestra CN (1990) Sedimentary paleoenvironments of fossil platyrrhine localities, Miocene Pinturas Formation, Santa Cruz Province, Argentina. J Hum Evol 19:87–119

Caperton Morton M (2013) Setting sail on unknown seas: the past, present and future of species rafting. http://www.earthmagazine.org/article/setting-sail-unknown-seas-past-present-and-future-species-rafting

Cartelle C, Hartwig WC (1996) A new extinct primate among the Pleistocene megafauna of Bahia, Brazil. Proc Natl Acad Sci U S A 93:6405–6409

Chatterjee HJ, Ho SYW, Barnes I, Groves C (2009) Estimating the phylogeny and divergence times of primates using a supermatrix approach. BMC Evol Biol 9:259. https://doi.org/10.1186/1471-2148-9-259

Ciochon RL, Chiarelli AB (1980) Paleobiogeographic perspectives on the origin of the Platyrrhini. In: Ciochon RL, Chiarelli AB (eds) Evolutionary biology of the new world monkeys and continental drift. Planum Publishing Corporation, New York, pp 459–493

Cooke SB (2011) Paleodiet of extinct platyrrhynes with emphasis on the Caribbean forms: three-dimensional geometric morphometrics of mandibular second molars. Anat Rec 294:2073–2091

Cooke SB, Rosenberger AL, Turvey S (2011) An extinct monkey from Haiti and the origins of the Greater Antillean primates. Proc Natl Acad Sci U S A 108(7):2699–2704

De Queiroz A (2005) The resurrection of oceanic dispersal in historical biogeography. Trends Ecol Evol 20(2):68–73

De Queiroz A (2014) The monkey's voyage: how improbable journeys shaped the history of life. Basic, New York

De Queiroz A (2016) The monkey's voyage. Blog. https://monkeysvoyage.wordpress.com/tag/panamacebus/

Delson E, Rosenberger AL (1984) Are there any anthropoid primate "living fossils"? In: Eldredge N, Stanley S (eds) Casebook on living fossils. Plenum, New York, pp 50–61

Fleagle JG (1989) New Fossil Platyrrhines from Santa Cruz Province, Argentina. Folia Primatol 48(1-2):65–77

Fleagle JG (1990) New fossil platyrrhines from the Pinturas Formation, southern Argentina. J Hum Evol 19:61–85

Fleagle JG (2013) Primate adaptation and evolution, 3rd edn. Academic Press, New York

Fleagle JG, Bown TM (1983) New primate fossils from late Oligocene (Colhuehuapian) localities of Chubut Province, Argentina. Folia Primatol 41:240–266. https://doi.org/10.1159/000156135

Fleagle JG, Kay RF (1989) The dental morphology of *Dolichocebus gaimanensis*, a fossil monkey from Argentina. Am J Phys Anthropol 78(2):221
Fleagle JG, Tejedor MF (2002) Early platyrrhines of southern South America. Cambridge Studies. In: Hartwig WC (ed) The Primate Fossil Record. Cambridge University Press, Cambridge, pp 161–173
Fleagle JG, Powers DW, Conroy GC, Watters JP (1987) New fossil platyrrhines from Santa Cruz province, Argentina. Folia Primatol 48:65–77
Fleagle JG, Brown TM, Swisher C, Buckley G (1995) Age of the Pinturas and Santa Cruz formation. Actas VI Congreso Argentino de Paleontología Biostratigraffa, 129–135
Flynn JJ, Wyss AR, Charrier R, Swisher CC (1995) An early Miocene anthropoid skull from the Chilean Andes. Nature 373:603–607
Flynn JJ, Guerrero J, Swisher CC III (1997) Geochronology of the Honda Group. In: Kay RF, Madden RH, Cifelli RL, Flynn JJ (eds) Vertebrate paleontology in the neotropics: the Miocene fauna of La Venta, Colombia. Smithsonian Institution Press, Washington, DC, pp 44–59
Flynn JJ, Charrier R, Croft DA, Wyss AR (2012) Cenozoic Andean faunas: shedding new light on South American mammal evolution, biogeography, environments, and tectonics. In: Patterson BD, Costa LP (eds) Bones, clones, and biomes. University of Chicago Press, Chicago, pp 51–75
Ford SM (1980) Affinities and interpretations of the primate fossil postcranials from South America. Am J Phys Anthropol 52:227
Ford SM (1990a) Locomotor adaptations of fossil platyrrhines. J Hum Evol 19:141–173
Ford SM (1990b) Platyrrhine evolution in the West Indies. J Hum Evol 19:237–254
Ford SM, Morgan GS (1986) A new ceboid femur from the late Pleistocene of Jamaica. J Vertebr Paleontol 6(3):281–289
Frailey CD, Campbell KE Jr (2004) Paleogene rodents from amazonian Peru: the Santa Rosa local fauna. In: Campbell KE Jr (ed) The Paleogene mammalian fauna of Santa Rosa, amazonian Peru. Natural History Museum of Los Angeles County, Los Angeles, pp 71–130
Fulwood E, Boyer DM, Kay RF (2016) Stem members of Platyrrhini are distinct from catarrhines in at least one derived cranial feature. J Hum Evol 100:16–24
Gebo DL (2002) Adapiformes: phylogeny and adaptation. In: Hartwig WC (ed) The primate fossil record. Cambridge University Press, Cambridge, pp 21–43
Glazko GV, Nei M (2003) Estimation of divergence times from major lineages of primate species. Mol Biol Evol 20(3):424–434
Goin FJ, Candela AM (2004) New Paleogene marsupials from the Amazon basin of eastern Peru. In: Campbell KE Jr (ed) The Paleogene mammalian fauna of Santa Rosa, amazonian Peru. Natural History Museum of Los Angeles County, Los Angeles, pp 15–60
Guerrero J (1997) Stratigraphy, sedimentary environments, and the Miocene uplift of the Colombian Andes. In: Kay RF, Madden RH, Cifelli RL, Flynn JJ (eds) Vertebrate paleonotology in the neotropics. Smithsonian Institution Press, Washington, DC, pp 15–43
Halenar LB (2011) Reconstructing the locomotor repertoire of *Protopithecus brasiliensis*. II. Forelimb morphology. Anat Rec 294:2048–2063
Halenar LB, Rosenberger AL (2013) A closer look at the "*Protopithecus*" fossil assemblages: new genus and species from Bahia, Brazil. J Hum Evol 65:374–390
Halenar LB, Cooke SB, Rosenberger AL, Rímoli R (2017) New cranium of the endemic Caribbean platyrrhine, *Antillothrix bernenesis*, from La Altagracia Province, Dominican Republic. J Hum Evol 106:133–153
Hartwig WC, Cartelle C (1996) A complete skeleton of the giant South American primate *Protopithecus*. Nature 381:307–311
Hartwig WC, Meldrum J (2002) Miocene platyrrhines of the northern Neotropics. In: Hartwig WC (ed) The primate fossil record. Cambridge University Press, Cambridge, pp 175–188
Hedges SB (1996) The origin of the west Indian amphibians and Reptiles. In: Powell R, Henderson RW (eds) Contributions to West Indian herpetology: a tribute to Albert Swartz. SSAR, Ithaca, pp 95–128

Hedges SB (2001) Biogeography of the West Indies: an overview. In: Woods CA, Sergile FE (eds) Biogeography of the West Indies: patterns and perspectives. CRC Press, New York, pp 15–33
Hedges SB (2006) Paleogeography of the Antilles and origin of West Indian terrestrial vertebrates. Ann Mo Bot Gard 93:231–244
Hershkovitz P (1974) A new genus of late Oligocene monkey (Cebidae, Platyrrhini) with notes on postorbital closure and platyrrhine evolution. Folia Primatol 21(1):1–35
Hodgson JA, Sterner KN, Matthews LJ, Burrell AS et al (2009) Successive radiations, not stasis, in the South American primate fauna. Proc Natl Acad Sci U S A 106(14):5534–5539
Hofstetter MR (1969) Un primate de l'Pliocène inférieur sudamericain: *Branisella boliviana* gen. et sp. Nov. C R. Acad Sci Paris 269:434–437
Hofstetter MR (1980) Origin and deployment of New World monkeys emphasizing the southern continents route. In: Chichon AB, Chiarelli AB (eds) Evolutionary biology of the New World monkeys and continental drift, Advances in primatology. Plenum Press, New York, pp 103–122
Horovitz I, MacPhee DE (1999) The quaternary Cuban platyrrhine Paralouatta varonai and the origin of Antillean monkeys. J Hum Evol 36(1):33–68
Houle A (1999) The origin of Platyrrhines: an evaluation of the Antarctic scenario and the Floating Island model. Am J Phys Anthropol 109:541–559
Irvine RDG (2014) Deep time: an anthropological problem. Soc Anthropol 22(2):157–172
Iturralde-Vinent M, MacPhee RDE (1999) Paleogeography of the Caribbean region: implications for Cenozoic biogeography. Bull Am Mus Nat Hist 238:1–95
Jaeger J-J, Beard KC, Chaimanee Y, Salem M, Benammi M, Hlal O, Coster P, Bilal AA, Duringer P, Schuster M, Valentin X, Marandat B, Marivaux L, Métais E, Hammuda O, Brunet M (2010) Late middle Eocene epoch of Libya yields earliest known radiation of African anthropoids. Nature 467:1095. https://doi.org/10.1038/nature09425
Kay RF (1990) The phyletic relationships of extant and fossil Pitheciinae (Platyrrhini, Anthropoidea). J Hum Evol 19:175–208
Kay RF (1994) "Giant" tamarin from the Miocene of Colombia. Am J Phys Anthropol 95:333–353
Kay RF (2010) A new primate from the early Miocene of Gran Barranca, Chubut province, Argentina: paleoecological implications. In: Madden RH, Carlini AA, Vucetich MG, Kay RF (eds) The paleontology of Gran Barranca. Cambridge University Press, Cambridge, pp 220–236
Kay RF (2015) Biogeography in deep time – what do phylogenetics, geology, and paleoclimate tell us about early platyrrhine evolution? Mol Phylogenet Evol 82:358–374
Kay RF, Cozzuol MA (2006) New platyrrhine monkeys from the Solimoes Formation (Late Miocene Acre State, Brazil). J Hum Evol 50:673–686
Kay RF, Fleagle JG (2010) Stem taxa, homoplasy, long lineages, and the phylogenetic position of Dolichocebus. J Hum Evol 59:218. https://doi.org/10.1016/j.jhevol.2010.03.002
Kay RF, Frailey CD (1993) Large fossil platyrrhines from the Rio Acre fauna, Late Miocene, western Amazonia. J Hum Evol 25:319–327
Kay RF, Madden RH (1997) Mammals and rainfall: paleoecology of the middle Miocene at La Venta (Colombia, South America). J Hum Evol 32:161–199
Kay RF, Meldrum DJ (1997) A new small Platyrrhine and the phyletic position of Callitrichinae. In: Hartwig WC 2002. The primate fossil record. Cambridge University Press, Cambridge, pp 435–458
Kay RF, Williams BA (1995) Recent finds of monkeys from the Oligocene/Miocene of Salla, Bolivia. Am J Phys Anthropol 20(Suppl):124
Kay RF, Madden RH, Plavcan JM, Cifelli RL, Diaz JG (1987) *Stirtonia victoriae*, a new species of Miocene Colombian primate. J Hum Evol 16:173–196
Kay RF, Madden RH, Cifelli RL, Flynn JJ (1997) Vertebrate paleontology in the Neotropics. Smithsonian Institution Press, Washington, DC
Kay RF, Johnson D, Meldrum DJ (1998) A new pitheciin primate from the middle Miocene of Argentina. Am J Primatol 45:317–336
Kay RF, Williams BA, Anaya F (2002) The adaptations of *Branisella boliviana*, the earliest South American monkey. In: Plavean et al (eds) Reconstructing behavior in the primate fossil record. Kluwer Academic/Plenum Publishers, New York, pp 339–370

Kay RF, Campbell VM, Rossie JB, Colbert MW, Rowe TB (2004) Olfactory fossa of *Tremacebus harringtoni* (Platyrrhini, Early Miocene, Sacanana, Argentina): implications for activity pattern. Anat Rec A 281A:1157–1172

Kay RF, Fleagle JG, Mitchell TRT, Colbert M, Brown T, Powers DW (2008) The anatomy of *Dolichocebus gaimanensis*, a stem platyrrhine monkey from Argentina. J Hum Evol 54:323–382

Kay RF, Hunt KD, Beeker CD, Conrad GW, Johnson CC, Keller J (2011) Preliminary notes on a newly discovered skull of the extinct monkey *Antillothrix* from Hispaniola and the origin of the Greater Antillean monkeys. J Hum Evol 60:124–128

Lathrap DW (2010) The antiquity and importance of long-distance trade relationships in the moist tropics of pre-Columbian South America. World Archaeol 5(2):170–186

Lavocat R (1980) The implications of rodent paleontology and biogeography to the geographical sources and origin of platyrrhine primates. In: Ciochon RL, Chiarelli AB (eds) Evolutionary biology of the new world monkeys and continental drift. Plenum Press, New York, pp 93–102

Luchterhand K, Kay RF, Madden RH (1986) *Mahanico hershkovitzi*, gen. et sp. nov., un primate du Miocène moyen d'Amérique du Sud. Comptes rendus de l'académie des sciences, Série 2, Mécanique, Physique, Chimi, Sciences de lúnivers. Sci Terre 303(19):1753–1758

Lund P (1840) View of the fauna of Brazil, previous to the last geological revolution. Charlesworth's Mag Nat Hist 4: 1–8; 49–57; 105–112; 153–161; 207–213; 251–259; 307–317; 373–389

MacFadden BJ (1990) Chronology of Cenozoic primate localities in South America. J Hum Evol 19:7–21

MacFadden BJ, Campbell KE Jr, Cifelli RL, Siles O et al (1985) Magnetic polarity stratigraphy and mammalian fauna of the Deseadan (Late Oligocene-Early Miocene) Salla Beds of northern Bolivia. J Geol 93(3):223–250

MacPhee RDE (2005) "First" appearances in the Cenozoic land-mammal record of the Greater Antilles: significance and comparison with South American and Antarctic records. J Biogeogr 32:551–564

MacPhee RDE (2009) *Insulae infortunatae*: establishing a chronology for late Quaternary mammal extinctions in the West Indies. In: Haynes G (ed) American megafaunal extinctions at the end of the Pleistocene. Springer, New York, pp 169–193

MacPhee RDE, Fleagle J (1991) Postcranial Remains of *Xenothrix mcgregori* (Primates, Xenotrichidae) and Other Late Quaternary Mammals from Long Mile Cave, Jamaica. Bull Am Mus Nat Hist 206:287–321

MacPhee R, Hoffmann M (2008) *Xenothrix mcgregori*. In: IUCN 2008. IUCN Red List of Threatened Species. http://www.iucnredlist.org/details/136515/0. Downloaded on 3 Jan 2009

MacPhee RDE, Horovitz I (2002) Extinct Quaternary platyrrhines of the Greater Antilles and Brazil. In: Hartwig WC (ed) The primate fossil record. Cambridge University Press, Cambridge, pp 189–200

MacPhee RDE, Horovitz I (2004) New craniodental remains of the Quaternary Jamaican monkey *Xenothrix mcgregori* (Xenotrichini, Callicebinae, Pitheciidae) with a reconsideration of the *Aotus* hypothesis. Am Mus Novit 3434:1–55

MacPhee RDE, Iturralde-Vinent M (1994) First Tertiary land mammal from greater Antilles: an early Miocene sloth (Xenarthra, Magalonychidae) from Cuba. Am Mus Novit 3094:1–13

MacPhee RDE, Iturralde-Vinent M (1995) Origin of the greater Antillean land mammal fauna: new Tertiary fossils from Cuba and Puerto Rico. Am Mus Novit 3141:1–31

MacPhee RDE, Meldrum J (2006) Postcranial remains of the extinct monkeys of the Greater Antilles, with Evidence for semiterrestriality in *Paralouatta*. Am Mus Novit 3516:1–65

MacPhee RDE, Rivero de la Calle M (1996) Accelerator mass spectrometry 14C age determination for the alleged "Cuban spider monkey", Ateles (=*Montaneia*) anthropomorphus. J Hum Evol 30:89–94

MacPhee RDE, Woods CA (1982) A new fossil Cebine from Hispaniola. Am J Phys Anthropol 58:419–433

MacPhee RDE, Horovitz I, Arredondo O, Vasquez OJ (1995) A new genus for the extinct Hispaniolan monkeys *Saimiri bernensis* Rimoli, 1977, with notes on its systematic position. Am Mus Novit 3134:1–21

Matsui A, Hasegawa M (2012) Molecular phylogeny and evolution in primates. In: Hirai H et al (eds) Post-genome biology of primates. Springer, New York
Meldrum DF, Kay RF (1997a) *Nuciruptor rubricae*, a new pitheciin seed predator from the Miocene of Colombia. Am J Phys Anthropol 102:407–427
Meldrum DF, Kay RF (1997b) Postcranial skeleton of Laventan platyrrhines. In: Kay RF, Madden RH, Cifelli RL, Flynn JJ (eds) Vertebrate paleontology in the neotropics. Smithsonian Institution Press, Washington, DC, pp 459–472
Miller ER, Simons EL (1997) Dentition of *Proteopithecus sylviae*, an archaic anthropoid from the Fayum, Egypt. Proc Natl Acad Sci U S A 94:13760–13764
Nakatsukasa M, Takai M, Setoguchi T (1997) Functional morphology of the postcranium and locomotor behavior of Neosaimiri fieldsi, a Saimiri-like Middle Miocene platyrrhine. Am J Phys Anthropol. https://doi.org/10.1002/(SICI)1096-8644(199704)102:4<515::AID-AJPA7>3.0.CO;2-Q
Nei M, Glazko GV (2002) Estimation of divergence times for a few mammalian and several primate species. Am Genet Assoc 93:157–164
Perez SI, Tejedor MF, Novo NM, Aristide L (2013) Divergence times and the evolutionary radiation of new world monkeys (Platyrrhini, Primates): an analysis of fossil and molecular data. PLoS One 8(6):1–16
Perry JMG, Kay RF, Vizcaíno SF, Bargo MS (2014) Oldest known cranium of a juvenile New World monkey (Early Miocene, Patagonia, Argentina): implications for the taxonomy and the molar eruption pattern of early platyrrhines. J Hum Evol 74:67–81
Poux P, Chevret P, Huchon D, De Jong WW, Douzery EJP (2006) Arrival and diversification of platyrrhine primates in South America. Syst Biol 55(2):228–244
Redford KW (1992) The empty forest. Bioscience 42(6):412–422
Rimoli R (1977) Una nueva especie de monos (Cebidae: Saimirinae: Saimiri) de la Hispaniola. Cuad de CENDIA, Univ. Aut. Santo Domingo 242: 5–14
Rivero M, Arredondo O (1991) *Paralouatta varonai*, a new Quaternary platyrrhine from Cuba. J Hum Evol 21:1–11
Rosenberger AL (1977) *Xenothrix* and ceboid phylogeny. J Hum Evol 6:461–481
Rosenberger AL (1979) Cranial anatomy and implications of *Dolichocebus*, a Late Oligocene ceboid primate. Nature 279:416–418
Rosenberger AL (1981) A mandible of *Branisella boliviana* (Platyrrhini, Primates) from the Oligocene of South America. Int J Primatol 2:1–7
Rosenberger AL (2002) Platyrrhine paleontology and systematics: the paradigm shifts. In: Hartwig W (ed) The primate fossil record. Cambridge University Press, Cambridge, pp 151–159
Rosenberger AL (2011) Evolutionary morphology, platyrrhine evolution, and systematics. Anat Rec 294:1955–1974
Rosenberger AL, Tejedor MF (2013) The misbegotten: long lineages, long branches and the interrelationships of *Aotus*, *Callicebus* and the saki-uakaris. In: Veiga LM, Barnett AA, Ferari SF, Norconk MA (eds) Evolutionary biology and conservation of titis, sakis and uacaris. Cambridge University Press, Cambridge, pp 13–22
Rosenberger AL, Setoguchi T, Shigehara N (1990) The fossil record of callitrichinae primates. J Hum Evol 19:209–236
Rosenberger AL, Setoguchi T, Hartwig WC (1991) *Laventiana annectens*, new genus and species: fossil evidence for the origins of Callitrichinae New World monkeys. Proc Nati Acad Sci U S A 88:2137–2140
Rosenberger AL, Halenar LB, Cooke SB, Hartwig WC (2008) Morphology and evolution of the spider monkey, genus *Ateles*. In: Campbell C (ed) Spider monkeys: behavior, ecology and evolution of the genus *Ateles*. Cambridge University Press, New York
Rosenberger AL, Tejedor MF, Siobhán BC, Pekar S (2009) Chapter 4: Platyrrhine ecophylogenetics in space and time. In: Garber PA, Estrada A, Bicca-Marques JC, Heymann EW, Strier KB (eds) South American primates: comparative perspectives in the study of behavior, ecology, and conservation. Springer, New York, pp 69–113

Rosenberger AL, Cooke SB, Rímoli R, Ni X, Cardoso L (2010) First skull of Antillothrix bernensis, an extinct relict monkey from the Dominican Republic. Proc R Soc B. https://doi.org/10.1098/rspb.2010.1249

Rosenberger AL, Cooke SB, Rímoli R, Ni X, Cardoso L (2011) First skull of *Antillothrix bernensis*, an extinct relict monkey from the Dominican Republic. Proc R Soc B 278:67–74

Rosenberger AL, Klukkert ZS, Cooke SB, Rimoli R (2013) Rethinking *Antillothrix*. The mandible and its implications. Am J Primatol 75:825–836

Rosenberger AL, Cooke SB, Halenar LB, Tejedor MF, Hartwig WC, Novo NM, Muñoz-Saba Y (2015a) Fossil Alouattines and the origins of *Alouatta*: Craniodental diversity and interrelationships. In: Kowalewski MM, Garber PA, Cortes-Ortiz L, Urbani B, Youlatos D (eds) Howler monkeys: behavior, ecology and conservation. Springer, New York, pp 21–54

Rosenberger AL, Pickering R, Green H, Cooke SB et al (2015b) 1.32 ± 0.11 Ma age for underwater remains constrain antiquity and longevity of the Dominican primate *Antillothrix bernensis*. J Hum Evol 88(2015):85–96

Rusconi C (1935) Las especies de primates del oligoceno de Patagonia (gen. *Homunculus*). Ameghiniana 1:39–126

Schrago CG, Russo CAM (2003) Timing the origin of new world monkeys. Mol Biol Evol 20(10):1620–1625

Schrago CG, Mello B, Soares ER (2013) Combining fossil and molecular data to date the diversification of New World primates. J Evol Biol 26:2438–2446

Sears KE, Finarelli JA, Flynn JJ, Wyss AR (2008) Estimating body mass in New World "monkeys" (Platyrrhini, Primates), with a consideration of the Miocene platyrrhine, *Chilecebus carrascoensis*. Am Mus Novit 3619:1–29

Setoguchi T (1985) *Kondous laventicus*, a new ceboid primate from the Miocene of the La Venta, Colombia, South America. Folia Primatol 44(2):96–101

Setoguchi T, Rosenberger AL (1985a) Some new ceboid primates from La Venta, Colombia, South America. Mem VI Congr Latinoam Geol 1:187–198

Setoguchi T, Rosenberger AL (1985b) Miocene marmosets: first fossil evidence. Int J Primatol 6(6):615–625

Setoguchi T, Rosenberger AL (1987) A fossil owl monkey from La Venta, Colombia. Nature 326:692–694

Simons EL (1997) Preliminary description of the cranium of *Proteopithecus sylviae*, an Egyptian late Eocene anthropoidean primate. Proc Natl Acad Sci U S A 94:14970–14975

Stahl P (2003) Pre-Columbian andean animal domesticates at the edge of empire. World Archaeol 34(3):470–483

Steadman DW, Takano OM (2013) A late-Holocene bird community from Hispaniola: refining the chronology of vertebrate extinction in the West Indies. The Holocene 23(7):921–935. https://doi.org/10.1177/0959683613479683

Stirton RA (1951) Ceboid monkeys from the Miocene of Colombia. Bull Calif Publ Geol Sci 28:315–356

Takai M (1994) New specimens of *Neosaimiri fieldsi* from La Venta, Colombia: middle Miocene ancestor of the living squirrel monkeys. J Hum Evol 27:329–360

Takai M, Anaya F (1996) New specimens of the oldest fossil platyrrhine, *Branisella boliviana*, from Salla, Bolivia. Am J Phys Anthropol 99:301–317

Takai M, Anaya F, Shigehara N, Segotuchi T (2000) New fossil materials of the earliest New World monkey, *Branisella boliviana*, and the problem of platyrrhine origins. Am J Phys Anthropol 111:263–281

Takai M, Anaya F, Suzuki H, Shigehara N, Setoguchi T (2001) A new platyrrhine from the middle Miocene of La Venta, Colombia and the phyletic position of Callicebinae. Anthropol Sci 109(4):289–307

Tejedor MF (1987) La denticion de *Callicebus* y el morfotipo ancestral de los platirrinos. Neotrop Primates 5:43–46

Tejedor MF (2002) Primate canines from the early Miocene Pinturas formation, southern Argentina. J Hum Evol 43:127–144

Tejedor MF (2005a) New specimens of *Soriacebus adrianae* Fleagle, 1990, with comments on pitheciin primates from the Miocene of Patagonia. Ameghiniana 42(1):249–251

Tejedor MF (2005b) New fossil platyrrhine from Argentina. Folia Primatol 76(3):146–150

Tejedor MF (2008) The origin and evolution of neotropical primates. Arch Mus Nac, Rio de Janeiro 66:251–269

Tejedor MF (2013) Sistemática, evolución y paleobiogeografía de los primates Platyrrhini. Rev Mus La Plata Sec Zool 20(176):20–39

Tejedor MF, Muñoz Y (2012) La sistemática de los platirrinos y el registro fósil. In: Defler TR, Stevenson PR, Bueno ML, Guzman D (eds) Primates Colombianos en Peligro de Extinción. Asociación Primatológica Colombiana, Bogotá

Tejedor MF, Rosenberger AL (2008) A neotype for *Homunculus patagonicus* Ameghino, 1891, and a new interpretation of the taxón. PaleoAnthropology 2008:68–82

Tejedor MF, Tauber AA, Rosenburger AL, Swisher CC III, Palacios ME (2006) New primate genus from the Miocene of Argentina. PNAS 103(14):5437–5441

Tejedor MF, Rosenberger AL, Cartelle C (2008) Nueva especie de *Alouatta* (Primates, Atelinae) del Pleistoceno Tardío de Bahía, Brasil. Ameghiniana (Rev Asoc Paleontol Argent) 45(1):247–251

Williams EE, Koopman KF (1952) West Indian fossil monkeys. Am Mus Novit 1546:16 pp

Wolff R (1984) New specimens of the primate *Branisella boliviana* from the early Oligocene of Salla, Bolivia. J Vertebr Paleontol 4(4):570–574

Yang Z, Yoder AD (2003) Comparison likelihood and Bayesian methods for estimating divergence times using multiple gene loci and calibration points with application to a radiation of cute-looking mouse lémur species. Syst Biol 52:705–716

Yoder AD, Yang Z (2004) Divergence dates for Malagasy lemurs estimated from multiple gene loci. Mol Ecol 13(4):757–773

Chapter 9
An Antarctic Mammalian Community

9.1 Introduction

Until about 30–36 Ma ago, South America and Antarctica were in intimate contact via a wide connection of the Antarctic Peninsula to southern Patagonia (the Weddellian Isthmus), so that fauna was able to move freely back and forth between the two continents (Shen 1995, 1998). In the Late Cretaceous or very early Paleocene, South American marsupials were able to reach Australia, before 64 Ma when the South Tasmanian Rise became submerged and the last land bridge to Australia disappeared and the continent continued to rift away from Antarctica (Woodburne and Case 1996). During Cretaceous and Paleocene times, temperatures in Antarctica were much milder and did not begin to degenerate until mid-Eocene times when the world temperature began to decrease from the high point of Late Paleocene to Early Eocene world temperatures that for a time had produced a tropical world climate (Figs. 9.1, 9.2, 9.3, 9.4, and 9.5).

9.2 Peninsular Connection of South America to Antarctic

The tip of South America was connected to Antarctica by an irregularly narrow land bridge (the Weddellian Isthmus), somewhat like the Isthmus of Panama of about 100–200 km in width and 700–900 km in length. This land bridge persisted from Early Cretaceous times to the end of the Eocene or the beginning of the Oligocene, lasting about 35 million years (Shen 1995, 1998) when a deep-sea rupture apparently occurred, allowing the establishment of the circumpolar current. The Weddellian Isthmus allowed a continuous flora and fauna to exist from southern South America to Antarctica and (in some cases) to Australia, although the istmus

T. Defler, *History of Terrestrial Mammals in South America*, Topics in Geobiology 42, https://doi.org/10.1007/978-3-319-98449-0_9

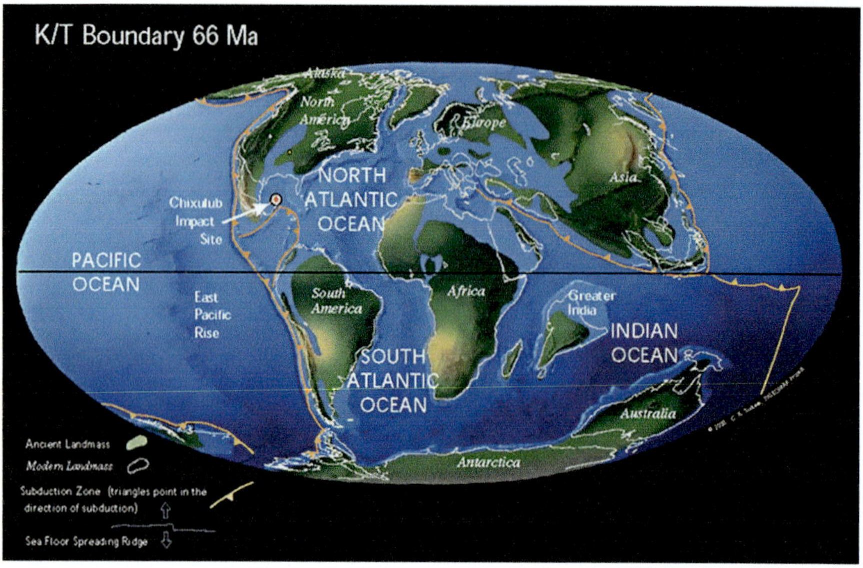

Fig. 9.1 Relationship of continents in the Early Paleocene 66 Ma showing Chicxulub (Chixulub on this map) zone, which impacted earth roughly 65.5 million years ago and caused the extinction of the nonavian dinosaurs among other groups. (By C. R. Scotese, Paleomap Project)

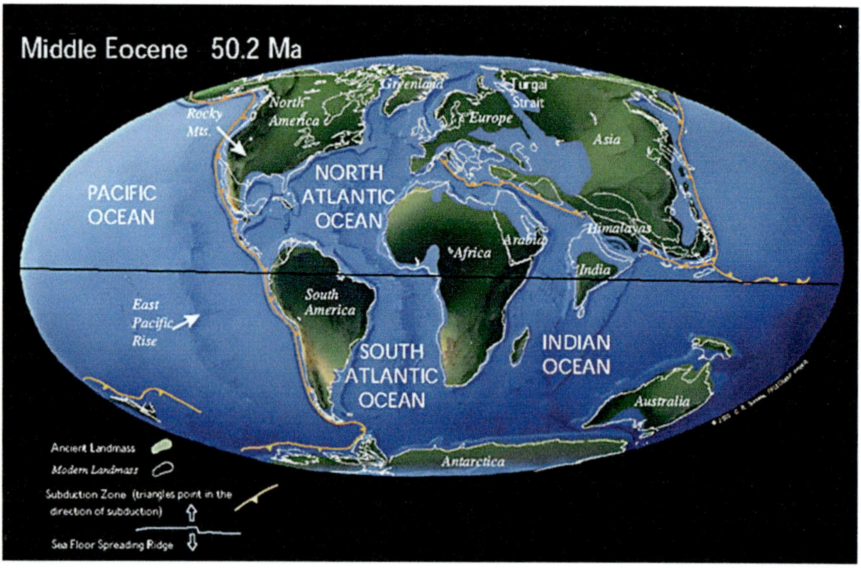

Fig. 9.2 Middle Eocene and the existence of the La Meseta fauna in Antarctica. By this time the Antarctic-Australian land connection had been broken, although a connection continued between southern South America and Antarctica until around 30 Ma. (By C. R. Scotese, Paleomap Project)

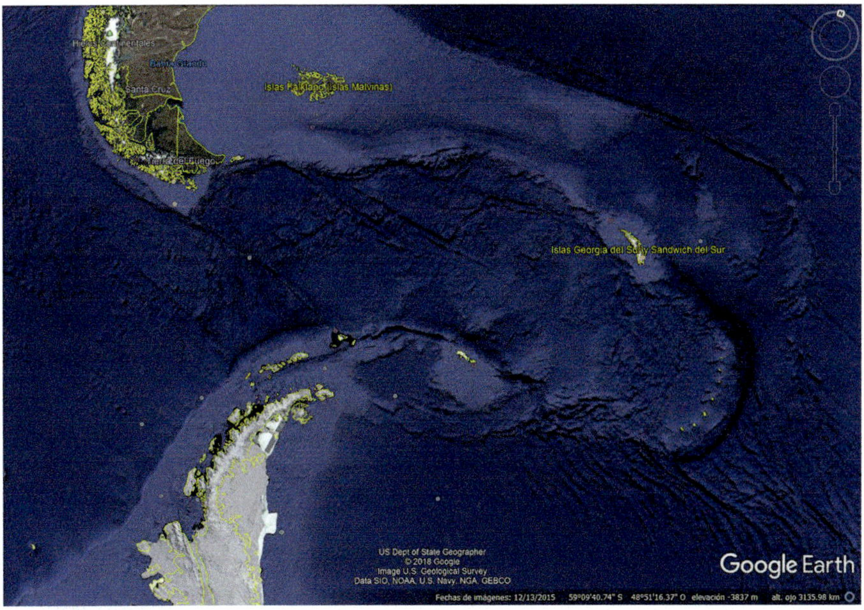

Fig. 9.3 The underwater topography that connects the southernmost point of South America with the Antarctic Peninsula. This topography was above water earlier than about 30 Ma and formed a barrier to ocean currents. (Image Lansat. Google Earth Image courtesy U.S. Geological Survey)

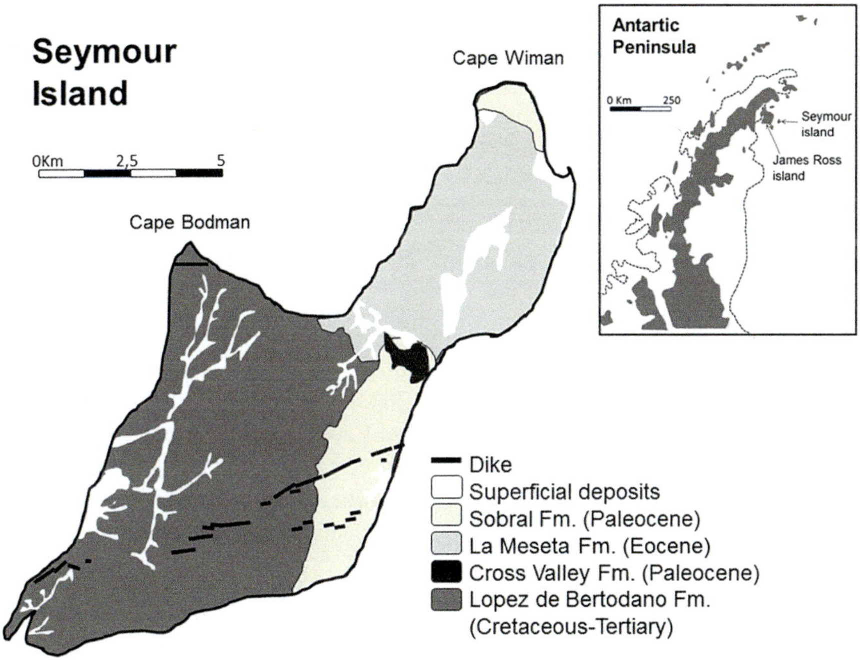

Fig. 9.4 Location of Seymour (Marambio) Island in relation to the Antarctic Peninsula. (Based on Bond et al. 2011 and Google maps, by Diego Casallas, Applied Biodiversity Foundation)

Fig. 9.5 Seymour-Marambio Island, Antarctica, and the location of the Argentinian base Antarctica-Marambio, Argentina's largest Antarctic scientific and military base plus a good airstrip

was possibly covered by a shallow sea from about 50 Ma, obstructing the entrance of notoungulates into the Antarctic and facilitating the development of endemism in the Antarctic metatheres (Reguero et al. 2014).

9.2.1 Vegetation of the Antarctic Peninsular Forest

In the Early–Late Eocene, now tentatively dated at about 51.6–47.89 Ma, the fauna of the peninsula existed in a temperate climate as a result of the most extreme warming event known in the history of the planet (Shellito et al. 2003; Weijers et al. 2007; Krause et al. 2017)). At that time (Early to mid-Eocene), the peninsular forest was a biome equivalent to the present Valdivian rainforest of Chile, a mixed forest of neotropical angiosperms of at least 36 species growing alongside plants characteristic of cooler temperatures and including podocarp and araucarian conifers alongside southern beech, other gymnosperms, and at least three ferns (Romero 1986; Case 2006; Cantrill 2012; Poole and Cantrill 2006).

There were also forests on the main part of the Antarctic continent. Assumedly these forests also sheltered a fauna similar to what we know about the Eocene Antarctic Peninsula and afterward. As the high temperatures began to decrease, the continental vegetation degenerated, becoming scarce, and then extinct. Forests

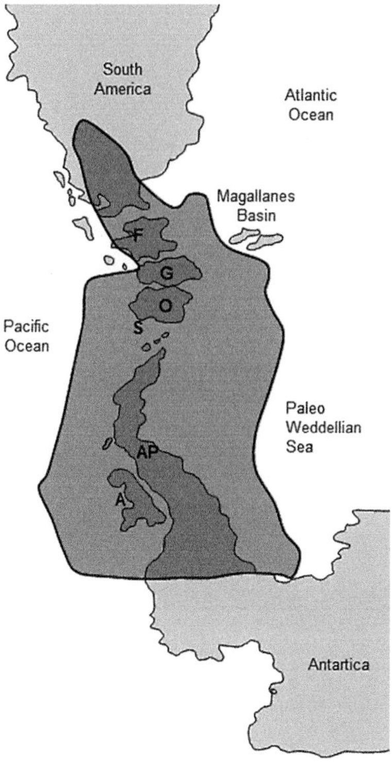

Fig. 9.6 The land bridge Isthmus of Scotia (based on Shen 1995) or Weddellian Isthmus (Reguero et al. 2014) that connected South America to Antarctica from Early Cretaceous times to the end of the Eocene (Shen 1998). (**a**) Early Cretaceous; (**b**) Late Cretaceous; (F) Tierra del Fuego Island; (G) South Georgia Island; (S) South Shetland Islands: (AP) Antarctic Peninsula; (A) Alexander Island. (By Diego Casallas, Applied Biodiversity Foundation)

existing after the Middle Eocene, when temperatures had begun to cool, were dominated by ferns *Nothofagus*, *Podocarpus*, and *Araucaria* in a mesophytic mixed forest (Barrett 1999; Case 1988, 2006; Gandolfo et al. 1998; Jamieson and Sugden 2008). These are the forests where we have some fossil knowledge of the fauna that included marsupials, ratite birds, ornithorhynchus (platypus), gondwanatherians, primitive xenarthrans, astrapotheres, litopterns (or condylarthras), and other groups (Fig. 9.6).

We know a lot about this southern temperate Antarctic forest (after the Middle Eocene) especially because of the discovery and analysis of fossils on Seymour Island. A rich Eocene fauna and flora has been recovered from fossils of La Meseta Formation (49–51 Ma) and from various subunits of the La Meseta Formation. These corresponded to the Middle Eocene age for the majority of the fossil recoveries, when temperatures had begun to decline in the world from the Early Eocene climatic optimum.

This is the only Cenozoic land vertebrate fauna known for the Antarctic. During early mid-Eocene, the world temperatures had reached the highest point in the entire Cenozoic (the Eocene Optimum). Argentina was subtropical (Fig. 9.7). In the Middle–Late Eocene, the climate was beginning to cool, and the forest was dominated by *Nothofagus* and three species of fern.

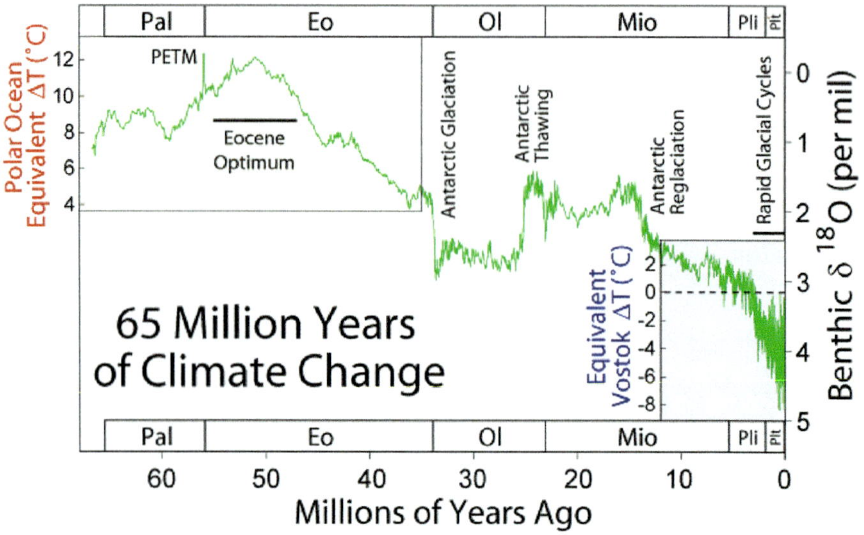

Fig. 9.7 The course of temperature change during the Cenozoic which allowed a rich fauna and flora to flourish in Antarctica during the Eocene. Temperature changes (green line) are based on a compilation of oxygen isotope measurements ($\delta^{18}O$) on benthic foraminifera published by Zachos et al. (2001). This can be converted into temperature changes (left axis). Temperature changes have been standardized by the observation that the oxygen isotope measurements of Lisiecki and Raymo (2005) are tightly correlated to temperature changes at Vostok as established by Petit et al. (1999). Present day is indicated by 0. (By R. A. Rohde for Global Warming Art)

Other flora was Podocarpaceous (*Dacrycarpus*, Araucariaceae (*Araucaria nathrosti*), Proteaceae (*Knightliophyllum andreae*), and the families Dilleniaceae, Myricaceae, Myrtaceae, Lauraceae, and Grossulariaceae). This was a mixed mesophytic forest, a seasonal cool-temperate rainy climate (Case 1988, 2006; Gandolfo et al. 1998) not dissimilar to some Eocene floras sampled in southern Patagonia. Based on this flora, the paleotemperature was probably about 11–13 °C, and the mean of the coldest month was probably between −3 °C and 2 °C (Dingle and Lavelle 1998a, b, 2000; Dingle et al. 1998). It will be very exciting to find evidence of the mammalian fauna corresponding to the height of the Pliocene-Eocene climatic optimum and the Early Eocene climatic optimum which existed before the cool temperate *Nothofagus* dominated rain forest that was the habitat of the only known ancient mammalian community of La Meseta.

9.3 Seymour Island (La Meseta) Fauna

The La Meseta fauna was intimately associated with the southern parts of South America. It included 16 or more taxa of terrestrial mammals. These included ten species of metatherians, including the families Polydolopidae, Microbiotheriidae (the group that gave rise to Australian marsupials), Derorhynchidae, and Prepidolopidae. It also included one species of gondwanatheria, one primitive xenarthran, and perhaps two species of astrapotheres (Hooker 1992; Goin et al. 1999, 2006; Case 2006) and one litoptern ungulate (Vizcaíno et al. 1997). Of course there were many birds and sea mammals as well; the birds included ratites (relatives of rheas and ostrich) as well as primitive penguins. The quality of the fossils is not good and except for the birds the majority are teeth that have been reworked by shallow seas. All of these animals must have been adapted to a climatic regimen that included several months of darkness and crepuscular conditions because of the high latitude involved (Seymour island is located at 64°14′S, 56°37′W).

A reconstruction of the environment and a vertebrate assemblage from the Middle to Late Eocene of Antarctic Peninsula, Seymour Island would include the following animals and plants: (1) Sparnotheriodontidae gen. et sp. nov. (condylarth-litoptern), (2) *Anthropornis nordenskjoeldi* (penguin), (3) *Delphinornis larseni* (penguin), (4) Gondwanathere sudamericid, (5) *Polydolops* (metatherian-marsupiales) (6) *Lyreidus antarcticus* (crab), (7) *Cucullaea* (bivalve), (8) *Eutrephoceras* (nautiloid), (9) Ratitae bird, (10) *Trigonostylops* sp. (astrapothere), (11) *Marambiotherium glacialis* (marsupial - Microbiotheriidae), (12) Didelphimorphia - marsupiales (12) Sloth, (13) Polyborinae indet (falconid bird), (14) *Araucaria*, and (15) *Nothofagus* (Southern beech) (Reguero et al. 2002; Reguero 2016).

9.3.1 *The Metatherians*

So far there have been five families of metatherians described, all of which were probably frugivorous/insectivorous. These eight species of mammals represent the most diverse group of mammals so far found from La Meseta. They include two species of Polydolopidae (common during the Early and mid-Eocene of Patagonia), *Polydolops dailyi* and *P. seymouriensis*), Microbiotheriidae? (1 species, *Marambiotherium glacialis*), Woodburnodontidae, *Woodburnodon casei*, Derorhynchidae (3 species, *Derorhynchus minutus, Pauladelphys juanjoi),* and an undetermined genus and species, an indeterminate family, sp. *Xenostylus peninsularis*, Prepidolopidae (1 species) *Perrodelphys coquinense*, and a possible marsupial whose taxonomic status is indeterminate (Goin et al. 1999; Chornogubsky et al. 2009; Reguero et al. 2013; Vizcaíno et al. 1988; Woodburne and Zinsmeister 1982). Finally a new species of marsupial of what is perhaps Glasbiidae has been reported (Reguero et al. 2016) (Fig. 9.8).

The endemism of several of these marsupials (prepidolopid and one derorhynchid marsupials) suggest there was some sort of isolating barrier, probably climatic, between La Meseta and Patagonia (Reguero et al. 2002, 2013). Others show close affinity to Patagonian fauna. In fact the polydolopid marsupials may have evolved as part of the *Nothofagus* flora (Reguero et al. 1998). Medium- to large-sized carnivore marsupials are lacking on the Antarctic Peninsula. These La Meseta marsupials were likely insectivores, frugivorous, and omnivores (Goin et al. 1999).

Fig. 9.8 *Polydolops* sp. Two species of these marsupials are known from La Meseta fauna, and they are the most abundant mammal from the Eocene of the Antarctic. (By William Stout, *Frozen in Time*)

9.3.2 Gondwanatheria

The gondwanatherian belongs in the family Sudamericidae and is closely related to *Sudamerica ameghinoi*, although it is more derived. Gondwanatherians are known for the Cretaceous and Early Paleocene of Patagonia, Madagascar, and India. This is the last (and most recent) gondwanatherian to be found, and since they had disappeared in Patagonia by this time, it seems to be evident for some sort of isolating barrier during the Eocene, probably some sort of water barrier (Reguero et al. 2002; Goin et al. 2007).

9.3.3 Xenarthra?

A supposed xenarthran fossilized caniform tooth was first considered to be a primitive sloth/anteater; they had not yet separated into distinctive lineages. This is the earliest record for a known Pilosa (anteaters and sloths (Tardigrada)) (Carlini et al. 1990; Vizcaíno and Scintilato-Yané 1995). However, MacPhee and Reguero (2010) have recently disagreed that it may or may not be xenarthran and state that this fossil is best considered to be a mammal *incertae sedis* (Reguero 2016).

9.3.4 *Astrapotheria*

Two astrapotherians were identified in the La Meseta fauna, one in the Trigonostylopidae family, according to Bond et al. (2011) but not belonging to the genus *Trigonostylops* (Marenssi et al. 1994; Case 2006; Bond et al. 2011). Another astrapothere has been identified as a new genus and species (*Antarctodon sobrali*) from the Astrapotheriidae, but the description of the new astrapothere and comparisons to all other previously collected materials concluded that *Antarctodon* is a non-trigonostylopid astrapothere (Bond et al. 2011) (Fig. 9.9).

9.3.5 *Youngest and Oldest Record of Ungulate from Eocene*

The M3 molar from the La Meseta is the oldest evidence of mammals from the Late Eocene (also proposed as Early Oligocene). It has been analyzed as a liptotern ungulate *Notiolofos arquinotiensis* of the family Sparnotheriodontidae (Vizcaíno et al. 1997; Bond et al. 1990, 2006, 2014; Gelfo et al. 2015).The ungulate's presence

Fig. 9.9 *Antarctodon sobrali*, La Meseta astrapothere, needed to be adapted to temporal winter conditions, which would have included sunless and crepuscular weeks and some seasonal freezing temperatures. (By Zimices or Julián Bayona). In contrast to this image, because of the climate, it probably had a hairy body

was accompanied by the beginning of glaciation in Antarctica and a cool-temperate *Nothofagus*-dominated forest on the Antarctic Peninsula (Case 1988). The tooth of the sparnotheriodontid ungulate *Notiolofos* was from the most recent subunit of La Meseta which was probably Late Eocene (37.8 Ma) or even Early Oligocene (34.2 Ma) and was a forest-adapted browser. It probably weighed about 395–400 kg and was the largest (and most common) Antarctic herbivore at the time. Corresponding plant fossils indicate that there was still a *Nothofagus*-dominated forest as during the Middle Eocene community. The tooth suggests a browsing herbivore (Vizcaíno et al. 1997).

A lower right molar was found of apparently the same species of litoptern, *Notiolofos arquinotiensis*, from Early Eocene deposits of Seymour Island, making it and an accompanying eutherian fossil phalanx (unidentified as to group) the oldest mammalian material found from Antarctica at about 55.3 Ma. If this fossil specimen is truly *N. arquinotiensis*, it indicates a very broad time range spanning as it does the oldest La Meseta to the youngest material for the species at 37.8 Ma (at least spanning 17.5 Ma, much longer than most mammalian species) and suggests a considerable evolutionary stasis for the species. So it is appropriate to consider that the physical conditions in West Antarctica remained the same during most of the Eocene (Gelfo 2014) despite evidence that this time frame represented a considerable ecological change from warm times around 55 Ma to a cool, *Nothofagus* forest at 37.4 Ma (Gelfo 2016) (Fig. 9.10).

A second species of *Notiolofos*, *N. regueroi*, has recently been described, smaller than *N. arquinotiensis*; it possibly massed at around 25–58 kg. This litoptern was also a browser in the local forests and existed sympatrically with the larger species at least from about 53 to 49 Ma during the Early and mid-Eocene (Gelfo et al. 2017).

Fig. 9.10 *Notiolofos arquinotiensis*, the archaic litoptern from Las Mesetas, the largest and one of the most common mammals in the La Meseta fauna. (Based on Bond et al. 2006, by Marie Giraud-López, Grupo EEMN Universidad Nacional de Colombia)

9.4 Comparison to Early Eocene Southern Patagonian Fauna

Case (2006) compares the Seymour Island La Meseta vertebrate fauna with Early Eocene Patagonian fauna (Casamayoran) and finds that the body size distribution in the two is quite different. On the one hand, the paleofauna of La Meseta is U-shaped with several small and several large species with an absence of medium-sized animals. On the other hand, the size distribution in Patagonia in the Early Eocene is normally distributed and much less diverse than further to the south. The La Meseta fauna existed in a temperate climate that was highly seasonal and mean winter temperatures of between −3° and −2 °C. The fauna reflects high latitude faunas where both small and large mammals have evolutionary tactics of either physiological adaptation or small surface area-to-volume ratios to cope with the cold, neither of which is available to medium-sized mammals. In contrast, the southern parts of Patagonia were subtropical in the Late Paleocene and Early Eocene with temperatures 10°–15.6 °C, permitting higher diversity and an equitable size distribution. Curiously, although notoungulates were very common in the Eocene mammalian fauna of Patagonia, none have been found in La Meseta, although a litoptern was common. Also, although small- to medium-sized carnivores (Sparassodonta) were common in Patagonia, none have been discovered in La Meseta (Vizcaíno et al. 1998).

9.5 Connection to South America, Disconnection

Southern Patagonian flora was a rich mixture of over 100 species identified with 88 species of dicots (flowering plants) and the rest monocots (angiosperms, ferns, conifers, cycads, and ginkgoes) (Wilf et al. 2003). In contrast the flora of La Meseta on Seymour Island was dominated by *Nothofagus*, podocarps, and araucarian conifers (Reguero et al. 2002), and this forest persisted after the climate had begun to cool.

The land bridge between South America and Antarctica lasted until about 30–35 Ma until it became interrupted, allowing a circumpolar current to become established. This circumpolar current locked in the ever colder temperatures, allowing Antarctica to begin developing glaciers around 34 Ma, which persisted throughout most of the Oligocene. By the end of the Eocene or the beginning of the Oligocene, the La Meseta fauna probably no longer existed because of the increasing cold.

References

Barrett P (1999) Antarctic climate history over the last 100 million years. Terra Ant Rep 3:53–72

Bond M, Pascual R, Reguero MA, Santillana SN, Marenssi SA (1990) Los primeros "ungulados" extinguidos sudamericanos de la Antártida. Ameghiniana 26:240

Bond M, Reguero MA, Vizcaino SF, Marenssi SA (2006) A new "South American ungulate" (Mammalia: Litopterna) from the Eocene of the Antarctic Peninsula. In: Francis JE, Pirrie

D, Crame JA (eds) Cretaceous-Tertiary high-latitude palaeoenvironments: James Ross basin, Antarctica. Geol Soc SP, London 258: 177–186

Bond M, Kramarz A, MacPhee RDE, Reguero M (2011) A new astrapothere (Mammalia, Meridiungulata) from La Meseta Formation, Seymour (Marambio) Island, and a reassessment of previous records of Antarctic astrapotheres. Am Mus Novit 3718:1–16

Bond M, Reguero MA, Vizcaíno SF, Marenssi SA (2014) A new "South American ungulate" (Mammalia: Litopterna) from the Eocene of the Antarctic Peninsula. In: Francis JE, Pirrie D, Crame JA (eds) Cretaceous-Tertiary high-latitude palaeoenvironments, James Ross basin, Antarctica. Geol Soc SP, London 258:163–176

Cantrill DJ (2012) The vegetation of Antarctica through geological time. Cambridge University Press, New York

Carlini, AA, Pascual R, Reguero MA, Scillato Yané GJ, Tonni EP, Vizcaíno SF (1990) The first paleogene land placental mammal from Antarctica: its paleoclimatic and paleobiogeographic bearings. In: IV international congress of systematic and evolutionary biology, Maryland, Abstracts, p 325

Case JA (1988) Paleogene floras from Seymour island, Antarctica Peninsula. In: Feldmann RM, Woodburne MO (eds) Geology and palaeontology of Seymour island, Antarctic peninusula. Geol Soc Am Mem 198:523–530

Case JA (2006) The late middle Eocene terrestrial vertebrate fauna from Seymour Island: the tails of the Eocene Patagonian size distribution. In: Francis JE, Pirrie D, Crame JA (eds) Cretaceous-Tertiary high-latitude palaeoenvironments: James Ross basin, Antarctica. Geol Soc SP, London 258: 177–186

Chornogubsky L, Goin FJ, Reguero M (2009) A reassessment of Antarctic polydolopid marsupials (Middle Eocene, La Meseta formation). Antarct Sci 21(3):285–297

Dingle R, Lavelle M (1998a) Antarctic Peninsula cryosphere: early Oligocene (c. 30 Ma) initiation and a revised glacial chronology. J Geol Soc Lond 155:433–437

Dingle R, Lavelle M (1998b) Late Cretaceous-Cenozoic climatic variations of the northern Antarctic Peninsula: new geochemical evidence and review. Palaeogeogr Palaeoclimatol Palaeoecol 107:79–101

Dingle RV, Lavelle M (2000) Antarctic peninsula Late Cretaceous-Early Cenozoic palaeoenvironments and Gondwana palaeogeographies. J Afr Earth Sci 31(1):91–105

Dingle R, Marenssi S, Lavelle M (1998) High latitude Eocene climate deterioration: evidence from the northern Antarctic Peninsula. J S Am Earth Sci 11:571–579

Gandolfo MA, Marenssi SA, Santillana SN (1998) Flora y paleoclima de la Formación La Meseta (Eocene medio), isla Marambio (Seymour), Antártida. In: Casado S (ed) Paleógeno de América del sur y de la península Antártica. Asoc Pal Argent Publ 5:155–162

Gelfo JN (2014) Implicatieons of the evolutionary statis of *Notiolofos arquinotiensis* (Mammalia) Eocene of Seymour Island, Antarctica. Symposium: cretaceous-tertiary palaeobiogeographic connections with Antarctica. 4th international palaeontological congress, Mendoza

Gelfo JN (2016) Considerations about the evolutionary stasis of *Notiolofos arquinotiensis* (Mammalia: Sparnotheriodontidae), Eocene of Seymour Island, Antarctica. Ameghiniana 53(3):316–332

Gelfo JN, Mörs T, Lorente M, López GM, Reguero M (2015) The oldest mammals from the Antarctic, early Eocene of the La Meseta Formation, Seymour Island. Palaeontology 58(1):101–110

Gelfo JN, López GM, Santillana SN (2017) Eocene ungulate mammals from West Antarctica: implications from their fossil record and a new species. Antarct Sci 29(5):445–455

Goin FJ, Case JA, Woodburne MO, Vizcaino SF, Reguero MA (1999) New discoveries of "opossum-like" marsupials from Antarctica (Seymour Island, Middle Eocene). J Mamm Evol 6(4):335–365

Goin FJ, Reguero MA, Pascual R, Von Koenigswald W, Woodburne MO, Case JA, Marenssi SA, Vieytes C, Vizcainso SF (2006) First gondwanatherian mammal from Antarctica. In: Francis JE, Pirrie D, Crame JA (eds) Cretaceous-Tertiary high-latitude palaeoenvironments: James Ross basin, Antarctica. J Geol Soc Lond 258:135–162

Goin FJ, Zimicz N, Reguero MA, Santillana SN, Marenssi SA, Moly JJ (2007) New marsupial (Mammalia) from the Eocene of Antarctica, and the origins and affinities of the Microbiotheria. Rev Asoc Geológ Argent 62(4):597–603

Hooker JJ (1992) An additional record of a placental mammal (Order Astrapotheria) from the Eocene of West Antarctica. Antarct Sci 4(1):107–108

Jamieson SSR, Sugden DE (2008) Landscape evolution of Antarctica. In: Cooper AK, Barrett PJ, Stagg H, Storey B, Stump E, Wise W (eds) Antarctica: a keystone in a changing world. Proceedings of the 10th International Symposium on Antarctic Earth Sciences. The National Academies Press, Washington, DC, pp 39–54

Krause JM, Clyde WC, Ibañez-Mejía M, Schmitz MD, Barnum T, Bellosi ES, Wilf P (2017) New age constraints for early Paleogene strata of Central Patagonia, Argentina: implications for the timing of South American Land Mammal Age. Geol Soc Am Bull 129(7/8):886–903

Lisiecki LE, Raymo ME (2005) A Pliocene-Pleistocene stack of 57 globally distributed benthic δ18O records (PDF). Paleoceanography 20; 1003. https://doi.org/10.1029/2004Ä001071

MacPhee RDE, Reguero MA (2010) Reinterpretation of a middle Eocene record of Tardigrada (Pilosa, Xenarthra, Mammalia) from La Meseta formation, Seymour Island, West Antarctica. Am Mus Novit 3698:1–21

Marenssi SA, Reguero MA, Santillana SN, Vizcaino SF (1994) Eocene land mammals from Seymour Island, Antarctica: palaeobiogeographical implications. Antarct Sci 6(1):3–15

Petit JR, Jouzel J, Raynaud D, Barkov NI, Barnola JM, Basile I, Bender M, Chappellaz J, Davis J, Delaygue G, Delmotte M, Kotlyakov VM, Legrand M, Lipenkov V, Lorium C, Pépin L, Ritz C, Saltzman E, Stieveard M (1999) Climate and atmospheric history of the past 420,000 years from the Vostok Ice Core, Antarctica. Nature 399:429–436

Poole I, Cantrill DJ (2006) Cretaceous and Cenozoic vegetation of Antarctica integrating the fossil wood record. In: Francis JE, Pirrie D, Crame JA (eds) Cretaceous-Tertiary high-latitude palaeoenvironments; James Ross basin, Antarctica. J Geol Soc Lond 258:63–81

Reguero MA (2016) Evolution of the Cenozoic mammals from Antarctica. In: Agnolin FL, Lio GL, Brissón Egli F, Chimento NR, Novas FE (eds) Historia evolutiva y paleobiogeográfica de los vertebrados de América del Sur. MACN-Museo Argentino de Ciencias Naturales "Bernardino Rivadavia" e Instituto Nacional de Investigación de las Ciencias Naturales, Buenos Aires, pp 403–412

Reguero MA, Vizcaíno SF, Goin FJ, Marenssi SA, Santillana SN (1998) Eocene high-latitude terrestrial vertebrates from Antarctica as biogeographic evidence In: Casadio S (ed) Paleógeno de América del Sur y de la península Antártica. Asoc Pal Argent Publ, 5:185–198 (cited in Chornobubsky et al. 2009)

Reguero MA, Marenssi A, Santillana SN (2002) Antarctic Peninsula and South America (Patagonia) Paleogene terrestrial faunas and environments: biogeographic relationships. Palaeogeogr Palaeoclimatol Palaeoecol 179:189–210

Reguero MA, Goin F, Acosta Hopitaleche C, Dutra T, Marenessi S (2013) Late Cretaceous/Paleogene West Antarctica terrestrial biota and its intercontinental affinities. Springer, New York

Reguero MA, Gelfo JN, López GM, Bond M et al (2014) Final Gondwana breakup: The Paleogene South American native ungulates and the demise of the South America-Antarctica land connection. Glob Planet Chang 123:400–413

Reguero MA, Goin FJ, Abello MA, Chornogubsky L, Gelfo JN (2016) New metatherian mammal from early Eocene levels of the la Meseta Formation, Antarctic Peninsula. Open Science Conference SCAR 2016, Kuala Lumpur, Malaysia. Abstracts

Romero EJ (1986) Paleogene phytogeography and climatology of South America. Ann Mo Bot Gard 73:449–461

Shen Y (1995) A paleoisthmus between southern South America with the Antarctic Peninsula during Late Cretaceous and Early Tertiary. VII International Symposium on Antarctic Earth Sciences, Abstracts. Siena

Shen Y (1998) A paleoisthmus linking southern South America with the Antarctic Peninsula during Late Cretaceous and Early Tertiary. Sci China Ser D Earth Sci 41(3):225–229

Shellito CJ, Sloan LC, Huber M (2003) Climate model sensitivity to atmospheric CO_2 levels in the Early-Middle Paleogene. Palaeogeogr Palaeoclimatol Palaeoecol 193:113–123
Vizcaíno SF, Carlini AA, Reguero MA (1988) Primer registro de un marsupial Didelphimorphia en Antártida. Su implicancia biogeográfica. In: Quiroga JC, Cione AL (eds) 5th Jornadas Argentinas de Paleontología Vertebrados, Abstracts. Universidad de La Plata, La Plata, pp 30–31
Vizcaíno SF, Scintilato-Yané GJ (1995) An Eocene tardigrade (Mammalia, Xenarthra) from Seymour Island, West Antarctica. Antarct Sci 7:407–408
Vizcaíno SF, Bond M, Reguero MA, Pascual R (1997) The youngest record of fossil land mammals from Antarctica; its significance on the evolution of the terrestrial environment of the Antarctic Peninsula during the late Eocene. J Paleontol 71(2):348–350
Vizcaíno SF, Reguero MA, Goin FJ, Tambussi CP, Noriega JI (1998) Community structure of Eocene terrestrial vertebrates from Antarctic Peninsula. In: Casadio S (ed) Paleóg Amér Sur Penins Antártida, Asociación Paleontológica Argentina. Publicación Especial, vol 5, pp 177–183
Weijers JWH, Schouten S, Sluijs A, Brinkhuis H, Damste JSS (2007) Warm arctic continents during the Paleocene-Eocene thermal maximum. Earth Planet Sci Lett 261:230–238
Wilf P, Cuneo NR, Johnson KR, Hicks JF, Wing SL, Obradovich JD (2003) High plant diversity in Eocene South America: evidence from Patagonia. Science 300:122–125
Woodburne MO, Zinsmeister WJ (1982) Fossil land mammal from Antarctica. Science 218:284–286
Woodburne MO, Case JA (1996) Dispersal, vicariance, and the Late Cretaceous to Early Tertiary land mammal biogeography from South America to Australia. J Mamm Evol 3(2):121–161
Zachos J, Pagani M, Sloan L, Thomas E et al (2001) Trends, rhythms, and aberrations in global climate 65 Ma to present. Science 292(5517):686–693. https://doi.org/10.1126/science.1059412

Chapter 10
La Venta: A Miocene Mammalian Community from Colombia

10.1 Introduction

La Venta is one of the most important fossil sites in South America for mammals of the Middle Miocene because of its high diversity and detailed paleoecological information, information that is not usually available for fossil sites. The place represents a narrow point in the Miocene of only 11.8–13.5 million years (Flynn et al. 1997). This is one of the youngest and most detailed faunas known before the late Miocene-Pliocene invasion (9.5 Ma–recent) of mammals that flooded into South America from the north, when the two continents established a land bridge allowing the so-called Great American Biotic Interchange. Also very important is that this faunal community also had little in common with the co-occurring faunas of Patagonia, thus illustrating the existence of ecological barriers.

Today, La Venta is located in the upper Magdalena River Valley, north of the city of Neiva, Huila. Currently the region is dry, semiarid scrubland, located as it is in a rain shadow of the Eastern Cordillera, but in the mid-Miocene, the area was a lowland (500 m) tropical region with scattered forest and savannah (Kay and Madden 1997a, b). The region was first collected by Brother Ariste Joseph of the Christian Brothers of La Salle, accompanied by Brother Nicéforo María, who became well-known in Colombia for his zoological collections. The two found many fossils and collected them, taking the material to Bogotá for further study. Brother Ariste sent specimens to J. B. Reside at the US Geological Survey. The material included the type specimens for a caiman and a notoungulate (*Metaxytherium ortegense*). Later, José Royo y Gómez collected many fossils in 1940 while he was surveying Huila department for mineral resources. The field notes of Royo y Gómez along with field notes and collections of the American R. A. Sirton greatly contributed to the fame of the La Venta fossil beds.

The most productive fossil beds are named The Honda Group, dated at 13.5–12.9 million years. They represent a new South American Land Mammal Age, the Laventan (Laventanense in Spanish) (Madden et al. 1997a, b). The Honda Group is

T. Defler, *History of Terrestrial Mammals in South America*, Topics in Geobiology 42, https://doi.org/10.1007/978-3-319-98449-0_10

divided into two principle parts, the La Victoria Formation (13.5–12.9 Ma) and the Villavieja Formation (12.9–11.5 Ma). The most productive fossil beds are found at the border between the two major formations (called the Monkey Beds) and may represent only about 15,000 years, which gives us great insight into an actual Miocene community (Flynn et al. 1997; Madden et al. 1997). Deposits of this period of time represent an ecosystem that was a broad plain of around 100 m altitude that continued to increase in altitude throughout the Honda Group's formation and with various small to large, sinuous rivers that flowed to the east and southeast, eventually emptying into the great brackish wetlands that then exited to the north, into the proto-Caribbean somewhere around the present-day mouth of the Orinoco River. These rivers were bordered by riverine forests that left wide areas of open savanna comparable to today's eastern llanos. although more fertile. This was not a closed-canopy forest, as amply proved by the mammalian fauna that has been collected.

The Honda Group is broadly correlated to the Pebas Formation of eastern Peru and southern Colombia that showed that the wetlands were brackish at the time (Hoorn et al. 2010). The Honda deposits have been dated using both paleomagnetism and argon analysis of hornblende volcanic deposits and are comprised by two Formations: La Victoria (the older) and the Villavieja (the younger). The initiation of the uprising of the Eastern Cordillera defines the end of the La Victoria Formation (pebbly volcanic litharenite and variegated mudstones and sandstones at 13.5–12.9 Ma) and may date from 12.9 Ma at the beginning of the Villavieja Formation (mostly red and gray mudstones with minor layers of medium-grained volcanic litharenites as well as sandstones 12.9–11.5 Ma) (Guerrero 1997), although complete establishment of the barrier to the east probably dates to about 11.8 Ma, judging by the change in direction of the paleo-river courses to the north instead of the east.

As of the publication of *Vertebrate Paleontology in the Neotropics* in 1997, there had been 142 species of vertebrates described from La Venta, including 72 species of mammals taken from the Monkey Beds of La Victoria Formation that probably represent only about 15,000 years of deposits (Kay and Madden 1997a). Non-mammals included fishes, reptiles, and birds. This area was located on a sort of peninsula with the Cordillera Central to the west which originated as a persistent crustal uplift since the latest Cretaceous as well as the site of a Neogene volcanic arc (Butler and Schamel 1988). However, generalized mountain building can be documented for the Cordillera Central only from about 30–25 Ma (Mora et al. 2010). To the east lay an immense brackish-freshwater wetlands of swamp and lakes (Hoorn et al. 2010) (Fig. 10.1).

The landscape of the Villavieja and Victoria Formations was low relief and tropical, with an altitude near sea level during the beginning of La Victoria Formation, but gradually increasing in altitude to about 1000 m at the end of the Villavieja Formation. The area had many meandering rivers with oxbow lakes (Guerrero 1997) with drainage to the east until about 11.8 Ma when the Eastern Cordillera had closed off that drainage. Precipitation during the Monkey Beds deposition was an average of only about 1500–2000 mm, much lower than regions east of the Eastern Cordillera today at the same latitude (Kay and Madden 1997a, b). During the deposition of the latter part of La Victoria Formation, large rivers of as much as 8–10 m in depth

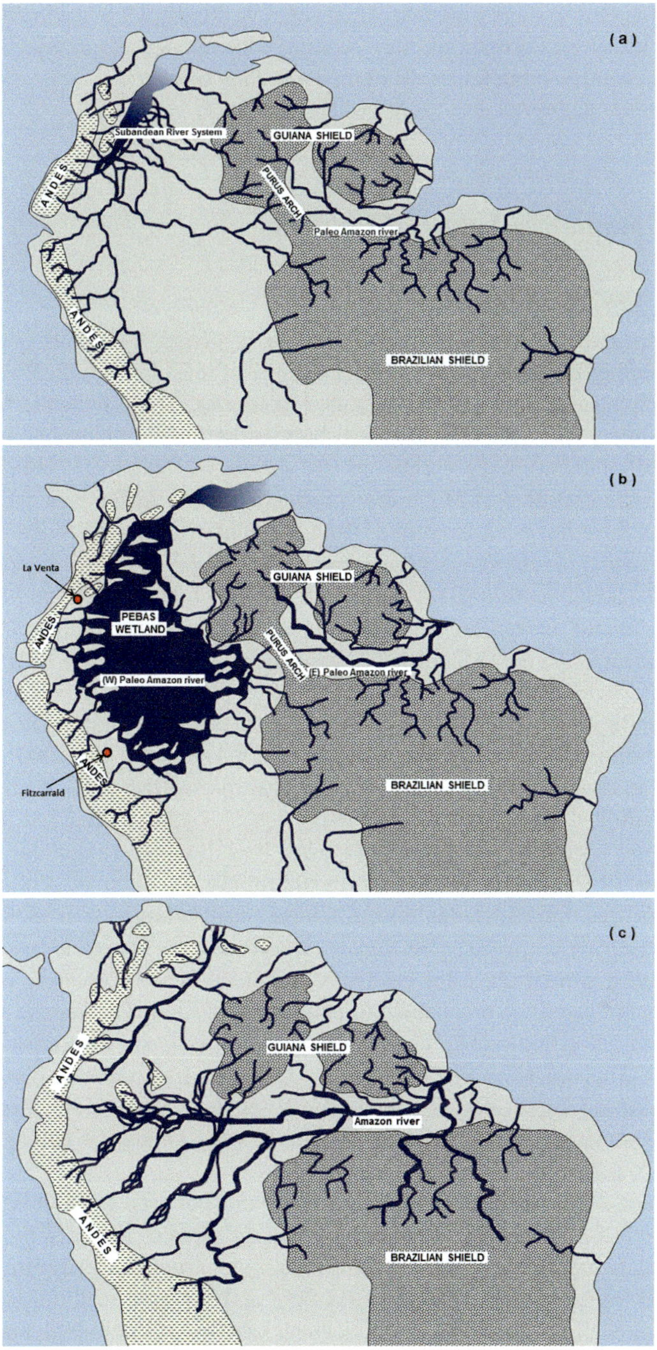

Fig. 10.1 Possible extent of mega-wetlands in the Middle Miocene and change in drainage patterns after about 11.8 Ma; (**a**) Early Miocene; (**b**) Middle Miocene; (**c**) Late Miocene. La Venta and Fitzcarrald identified with red dots defining extension of open habitat typical of broad area west of mega-wetlands. (Based on Hoorn et al. 2010; by Diego Casallas, Applied Biodiversity Foundation)

flowed to the east to empty into the Amazonian wetlands, while during the deposition of the Villavieja Formations, the direction of the rivers was variable, first toward the east and southeast but later toward the west, indicting the first appearance of the Cordillera Oriental by about 11.8 Ma (Guerrero 1997).

10.2 Mammalian Fauna of La Venta

10.2.1 Marsupialia and Metatheria

Not surprisingly La Venta had a large community of marsupials with about 15 taxa, 12 of which have been able to be classified to species. These included four orders and six families (Goin 1997). The usual large order Didelphimorphia and ancient family Didelphidae were represented by two species of *Thylamys* (*T. minutus* and *T. colombianus*), a species of *Micoureus laventicus*, and a Marmosini whose genus and species have not been identified. Both *Thylamys* (ten species) and *Micoureus* (woolly mouse opossums) (six species) still exist in the marsupial fauna of today. There is also an unidentifiable (to genus and species) Didelphinae. One Marmosinin and a Didelphinae are represented whose genus and species are undetermined.

The marsupial order Microbiotheria is also represented by a new species *Pachybiotherium minor* placed in the family Microbiotheridae. This genus is represented in Patagonia by two other species (Goin et al. 2010) and is one of the few faunal elements to be connected to southern South America (Goin 2003).

A third marsupial order Paucituberculata was discovered in La Venta with two different families. The species *Pithiculites chenche* (Abderitidae) and the species *Hondathentes cazador* (Palaeothentidae) have been described (Dumont and Bown 1997; Goin 1997). Although modern caenolestoids are faunivores, the Miocene members of this order included both frugivores and faunivores, as well as mixed feeders. Based on comparison to *Phalanger orientalis* a living cuscus from New Guinea with a similar shearing mechanism, *Pithiculites* possessed a dental morphology that allowed the breakdown of hard or tough food. This unique specialization permitted a varied diet so they can be viewed as dietary generalists (Dumont et al. 2000). This order was previously quite diverse with a wide distribution, but the order is currently represented by about seven species called shrew opossums (rat opossums or caenolestids) and found only in the high Andes and belonging to one family, the Caenolestidae (Patterson 2007).

The Sparassodonta, considered metatheres and not true opossums, are well-represented at La Venta with maybe seven species, most represented by teeth. The most complete predator found at La Venta is *Lycopsis longirostrus* a borhyaenid that is known from a complete skull and skeleton and was one of the largest known Prothylacyninae (Marshall 1977). This animal was about 35 cm high and weighed 15 kg and was probably primarily terrestrial but with some ability for grasping and climbing, probably hunting small rodents and marsupials as an ambush predator as well as being omnivorous. The animal probably used its forelimbs to manipulate

Fig. 10.2 *Dukecynus magnus* (Prothylacynidae) the largest predator found at La Venta. (By Roman Uchytel)

prey (Argot 2004). Some rodent bones were found in the animal's thorax and probably represented prey.

Dukecynus magnus (Fig. 10.2) was probably the largest predator in the La Venta fauna and weighed 68 kg was 2 m long and 60 cm high and was the largest Prothylacynidae known.

Other species of sparassodonts included *Hondadelphys fieldsi* (Hondadelphidae) and *Anachlysictis gracilis* (Thylacosmilidae) (Fig. 10.3) weighing around 18 kg, as well as a species whose genus and family are uncertain. *Hondadelphys fieldsi* (Fig. 10.4) may be the sister clade of the thylacosmilids, the group that includes the marsupial "cat" (*Thylacosmilus atrox*), so this species is closely related. *Anachlysictis gracilis* is closely related to *Thylacosmilus* and may be considered to be more primitive. Like *Thylacosmilus*, *Anachlysictis* possessed phalanges on its lower jaw which protected its smaller "saber" canines.

Three other taxa of Sparassodonta are known that cannot be identified to family. One could be a Thylacosmilidae or a borhyaenoid lineage that paralleled thylacosmilids. Two others are sparassodonts that cannot be assigned to a family.

10.2.2 Xenarthra

The Xenarthra found at La Venta include a rich diversity of Cingulata (the armadillos, glyptodonts), and they represent totally new genera from those found in Patagonia. Six species have been identified including three dasypodids (Dasypodidae—armadillos), one pampatherid (Pampatheridae), and two species of glyptodonts (Glyptodontidae).

Fig. 10.3 *Anachlysictis gracilis* (Thylacosmilidae) in a riverine forest. (By Roman Uchytel)

Fig. 10.4 *Hondathentes cazador*, a small Caenolestoidea La Venta predator. (By Zimeces or Julián Bayona)

10.2.2.1 Dasypodidae: Armadillos

The armadillos (*Nanoastegotherium prostatum*, *Anadasypus hondanus*, and *Pedrolypeutes praecursor*) are most likely forest species, and they are small, one reaching somewhere around 5 kg (Carlini et al. 1997). *Nanoastegotherium prostatum* weighed less than 2 kg and probably existed within the forest feeding on arthropods. It is the smallest armadillo found at La Venta. Interestingly, all fossil material for this species was found in what are thought to be crocodilian coprolites. Could this mean that the species was frequently aquatic or at least a swimmer?

Anadasypus hondanus was an omnivore similar in size to the living *Dasypus novemcinctus* (5 kg) and had at least six movable bands and was a precursor to this species and is considered to be the most ancient Dasypodini. Since it is the first Dasypodinae known, it may be ancestral to *Dasypus*. Anadasypus was probably omnivorous-insectivorous and lived in ecotonal areas of humid forest and grasslands.

Pedrolypeutes praecursor weighed less than 4 kg, fed on insects, and was *ecotonal* to dry forests and savanna. It is the oldest and most primitive Tolypeutinae known (Carlini et al. 1997; Castro 2015).

10.2.2.2 Pampatheriidae: Pampathere

The herbivorous pampatherid *Scirrotherium hondaensis* (Fig. 10.5) weighed around 40 kg and probably lived ecotonally in the border between savanna and forest. This species appears to have been common, judging from the many fossils that have been found. This is the earliest known pampatherid and shows the typical division of the carapace, separated into anterior and posterior bucklers that are separated by three rows of imbricating osteoderms. *Scirrotherium* has been widely reported as belonging to *Kraglievichia paranensis*, but this is mistaken, comparisons to osteoderms of *Kraglievichia*, showing that *Scirrotherium* osteoderms are distinct. Jaw biomechanics suggest that pampathere masticatory musculature was more powerful and more adapted for transverse movements than armadillos (De Iuliis et al. 2000). Much of their diet was apparently coarse vegetation, so they were primarily grazers, while

Fig. 10.5 The oldest record of an undoubted pampathere (*Scirrotherium hondaensis*) was found at La Venta. (By Marie Joelle Giraud, Grupo Evolución y Ecología de Mamíferos Neotropicales, Universidad Nacional de Colombia)

armadillos are omnivorous or insectivorous (Edmund and Theodor 1997; Vizcaíno et al. 1998).

10.2.2.3 Glyptodontidae: Glyptodont

Of the two glyptodonts *Neoglyptatelus originalis* and *Boreostemma* (formerly placed in the genus *Asterostemma*), the smallish *Neoglyptatelus* was probably ecotonal along the forest edges. This species was a primitive Glyptatelinae that existed after the apparent disappearance of the subfamily in southern parts of South America. The much larger *Boreostemma* (50–100 kg) probably inhabited the surrounding savannas. A second species of *Neoglyptatelus* (*N. sincelejanus*) has been described for northern Colombia in Sucre department, but the genus is unknown from southern parts of South America (Villarroel and Clavijo 2005).

Asterostemma gigantea and *Asterostemma acostae* (=*Boreostemma gigantea* and *B. acostae*) are both found in La Venta. Two other species are known from Venezuela. The genus is also known from southern parts of the continent, but further analysis has concluded that the *Asterostemma* from the north of the continent belong in a separate genus *Boreostemma*, thus preserving the separate identity of the northern fauna from the southern fauna (Carlini et al. 2008). This reclassification of the northern *Asterostemma* in *Boreostemma* also recognizes that the group belongs to the forest stages of the Glyptodontinae and thus establishes northern South America as the geographic origin of this group (Carlini et al. 2008).

All taxa of the La Venta Cingulata are new and only known from northern parts of South America. They are relatively distantly related to taxa from Patagonia. This lack of easy comparisons questions the utility of the SALMA system (Carlini et al. 1997).

10.2.2.4 Pilosa: Sloths

La Venta includes a rich fauna of sloths, including about nine species, and they probably inhabited both forest and open savanna. These include three different families, and, unlike communities in the south of the continent, the Mylodontidae is dominant, represented by three different species, one of which (*Pseudoprepotherium confusum*, Mylodontinae) (Fig. 10.6) was the largest member of the group and weighed perhaps 500 kg. *Pseudoprepotherium confusum* (Mylodontinae) was apparently a long-time resident of the region, as it has been found from both the La Victoria and Villavieja Formations. It was a savanna species that probably ate grasses.

The other two species are somewhat smaller but are also of the open savanna (McDonald 1997). Material identified at first as *Glossotheriopsis pascuali* (also belonging to the Mylodontinae), considered a more derived mylodontine is now recognized as a new taxon *Brievabradys laventensis*, while *Neonematherium flabellatum* is considered a Scelidotheriinae (Scillato-Yané 1976). *Brievabradys laven-*

Fig. 10.6 *Megadolodus molariformis* (Proterotheriidae) a common litoptern from La Venta. (By Zimices or Julián Bayona)

tensis was a small sloth the size of the living *Choloepus* and found in the very productive "Monkey Beds" (Villarroel 2000). This small species coexisted with the megatherid *Huilabradys* and the mylodontid *Pseudoprepotherium.*

The family Megalonychidae is represented by two species of sloth, a large species and a small species, neither of which has been identified to genus or species (Hirschfeld 1985; McDonald 1997).

In the Megatheriidae, there appear to be three species from the subfamily Nothrotheriinae: *Hapalops* sp., a well-known, small, and gracile genus that may be an ancestor of the arboreal *Bradypus* and *Choloepus*. This small sloth probably spent an appreciable amount of time in trees (White 1997). There are many small *Hapalops* described, and they may be a bridge along the origin and evolution of arboreal adaptations Pujos et al. (2016). For example, *Hapalops longiceps* was more or less the size of the terrestrial vermilingua *Myrmecophaga* (Scott 1903). Another even smaller species from La Venta belonging also to the Nothrotheriinae subfamily seems to belong to the genus *Eucholoeops*, also known from Argentina. Another small megatherid is recognized as a new genus and species *Huilabradys magdaleniensis* (McDonald 1997; Villarroel 1998). Although these are considered as part of the "ground sloths," they were small and probably spent a lot of time in the trees. Future studies of them may shed light on the evolution of arboreal sloths, the new genus and species *Huilabradys magdaleniensis* (Villarroel 1998). It is unclear whether the Nothrotheriinae belong to the Megatheridae or the Megalonychidae (McDonald 1997).

10.2.2.5 Vermilingua: Anteaters

Finally, an ancestor of the anteaters, *Neotamandua borealis,* was found that is considered to be an intermediate between the giant anteater (*Myrmecophaga tridactyla*) and the *Tamandua* although analysis suggests direct ancestry to *Myrmecophaga*

(Hirschfeld 1976). *Neotamandua* was intermediate in size between *Tamandua* and *Myrmecophaga* but its tail was more non-flexible as in *Myrmecophaga*. Although it was mostly arboreal, it seems to be a direct ancestor of *Myrmecophaga*. This was the first fossil anteater found for northern South America and adds to the various species described for this genera, all discovered in the southern parts of the continent (McDonald 1997; Croft 2016).

10.2.3 Meridiungulata: Ungulates

Various ungulates from La Venta further confirm the presence of open savanna around the forests (Madden et al. 1995). A surprise was the discovery of a Didolodontidae-like ungulate, previously only known from the Paleocene and Eocene in southern South America. *Megadolodus molariformis* (Fig. 10.7) has been collected from various sites in the Honda group, resulting in a skeleton and most of the anterior dentition. Predictions of body weight result in 60–80 kg, and further analysis has resulted in classifying this species as a Proterotheriidae litoptern that was probably omnivorous and used hard fruits as part of its diet. This animal probably foraged in forests, and in its habits, it was probably similar to living pigs (Cifelli and Villarroel 1997; Croft 2016).

Fig. 10.7 The macrauchenid *Theosodon* defending its young from two *Borhyaena tuberata*, (Borhyaenidae) predatory metatherans, indicated in a savannah habitat, although it was probably a forest animal. (By Charles Knight 1913. *A History of Land Mammals in the Western Hemisphere*)

10.2.3.1 Litopterns

Other Litopterna that have been discovered include three other species: *Prolicaphrium sanalfonensis* (Proterotheriidae), *Villarroelia toyotoi* (Proterotheriidae), and an unidentified genus and species of macrauchenid. These are based on less fossil material and cannot be confidently depicted although their low-crowned molars suggest a forest habitat. The three-toed macrauchenid guanaco-like *Theosodon* (Fig. 10.8) was widespread and also were present in Peru, Argentina, Bolivia, and Chile. There are about six species known, and they are about 2 m in length, weighing up to 95.6 kg. The low-crowned brachydont molars of these animals suggest a browsing habit rather than grazing, intimating a forest habitat rather than savanna (Cifelli and Guerrero 1997; Croft 2016).

10.2.3.2 Notoungulates and the Miocochilius Assembly Zoner

The most common fossil mammal found in La Venta area is a small typothere notoungulate *Miocochilius anomopodus* (Fig. 10.9). This animal was the size of a large cat (2–11 kg). It had large feet with three toes on the forefeet and two toes on the hind feet. The two of the front toes were dominant, and this rodent-like notoungulate was adapted to run (somewhat like a hare) in order to escape the several borhyaenid predators present.

Miocochilius was so common that an associated fauna has been named after the genus the Miocochilius Assemblage Zone (an ecological community?) defined by the presence of this species together with *Scleromys colombianus* (a rodent), *Granastrapotherium snorki* (an astrapothere), *Scirrotherium hondaensis* (a pampathere) at the beginning and the last appearance of *Miocochilius* along with *Scleromys schurmanni* (a rodent), *Prodolichotis pridiana* (a caviid rodent), *Olenopsis* (large spe-

Fig. 10.8 *Miocochilius anomopodus* (Interatheriidae) a very common notoungulate in La Venta, adapted to running from its predators. (By Marie Joelle Giraud, Grupo Evolución y Ecología de Mamíferos Neotropicales, Universidad Nacional de Colombia)

Fig. 10.9 *Huilatherium pluriplicatum* a leonteniid toxodont. (By Zimices or Julián Bayona)

Fig. 10.10 *Pericotoxodon platignathus* (Toxodontidae) was abundant in La Venta and was adapted to an abrasive diet. (By Marie Joelle Giraud, Grupo Evolución y Ecología de Mamíferos Neotropicales, Universidad Nacional de Colombia)

cies of rodent), *Xenastrapotherium kraglievichi* (an astrapothere), and *Pericotoxodon platignathus* (a toxodont ungulate). The zone also includes *Pedrolypeutes praecursor* (a tolypeutine armadillo), *Neoreomys huilensis* (a dasyproctid rodent), *Stirtonia* spp. (an ateline primate), *Huilatherium pluriplicatum* (a notoungulate), and *Glossotheriopsis pascuali* (a mylodontid glyptodont) (Madden et al. 1997).

10.2.3.3 Toxodont

A partial skull of a Toxodonta (Leontiniidae, *Huilatherium pluriplicatum*) (Fig. 10.10) and other material allows no conclusion as to the habitat of this toxodont other than the probability that it was a savanna-woodland grazer weighing about 800 kg.

Another large La Venta toxodont (Toxodontidae, *Pericotoxodon platignathus* (Dinotoxodontinae)) (Fig. 10.11) was about the same size as a Sumatran rhinoceros

Fig. 10.11 *Granastrapotherium snorki* an Astrapotherium weighing up to 3500 kg. This male exhibits only slightly curved lower canines, while females had more curved lower canines. Upper canines of males could reach 1 m in length (Johnson and Madden 1997). (By Roman Uchytel)

3.4 m long and 1.9 m tall and weighed also around 800 kg, very comparable to the black rhino (*Diceros bicornis*) or the Sumatran rhino (*Dicerorhinus sumatrensis*). The teeth suggested that it was a grazer of grass, like all toxodonts. This species also exhibited sexual dimorphism in tusks, which might suggest polygynous tendencies (Madden 1997; Villarroel and Colwell-Danis 1997).

10.2.3.4 Astrapotheres

By far the largest of the mammals of this community were three species of astrapotheres that have been described for La Venta: *Granastrapotherium snorki* (Fig. 10.12), *Xenastrapotherium kraglievichi*, and *Xenastrapotherium* sp. These animals probably reached 3000 kg or more and inhabited the savannas of the area. *Granastrapotherium snorki* was the largest mammal in this community (2500–3500 kg), being probably one third larger than *X. kraglievichi* which has been compared to a living black rhinoceros *Diceros bicornis* (900–1500 kg). Although *Granastrapotherium snorki* was described for the first time from La Venta, fossil material has been found from eastern Peru, extending the species into the Amazon region and from the local fauna of Fitzcarrald (Johnson and Madden 1997; Antoine et al. 2007). This material is the latest evidence of astrapotheres in South America and represents the largest assemblage of uruguaytheriine astrapotheres known. The partial skull and lower mandibles of a female have recently been described (Pardo Jaramillo 2010).

The genus *Xenastrapotherium* has been found in Venezuela, Colombia, Ecuador, Peru, as well as the upper Juruá in Brazil (Goillot et al. 2011). These were large grazers of the savannas, and all probably possessed proboscis that were a bit larger than that of tapir; probably they extended at least to the end of the long tusks (Johnson and Madden 1997). Whether the same species of *Granastrapotherium* and *Xenastrapotherium* were in eastern Peru as found in Colombia is not clear because of inadequate material. But fossils of these genera have been found in an arc from Venezuela, Colombia, and Peru and they probably represent a closely related Middle Miocene fauna that was found to the west of the Amazonian wetlands in a landscape that was made up of broken savanna—forest, somewhat similar to caatinga and more fertile than the llanos.

10.2.4 Rodents (Walton 1997)

The rodents of La Venta were made up of cavimorph rodents. Specifically the presence of Caviidae suggests a mixed vegetation with open areas and forest. The Caviidae had teeth adapted to an abrasive diet, which occurs in open habitat mostly in the form of grasses. This is considered to be one of the most diverse rodent faunas known from South America and has little in common with rodent faunas known from southern parts of South America.

The La Venta fauna fossils have not produced cricetid rodents; they did not arrive in South America until about 3.5–10 Ma years ago with the joining of the two continents. However, the caviomorph that have been found in La Venta total 5 families, 13 genera and 28 species.

10.2.4.1 Dinomyidae (Pacaranas and Relatives)

Five species are known from this family from La Venta that produced the largest rodents known, and in La Venta there are several stout rodents. Nevertheless, the smallest dinomyid known is also found in La Venta, *Scleromys schurmanni*, perhaps not exactly *Scleromys*, but it is recognized as the smallest Dinomyidae known so far at 1–2 kg. Also present was *Scleromys colombianus* weighed about 2.5–3.5 kg. Both of these "*Scleromys*" may actually be a new genus, although very similar to *Scleromys*.

A rodent similar to *Olenopsis* sp. 1 weighed 8–15 kg (the size of a pacarana), and another possible *Olenopsis* sp. 2 weighed 13–20 kg. These four species have been found southward to Argentina. These *Olenopsis*-like rodents are probably a new genus. Finally another genus, distinct from Olenopsis has been collected from the Monkey Beds but due to poor material it has not been named.

10.2.4.2 Caviidae (Cuys, Maras)

Three Caviidae are known: *Prodolichotis pridiana* (2–3.5 kg) is the most common found in La Venta, and two species from an unnamed genera are also known, the larger with a body weight of 1–2 kg, the smaller species weighed about 0.8–1.3 kg. These species of the same genus have not been given a name.

10.2.4.3 Erethizontidae (Puerçoespín)

1. ? *Steiromys* (2.5–4 kg) Two species of erethizontids are known from La Venta.
2. *Microsteiromys jacobsi* (200–300 g), The second erethizontid is the smallest porcupine known and weighted about 200-300 g.

10.2.4.4 Echimyidae (Spiny Rat)

Acarechimys is abundant in Argentina and is present in La Venta. *Acarechimys minutissimus* of La Venta is one of the smallest caviomorphs known.

1. *Acarechimys minutissimus* (90–150 g)
2. *Ricardo longidens* (150–250 g)
3. Gen. et sp. indet.
4. The Smallest Caviomorph Ever Located

10.2.4.5 Dasyproctidae (pacas, ñeques) is represented by three species in La Venta

1. *Neoreomys huilensis* (3.2–4.2 kg)
2. *Microscleromys paradoxalis*
3. *Microscleromys cribriphilus* (100–200 g)

10.2.5 Primates

The primates from La Venta represent the richest primate assemblage known in South America, and the comparatively close dates of the fossils allow us to view this assemblage as a primate community (Fleagle and Reed 1999). This community was discussed in more detail in Chap. 8 of this book. The group has been described from the La Venta deposits that capture a very short (geologically) and defined time span of around 12 Ma ago (11.5–13.5 Ma according to Flynn et al. 1997). They also represent the numerically richest ancient primate assemblage known for South America. These primates show very clear connections to modern primates, and many can even be placed into modern subfamilies, and though the fossil deposits are found in the upper Magdalena River, which is now a deep river valley between two branches of the northern Andes, at the time of the deposits of the fossils, the area was actually peripheral to the Amazon basin with no mountain barrier between the animals and the immense flatlands of the east. Fossils from La Venta are placed in the Middle Miocene, and they have been found in two geological formations, La Victoria (13.5–12.9 Ma) and Villavieja (12.9–11.5 Ma) (Guerrero 1997). These

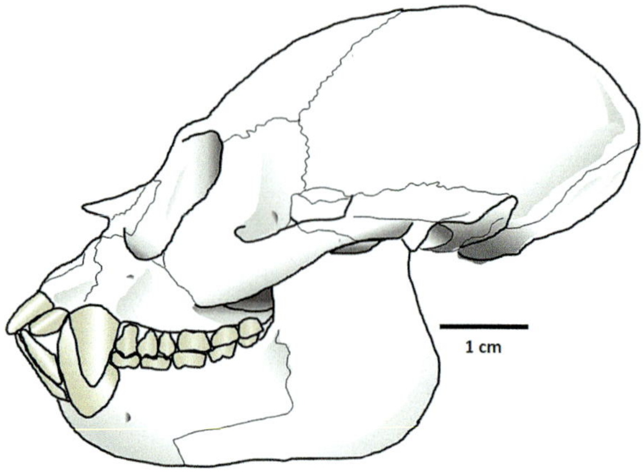

Fig. 10.12 *Cebupithecia sarmientoi.*

fossils represent at least 13 species of primates, 8 of which can be placed in a modern phylogenetic context as follow (Fleagle et al. 1997):

Two species of the atelid *Stirtonia* have been discovered (*S. tatacoensis* Stirton 1951; *S. victoriae* Kay et al. 1987), which are clearly related to *Alouatta* and are probably direct ancestors of this modern genus, dated about 12.6–13.7 Ma. Another atelid, Kondous laventicus is known from molars (Setoguchi 1985) and is probably closely related to *Ateles*.

Figure 10.12 The very famous and almost complete skeleton of *Cebupithecia sarmientoi* (Stirton 1951) (Fig. 10.12) is clearly a pithecine (Pitheciini) primate very similar to *Pithecia*. This animal would have weighed 2–3 kg (Diego Casallas-Pabon, Applied Biodiversity Foundation).

Nuciruptor rubricae (Meldrum and Kay 1997) is also clearly belongs to the pithecids but is midway in size between the modern genera *Pithecia* and *Callicebus* but it is unclear in what subfamily it should be placed (*Callicebus* has now been divided into three separate genera (Byrne et al. 2016)).

Neosaimiri fieldsi (Stirton 1951) is a small cebid that resembles *Saimiri* in many aspects and has even been suggested at belonging in the modern genus *Saimiri*. The species is now known from fragments of jaw, teeth, distal humerus, proximal ulna, proximal femur, distal tibia, calcaneus, and four tali and shares some traits with *Aotus* as well (Hartwig and Meldrum 2002; Rosenberger et al. 1991).

Laventiana annectens (Rosenberger et al. 1991) another cebid is a *Saimiri*-sized mandible closely related to *Saimiri* and *Neosaimiri* but with some differences. Some would include this species within *Neosaimiri* (Meldrum and Kay 1997; Nakatsukasa et al. 1997; Takai 1994).

Aotus dindensis (Setoguchi and Rosenberger 1987) (12.1–12.5 Ma) has been clearly placed within the modern genus *Aotus*, which clearly suggests the antiquity of this genus and may explain the very similar phenotype of the many species known today, since this is a highly evolved, mostly nocturnal primate genus, well-adapted to a nocturnal life.

Mohanamico hershkovitzi (Luchterhand et al. 1986) is a complete lower jaw, of about a 1 kg, primate with similarities to the callithrichine *Callimico*.

Unfortunately the primate fossils *Patasola magdalenae*, *Lagonimico conclutatus*, and *Micodon kiotensis* are not easily placed into a known group of modern primates (Kay and Meldrum 1997; Kay 1994; Setoguchi and Rosenberger 1985). *Patasola* and *Lagonimico* show some similarities to the callitrichines, and *Patasola* has a large third molar like *Callimico*. And *Lagonimico* (represented by a badly crushed skull and mandible) is somewhat larger than modern callitrichines. However, modern callitrichines are possibly dwarfed cebids (or callitrichids), and *Lagonimico* may come from a larger ancestor than present-day callitrichines. *Micodon kiotensis* might belong with the callitrichines, mostly based on its very small size, based on an isolated upper molar and upper incisor and lower premolar (Rosenberger et al. 1990).

Finally *Miocallicebus villaviejai* was described from a right maxillary fragment with three molars found from the Middle Miocene of La Venta and has been associated with the Callicebinae. This belongs with the *Callicebus* monkeys, though it is much larger than modern species of *Callicebus* (Takai et al. 2001).

An attempt to test the idea of evolutionary time and its impact on the evolution of ecological diversity and phylogenetic richness as suggested by Fleagle and Reed (1996) found that La Venta primate community was not significantly less ecologically diverse than modern communities, suggesting that ecological diversification occurred much earlier in platyrrhine evolution (Wheeler 2010).

A comparison of the Middle Miocene Laventan Quebrada Honda Fauna to that of La Venta finds very few taxa shared between the two, despite very similar ages. The Honda Fauna from southern Bolivia is much more similar to the slightly older and high-altitude fauna of Collón Cura in Chile than it is to La Venta in Colombia. This suggests that there were isolating mechanisms during the Middle Miocene between the low-latitude (Amazonian) faunas and Quebrada Honda, which extended at least to middle Perú and the local faunas of Fitzcarrald and that the use of SALMA's for correlation from low-latitude faunas and other parts of South America might not be useful (Croft 2007).

References

Argot C (2004) Functional adaptive analysis of the postcranial skeleton of a Laventan borhyaenoid, *Lycopsis longirostris* (Marsupialia, Mammalia). J Vertebr Paleontol 24(3):689–708

Antoine P-O, Salas-Gismondi R, Baby P, Benammi M, Brusset S, de Franceschi D, Espurt N, Goillot C, Pujos F, Tejada J, Urbina M (2007) The middle Miocene (Laventan) Fitzcarrald Fauna, Amazonian Peru. In: Días-Martínez E, Rábano I (eds) 4th European meeting on the palaeontology and stratigraphy of Latin America. Cuadernos del Museo Geominero. Instituto Geológico y Minero de España, Madrid, no. 8:19–24

Butler K, Schamel S (1988) Structure along the eastern margin of the central cordillera, upper Magdalena Valley, Colombia. J S Am Earth Sci 1(I):109–120
Byrne H, Rylands AB, Carneiro JC, Lynch JW et al (2016) Phylogenetic relationships of the New World titi monkeys (*Callicebus*): first appraisal of taxonomy based on molecular evidence. Frontiers Zool 13: https://doi.org/10.1186/s12983-016-0142-4
Carlini AA, Vizcaino SF, Scillato-Yané GJ (1997) Armored Xenarthrans: a unique taxonomic and ecologic assemblage. In: Kay RF, Madden RH, Cifelli RL, Flyn JJ (eds) Vertebrate paleontology in the neotropics. Smithsonian Institution Press, Washington, DC, pp 213–226
Carlini AA, Zurita AE, Scillato-Yané GJ, Sánchez R, Aguilera OA (2008) New glyptodont from the Codore Formation (Pliocene), Falcón State, Venezuela, its relationship with the Asterostemma problema, and the paleobiogeography of the Glyptodontinae. Palaeontol Z 82(2):139–152
Castro MC (2015) Sistemática y evolución de los armadillos Dasypodini (Xenarthra, Cingulata, Dasypodidae). Rev Mus La Plata 15(73):1–50
Cifelli RL, Guerrero J (1997) Litopterns. In: Kay RF, Madden RH, Cifelli RL, Flyn JJ (eds) Vertebrate paleontology in the neotropics. Smithsonian Institution Press, Washington, DC, pp 289–302
Cifelli RL, Villarroel C (1997) Paleobiology and affinities of *Megadolodus*. In: Kay RF, Madden RH, Cifelli RL, Flyn JJ (eds) Vertebrate paleontology in the neotropics. Smithsonian Institution Press, Washington, DC, pp 265–288
Croft DA (2007) The middle Miocene (Laventan) Quebrada Honda Fauna, southern Bolivia and a description of its notoungulates. Palaeontology 50(1):277–303
Croft DA (2016) Horned armadillos and rafting monkeys. Indiana University Press, Bloomington
De Iuliis G, Bargo MS, Vizcaíno SF (2000) Variation in skull morphology and mastication in the fossil giant armadillo *Pampatherium* spp. and allied genera (Mammalia: Xenarthra: Pampatheriidae), with comments on their systematics and distribution. J Vertebr Paleontol 20(4):743–754
Dumont ER, Bown TM (1997) New Caenolestoid marsupials. In: Kay RF, Madden RH, Cifelli RL, Flyn JJ (eds) Vertebrate paleontology in the neotropics. Smithsonian Institution Press, Washington, DC, pp 207–212
Dumont ER, Strait SG, Friscia R (2000) Abderitid Marsupials from the Miocene of Patagonia; an assessment of form, function, and evolution. J Paleontol 74(6):1161–1172
Edmund G, Theodor J (1997) A new giant pampatheriid armadillo. In: Kay RF, Madden RH, Cifelli RL, Flyn JJ (eds) Vertebrate paleontology in the neotropics. Smithsonian Institution Press, Washington, DC, pp 227–232
Fleagle JG, Reed KE (1996) Comparing primate communities: a multivariate approach. J Hum Evol 30:489–510
Fleagle JG, Reed KE (1999) Primate communities. Cambridge University Press, Cambridge
Fleagle JG, Kay RF, Anthony MRL (1997) Fossil New World monkeys. In: Kay RF, Madden RH, Cifelli RL, Flyn JJ (eds) Vertebrate paleontology in the neotropics. Smithsonian Institution Press, Washington, DC, pp 473–495
Flynn JJ, Guerrero J, Swisher CC III (1997) Geochronology of the Honda Group. In: Kay RF, Madden RH, Cifelli RL, Flyn JJ (eds) Vertebrate paleontology in the neotropics. Smithsonian Institution Press, Washington, DC, pp 44–59
Goin FJ (1997) New clues for understanding Neogene marsupial radiations. In: Kay RF, Madden RH, Cifelli RL, Flyn JJ (eds) Vertebrate paleontology in the neotropics. Smithsonian Institution Press, Washington, DC, pp 187–206
Goin FJ (2003) Early marsupial radiations in South America. In: Jones M, Dickman C, Archer M (eds) Pedators with pouches: the biology of carnivorous marsupiales. CSIRO Publishing, Collingwood, pp 30–42
Goin FJ, Tejedor MF, Abello MA, Martin GM (2010) Un Nuevo microbiotérido (Mammalia, Marsupialia, Microbiotheria) de la formación Pinturas (Mioceno temprano) de la provincia de Santa Cruz. Ameghiniana 47(1):117–122

Goillot C, Pierre-Olivier A, Tejada J, Pujos F, Salas Gismondi R (2011) Middle Miocene Urugyaytheriinae (Mammalia, Astrapotheria) from Peruvian Amazonia and a review of the astrapotheriid fossil record in northern South America. Geodiversitas 33(2):331–345

Guerrero J (1997) Stratigraphy, sedimentary environments, and the Miocene uplift of the Colombian Andes. In: Kay RF, Madden RH, Cifelli RL, Flyn JJ (eds) Vertebrate paleontology in the neotropics. Smithsonian Institution Press, Washington, DC, pp 1–12

Hartwig WC, Meldrum DJ (2002) Miocene platyrrhines of the northern Neotropics. In: Hartwig WC (ed) The primate fossil record. Cambridge University Press, Cambridge, pp 175–188

Hirschfeld SE (1976) A new fossil anteater (Edentata, Mammalia) from Colombia, S.A. and evolution of the Vermilingua. J Paleontol 50(3):419–432

Hirschfeld SE (1985) Ground sloths from the Friasian La Venta fauna, with additions to the pre-Friasian Coyaima fauna of Colombia, South America. Geol Sci Univ Calif Publ 28:1–91

Hoorn C, Wesselingh FP, Hovikoski J, Guerrero J (2010) The development of the Amazonian mega-wetland (Miocene; Brazil, Colombia, Peru, Bolivia). In: Hoorn C, Wesselingh F (eds) Amazonia: landscape and species evolution. Wiley-Blackwell, Oxford, pp 123–142

Johnson SC, Madden RH (1997) Uruguaytheriine astrapotheres of tropical South America. In: Kay RF, Madden RH, Cifelli RL, Flyn JJ (eds) Vertebrate paleontology in the neotropics. Smithsonian Institution Press, Washington, DC, pp 355–381

Kay (1994) "Giant" tamarin from the Miocene of Colombia. Am J Phys Anthr 95:333–353

Kay RF, Madden RH (1997a) Paleogeography and paleoecology. In: Kay RF, Madden RH, Cifelli RL, Flyn JJ (eds) Vertebrate paleontology in the neotropics. Smithsonian Institution Press, Washington, DC, pp 520–550

Kay RF, Madden RH (1997b) Mammals and rainfall: paleoecology of the middle Miocene at La Venta (Colombia, South America). J Hum Evol 32(2–3):161–199

Kay RF, Meldrum DJ (1997) A new small Platyrrhine and the phyletic position of Callitrichinae. In: Kay RF, Madden RH, Cifelli RL, Flyn JJ (eds) Vertebrate paleontology in the neotropics. Smithsonian Institution Press, Washington, DC, pp 435–458

Kay RF, Madden RH, Plavcan JM, Cifelli RL, Guerrero Diaz J (1987) *Stirtonia victoriae*, a new species of Miocene Colombian primate. J Hum Evol 16:173–196

Luchterhand K, Kay RF, Madden RH (1986) *Mahanico hershkovitzi*, gen. et sp. nov., un primate du Miocène moyen d'Amérique du Sud. CR Acad Sci Paris 303(2):1753–1758

Madden RH (1997) A new toxodontid notoungulate. In: Kay RF, Madden RH, Cifelli RL, Flyn JJ (eds) Vertebrate paleontology in the neotropics. Smithsonian Institution Press, Washington, DC, pp 335–354

Madden RH, Cifelli RL, Flynn JJ (1995) Vertebrate paleontology in the neotropics: the Miocene fauna of La Venta, Colombia. Smithsonian Institute Press, Washington, D.C.

Madden RH, Guerrero J, Kay RF, Flynn JJ, Swisher CC III, Walton AH (1997) The Laventan stage and age. In: Kay RF, Madden RH, Cifelli RL, Flyn JJ (eds) Vertebrate paleontology in the neotropics. Smithsonian Institution Press, Washington, DC, pp 499–519

Marshall LG (1977) A new species of *Lycopsis* (Borhyaenidae: Marsupialia) from the La Venta fauna (Late Miocene) of Colombia, South America. J Paleontol 51:633–642

McDonald HG (1997) Xenarthrans: Pilosans. In: Kay RF, Madden RH, Cifelli RL, Flyn JJ (eds) Vertebrate paleontology in the neotropics. Smithsonian Institution Press, Washington, DC, pp 233–245

Meldrum DJ, Kay RF (1997) Postcraneal skeleton of Laventan Platyrrhines. In: Kay RF, Madden RH, Cifelli RL, Flyn JJ (eds) Vertebrate paleontology in the neotropics. Smithsonian Institution Press, Washington, DC, pp 459–472

Mora A, Baby P, Roddaz M, Parra M, Brusset S, Hermoza W, Espurt N (2010) Tectonic history of the Andes and sub-Andean zones: implications for the development of the Amazon drainage system. In: Hoorn C, Wesselingh F (eds) Amazonia: landscape and species evolution. Wiley-Blackwell, New York, pp 38–60

Nakatsukasa M, Takai M, Setoguchi T (1997) Functional morphology of the postcranium and locomotor behavior of *Neosaimiri fieldsi*, a *Saimiri*-like middle Miocene platyrrhine. Am J Phys Anthropol 102:515–544

Pardo Jaramillo M (2010) Reporte de un Nuevo ejemplar de *Granastrapotherium snorki* en el valle superior del Magdalena, desierto de La Tatacoa, Huila, Colombia. Rev Acad Colomb Cienc 34(131):253–256

Patterson BD (2007) Family Caenolestidae Trouessart, 1898. In: Gardner AL (ed) Mammals of South America, Marsupials, xenarthrans, shrews, and bats, vol 1. University of Chicago Press, Chicago, pp 119–127

Pujos F, De Iuliis G, Cartelle C (2016) A paleogeographic overview of tropical fossil sloths: towards an understanding of the origin of extant suspensory sloths? J Mamm Evol 24(1):19–38

Rosenberger AL, Setoguchi T, Shigehara N (1990) The fossil record of callitrichinae primates. J Hum Evol 19:209–236

Rosenberger AL, Setoguchi T, Hartwig WC (1991) *Laventiana annectens*, new genus and species: fossil evidence for the origins of Callitrichinae New World monkeys. Proc Natl Acad Sci U S A 88(6):2137–2140

Scillato-Yané GJ (1976) El más antiguo Mylodontinae (Edentata, Tardigrada) conocido: *Glossotheriopsis pascuali* n. gen., n. sp., del "Colloncurense" (Mioceno superior) de la Provincia de Río Negro (Argentina). Ameghiniana 13:333–334

Scott WB (1903) Mammalia of the Santa Cruz beds. Vol. 5, paleontology, pts. 1-3, edentata. Reports Princeton University expeditions to Patagonia, 1896–1899. Princeton University, Stuttgart

Setoguchi T (1985) *Kondous laventicus*, a new ceboid primate frm the Miocene of La Venta, Colombia, South America. Folia Primatol 44:96–101

Setoguchi T, Rosenberger AL (1985) Miocene marmosets: first fossil evidence. Int J Primatol 6(6):615–625

Setoguchi T, Rosenberger AL (1987) A fossil owl monkey from La Venta, Colombia. Nature 22:692–694

Stirton RA (1951) Ceboid monkeys from the Miocene of Colombia. Bull Cal Publ Geolog Sci 28:315–356

Takai M (1994) New specimens of *Neosaimiri fieldsi* from La Venta, Colombia: a middle Miocene ancestor of the living squirrel monkeys. J Hum Evol 27(4):329–360

Takai M, Anaya F, Suzuki H, Shigehara N, Setoguchi T (2001) A new platyrrhine from the middle Miocene of La Venta, Colombia, and the phyletic position of Callicebinae. Anthropol Sci 109(4):289–307

Villarroel AC (1998) Los Nothrotheriinae (Megatheriidae, Gravigrada) del Miocene de La Venta (Colombia). descripción de *Huilabradys magdaleniensis*, nuevos género y especie. Rev Acad Colomb Cienc 22(85):497–506

Villarroel AC (2000) Un nuevo mylogontinae (Xenarthra, Tardigrada) en la fauna de La Venta, Mioceno de Colombia: el estado actual de la familia Orophodontidae. Rev Acad Colomb Cienc 24(90):117–127

Villarroel AC, Colwell-Danis J (1997) A new leontiniid notoungulate. In: Kay RF, Madden RH, Cifelli RL, Flyn JJ (eds) Vertebrate paleontology in the neotropics. Smithsonian Institution Press, Washington, DC, pp 303–318

Villarroel AC, Clavijo J (2005) Los mamíferos fósiles y las edades de las sedimentitas continentales del Neógeno de la Costa Caribe Colombiana. Revista de la Academia Colombiana de Ciencias 29(112):345–356

Vizcaíno SF, De Iuliis G, Bargo MS (1998) Skull Shape, Masticatory Apparatus, and Diet of Vassallia and Holmesina(Mammalia: Xenarthra: Pampatheriidae): When Anatomy Constrains Destiny. J Mamm Evol 5(4):291–322

Walton AH (1997) Rodens. In: Kay RF, Madden RH, Cifelli RL, Flyn JJ (eds) Vertebrate paleontology in the neotropics. Smithsonian Institution Press, Washington, DC, pp 392–409

Wheeler BC (2010) Community ecology of the middle Miocene primates of La Venta, Colombia: the relationship between ecological diversity, divergence time, and phylogenetic richness. Primates 51:131–138

White JL (1997) Locomotor adaptations in Miocene Xenarthrans. In: Kay RF, Madden RH, Cifelli RL, Flyn JJ (eds) Vertebrate paleontology in the neotropics. Smithsonian Institution Press, Washington, DC, pp 246–264

Chapter 11
Mammalian Invasion of the Caribbean Islands

11.1 Introduction

The Caribbean Islands are a zoogeographic puzzle. How could islands so far away from South America have come to have had a rich fauna whose obvious source was mostly from South America? Many of the islands form themselves into what we call the Antilles (the Lesser Antilles, an arc of small and not so small islands that curves away to the northeast of the Venezuelan coast and arcs back to the northwest to end just south of Hispaniola, Cuba, and Jamaica, the Greater Antilles) as well as Lucayan Archipelago, or Bahamas Archipelago, to the north of Cuba. Most of the Caribbean Islands (also called the West Indies) are ancient; they began consolidating in the Early Cretaceous with the formation of the Caribbean plate and through many tectonic forces and as a result of the meeting of the Pacific, Atlantic, and South American plates with the Caribbean plate. Geological forces included volcanic activity especially in the Lesser Antilles and accretions of uplifted material from nearby sources such as North and South America, Yucatan, the ocean floor (both Pacific and Atlantic), etc. These have continued their consolidation throughout the Cenozoic, resulting in mountains in Cuba, Jamaica, Hispaniola, and Puerto Rico (Donnelly 1985; Burk 1988). The consolidation of the Caribbean Islands has been intimately connected with the consolidation of Central America (Coates and Obando 1996) (Fig. 11.1).

Even though the evolution of the Caribbean Islands dates back to Cretaceous times, the geological and tectonic processes have involved immersion of the land masses and the nearby strike of the huge bolide Chicxulub 65 million years ago which resulted in the extinction of the dinosaurs. This event surely impacted the islands greatly with unprecedented tsunamis. However, most of the mammalian migrations into the islands appeared to have occurred later in the Cenozoic.

T. Defler, *History of Terrestrial Mammals in South America*, Topics in Geobiology 42, https://doi.org/10.1007/978-3-319-98449-0_11

Fig. 11.1 The West Indies and their geographic relation to South America. (From US Central Intelligence CIA World Factbook)

The West Indies and especially the Greater and Lesser Antilles possessed a striking mammalian fauna (much now extinct), and much of which clearly had a connection to South America, though there are important elements from Central and North America as well. For example, the mysterious West Indian solenodons have evolved from insectivores and have extremely ancient Late Cretaceous origins. Sloths, primates, and caviomorph rodents, all of whose relatives are South American, evolved distinct and endemic species on the islands. These mammals seem to have become extinct only in recent times, most probably due to the arrival of human beings.

About 90% of the West Indian fauna that existed 20,000 years ago has become extinct. The oldest evidence of humans in the Caribbean is the Casirimoid culture in Cuba and Hispaniola that dates back to 4500 BCE of people traveling from Central America (Wilson 2007). The oldest carbon date for cultural remains in Cuba is 5270 y BP (i.e., 3270 BC). But other peoples traveled up the island arc from South America. Many other dates for the arrival of humans are carbon-dated between 3270 and 4000 years before the present. Since the Caribbean fauna was innocent of predators, it would have been exceedingly easy for the human hunters to pick the animals off, as often animals evolving on islands without predators lose their cautious behaviors that are their defense in the face of novel dangers. Since island populations are usually limited in size, it wouldn´t have taken long to wipe them out. There are many other examples of islands in the world that lost great swaths of fauna when the hunters arrived. Especially studied with respect to this theme have been the faunas of Easter Island (Steadman 1995; Steadman et al. 1994), Hawaii (Olson and James 1984), New Zealand (Trotter and MCulloch 1984; Anderson 1984), and Madagascar (Goodman and Jungers 2014). Along with hunters, often other animals such as rats, mongoose, pigs, and dogs arrive to contribute to the pressures that result in extinction of an island fauna. Indeed, island faunas have been shown to be the most vulnerable to extinction.

11.2 West Indian Mammals (Greater and Lesser Antilles)

11.2.1 Sloths

Interestingly giant sloths arrived in North America around nine million years ago, long before the establishment of the Isthmus of Panama that allowed the Great American Biotic Interchange. However, new evidence suggests a path opened between North and South America (whether a land bridge or shallow waters) far earlier than traditionally accepted and as early as 9.5 MYA. But these animals arrived in the West Indies far earlier than when they arrived in North America. The earliest fossil sloth (unnamed) so far encountered is from Puerto Rico and has been dated from the Early Oligocene at around 30–31 MYA. In Cuba a record for the sloth *Imagocnus zazae* has been found from the Early Miocene as well (White and MacPhee 2001; MacPhee 2005). Dates for the presence of sloths have been found at least 1000 years after the arrival of human beings in Cuba (MacPhee et al. 2007). For Cuba, there are radiocarbon dates of about 5000 (3000 BC) years BP for a specimen of *Parocnus browni* and a date of about 6250 (4250 BC) years BP for a specimen of *Megalocnus rodens* (Steadman 2005). These are dates when we know that humans had arrived in Cuba, so it is probable that humans were hunting sloths (MacPhee et al. 2007). For Haiti there are at least two carbon dates of sloth remains (4500 BP and 4391 BP) from a time period overlapping the first presence of human beings (Steadman et al. 2005).

There are 18 known Caribbean species of sloths, according to a recent count (White and MacPhee 2001), all of which belong to one family (Megalonychidae). Surprisingly the family has not yet been detected for the dates when they seemed to enter the West Indies (perhaps 34–31 MYA ago) (MacPhee 2005). These include 8 Cuban species, 6–7 Hispaniolan species, 2 species found in Puerto Rico, and 1 species in both Grenada and Curaçao.

Caribbean sloths have experienced dwarfism in their evolution like many large mammals that are isolated on islands and are separated from larger, ancestral forms on the mainland. It is clear that body size decreases linearly with island age, although it is probable that other factors must influence this phenomenon such as resource limitation (Anderson and Hadley (2002). Caribbean sloths varied in size from the 200 kg Cuban *Megalocnus rodens* (the largest West Indian land mammal) to the smallest sloth, the 4 kg *Neocnus toupiti* (Steadman et al. 2005). Only two of these

Fig. 11.2 Recreation of *Megalocnus rodens*, alive in Cuba at least 6000 years BP. The adult probably reached a weight of 270 kg and was the largest ground sloth known from the Caribbean fauna. It was hunted by the first humans to arrive on the island and remains have been found in ancient human kitchen middens (based on a drawing of Paulo Couto 1979). This was probably one of the last ground sloths to survive. (By Marie Joelle Giraud, Grupo Evolución y Ecología de Mamíferos Neotropicales, Universidad Nacional de Colombia)

fossils are Tertiary, the rest are Quaternary ages. Many islands have not been studied for fossils, so this count of known species should increase in the future. A phylogenetic study of Caribbean sloths shows them to be diphyletic with origins from two sister groups. This suggests that there were two invasions into West Indies, not one (White and MacPhee 2001; MacPhee 2005) (Fig. 11.2).

11.2.2 Primates

At least five species of extinct platyrrhine primates have now been discovered in Cuba, Jamaica, and Hispaniola, and these are specialized, primitive, and difficult to associate with other neotropical primates from South America. They have been at one time or another associated with *Alouatta* (*Paralouatta*) (Rivero and Arredondo 1991; Rosenberger et al. 2011), *Saimiri* (*Xenothrix*) (Rímoli 1977), *Cebus* (*Antilothrix*) (Woods and Ottenwalder 1992), Callitrichidae (*Xenothrix*) (Williams and Koopman 1952; Ford and Morgan 1986), Pitheciidae (*Callicebus* ancestor) (all Caribbean species) (Horovitz 1999), and *Aotus* (Xenothrix) (Rosenberger 1977, 1978), pretty much covering all the known platyrrhine families. They are probably stem platyrrhines, unrelated to any living platyrrhine family (Kay et al. 2011) except as a sister group. Lately in a refreshing turn of events, Antillean primates have been recognized as stem platyrrhines that became established in the Greater Antilles in the Early Miocene (before the appearance of the families with which they have been identified) (Kay 2013). At one time the species *Montaneia anthropomorpha* was described from a tooth found in Cuba (Ameghino), but this was later recognized as an *Ateles* tooth, and later it was carbon-dated to no more than 500 years ago. Although Arredondo and Varona (1983) have argued that this species was an endemic Cuban species, it is well-known that pre-Colombian indigenous trading included much wildlife and that primates were commonly kept as pets. Unless further *Ateles* material is discovered in Cuba, the consensus is that the *Ateles* was introduced by humans.

Cuban primate discoveries have included *Paralouatta varonai* (Rivero and Arredondo 1991) in Cuba and *Paralouatta marianae* (McPhee and Meldrum 2006). *Paralouatta varonai* was found in association with Late Quaternary fauna in Western Cuba, while *P. marianae* was found in Early Miocene (17.5–18.5 MYA) deposits and so is the oldest Caribbean primate known, establishing an early date for the possible arrival of West Indian primates. *Paralouatta varonai* was initially thought to be similar to *Alouatta* (it was the size of modern-day *Alouatta*, and the skull was vaguely similar to *Alouatta*) (Rivero and Arredondo 1991; Rosenberger et al. 2011), but proximate analyses linked it to the other Caribbean fossils, establishing a Caribbean clade, according to Horovitz and MacPhee (1998) and showed that the teeth were very different from *Alouatta*. Thus, the authors decided, after a phylogenetic analysis of all the fossils known at the time, that all Caribbean primates were linked to some ancestral Callicebinae from South America (Horovitz and MacPhee 1998; Horovitz 1999). Meanwhile, cladistic analyses of postcranial fossils of *Paralouatta* suggested locomotion semi-terrestriallity in the genus, short digits for its size, and a lack of ability for leaping and slow climbing, not unsurprising for an ecosystem without predators (MacPhee and Meldrum 2006). However, a contrarian opinion places *Paralouatta* into the Alouattinae as an atelid, rather than being an ancient pitheciid (Rosenberger et al. 2015).

Xenothrix mcgregori was discovered in the form of several incomplete mandibles with molars, skull face, and some postcranial bones at several cave sites in Jamaica (Williams and Koopman 1952; MacPhee and Horovitz 2004). The holotype jaw was found in a rock overhang underneath ("10 inches") a kitchen midden that contained human bones and was found underneath the human remains. They were found by H. E. Anthony in 1920 and not published until 1952 (Williams and Koopman 1952). The age of this fossil material is thought to be latest Pleistocene to Late Holocene. The relationship of this primate to other platyrrhines has been posited as "cebid," "closest to *Callicebus*" (Rosenberger 1977, 2002) or *Tremacebus* and *Aotus* (Rosenberger 2002), based on a cladistic analysis primarily of morphological evidence of skulls and teeth. The animal weighed in at least 2 kg and at most 5 kg, and it had a dental formula similar to marmosets and tamarins (2.1.3.2). The molars were quite different from any living platyrrhine (Fleagle 2013). Because the fossil was so distinct from other cebid monkeys, Hershkovitz (1974) placed the monkey in a new family that he called Xenothricidae. Judging from the postcranial remains, this animal seemed to be very slow-moving, perhaps like a sloth, unlike any other primate (MacPhee and Horovitz 2004).

Rímoli (1977) first described a fossil maxillary fragment with P4-M2 found in the Dominican Republic as a *Saimiri*. It was possible to establish a radiometric date of 3860 + 135 years. But examination by Rosenberger (1978) quickly established that this primate was a new genus and that it had occupied Hispaniola for a very long time. A find of an Early Holocene mandibular fragment in Hispaniola was identified as the same species as *Saimiri bernensis* (MacPhee and Woods 1982). Much earlier Miller (1929) had found the distal end of a tibia that Ford (1980) thought might represent a very large tamarin, but this was not later possible to associate with the two mandibular fragments. MacPhee et al. (1995) established a new genus for this fossil, calling it *Antillothrix*, and he established a phylogenetic connection with *Paralouatta* and suggested *Callicebus* as a sister group, denoting this genus as the closest relatives. Horovitz (1999) agreed with this relationship in a phylogenetic analysis of all living and extinct platyrrhines.

Finally a spectacular skull of *A. bernensis* was discovered in 2009 in a submerged cave system in eastern Dominican Republic. It consists of a cranium with intact premolars and molars (which closely map the holotype), a femur and ulna, one vertebra, and five ribs. The living animal may have weighed about 4.7 kg (based on regression equations relating dental measures to adult body mass) (Rosenberger et al. 2011) and it was probably a typical arboreal quadruped (MacPhee et al. 2007). Rosenberger et al. (2011) at first identified this primate with a Cebidae, but the posterior discovery of more material in 2009 prompted them to change their conclusion and identify this taxon as a pithecid (Rosenberger et al. 2015).

Although there are some similarities to cebine monkeys, the dental characteristics suggest a soft fruit diet with insect but with few hard seeds (Kay et al. 2011). Although the ulna is comparative with *Cebus* monkeys of this size, the femur is thick and comparatively short unlike any other platyrrhines except, intriguingly, *Xenothrix* of Jamaica (Cooke et al. 2011). This may reflect feeding adaptations somewhat like a treesloth which cling to tree supports with the hindfeet while using the longer forelimbs to rake food in toward the mouth (Rosenberger et al. 2011).

Meanwhile Cooke et al. (2011) reported on and described a fossil maxillary and mandibular fragment with almost complete dentition found in southwest Haiti (nos. 7 and 8 in Fig. 11.3). This new species, *Insulacebus toussaintiana*, is clearly a dif-

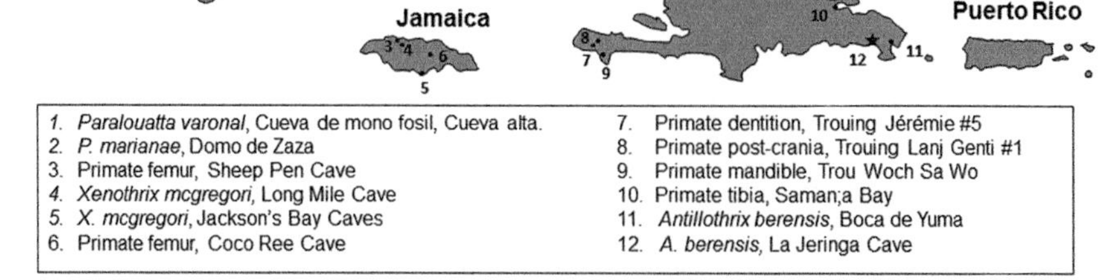

Fig. 11.3 Collection sites for Caribbean platyrrhine monkeys. (By Diego Casallas-Pabón, Applied Biodiversity Foundation)

ferent species from *X. mcgregori*, but they are related, according to the authors, to the modern genus *Aotus*, or they may be allied to *Callicebus*.

Kay et al. (2011) reports on a very well-preserved skull from another underwater cavern in eastern Dominican Republic that is clearly a specimen of *Antillothrix bernensis*, while Halenar et al. (2017) describe a further complete skull not far from the first reported specimen. Kay et al. (2011) had reported the bone arrangement of the pterion to be similar to Old World primates with the parietal plate contacting the zygomatic plate and separating the frontal from the sphenoid plate, but closer examination and comparison with the later specimen confirm that the parietal plates were in accordance with platyrrhine primates (Halenar et al. 2017). Most platyrrhines have the opposite arrangement of frontal and sphenoidal in contact and separating the zygomatic and the parietal. The early playrrhines *Tremacebus* and *Homunculus* also had a catarrhine-type arrangement. The consensus at the moment is that *Antillothrix*, *Xenothrix*, and *Insulacebus* belong to a single clade that is related to ancient pithecids, while Rosenberger et al. (2015) recognize *Paralouatta* as belonging to the alouattine atelid clade.

While MacPhee and coworkers argue that there has been a single ancestor for the West Indian primates, Rosenberger (2002) suggested that at least two clades were present in the Caribbean monkeys, one *Paralouatta* related to the Alouattinae (howler monkeys). Recent analysis of new specimens of *Antillothrix* suggests that the Caribbean primates might have had multiple origins (Halenar et al. 2017) and the latest analysis of Rosenberger et al. (2015) suggests that, indeed, *Antillothrix* was a stem pithecid (Fig. 11.3).

11.2.3 West Indian Rodents

West Indian rodents are made up of the families Capromyidae, Echimyidae, Heptaxodontidae, and Muridae. The first three families are neotropical, and the Muridae have invaded from the North.

Capromyidae

The capromyids are a family of rodents known as hutias. They are similar to cavies and are endemic to the West Indies. There are about 20 species, half of which are extinct. These animals have always been a food item for human beings, and hunting pressure and the loss of habitat have been the greatest risks. Molecular analysis shows that the Capromyidae have evolved either from the South American Echimyidae (spiny rats) or from the Myocastoridae (nutrias) as all three families are closely related (Fig. 11.4).

Heptaxodontidae

The Heptaxodontidae is a paraphyletic family that is used to classify various very large rodents, the largest in the Antilles. All are extinct, probably because of human predation, although no fossils have as yet been found associated with humans. The largest known rodent from the Caribbean was the 200 kg *Amblyrhiza inundata*, the

Fig. 11.4 A typical capromyid *Capromys pilorides*, the only common and widespread species left in the group. (By Jensflorian)

blunt-toothed giant hutia. About the size of an American black bear (*Ursus americanus*), it lived on the islands of Anguilla and San Martin.

The Heptaxodontidae or giant hutias are divided into two subfamilies, five genera, and six species. *Amblyrhiza inundata* is the only species of this genus. *Elasmodontomy obliquus* was a 13 kg monophyletic rodent that lived in Puerto Rico until the arrival of humans 2000 years ago. *Quemisia gravis* lived on Hispaniola. *Xaymaca fulvopulvis* was from Jamaica and *Clidomys osborni* was also found on Jamaica.

Cricetidae

The species *Megalomys desmarestii* or the giant rice rat was a cat-sized rodent that was common in Martinique until the end of the nineteenth century when the mongoose was introduced. This and widespread volcanic eruptions brought about its extinction. The Cricetidae invaded the Antilles from North America rather from South America. *Megalomys desmarestii* was the largest species known from this family of rodents (Fig. 11.5). Another large and now extinct species of rice rat was *Ekbletomys hypenemus*, similar in size to *Megalomys* and known from fossils in Antigua and Barbuda. Other extinct species were *Oryzomys antillarum* from Jamaica and *Oligoryzomys victus* from the Lesser Antilles. Several other unnamed rice rat fossils are also known from around the Atilles and now extinct, possibly because of introduced animals like the rat, mongoose, pig, cat, and dog.

Fig. 11.5 *Megalomys desmarestii*. (From Gervais, P., 1855, Histoire naturelle des mammiferes..., vol. 1)

Fig. 11.6 *Hyrachyus*, a primitive rhinoceros whose fossil has been found in Western Jamaica but which was probably rafted there from Central America. (By RB Horsfeld, from *Land Mammals of the Western Hemisphere*, 1913)

11.2.4 Ungulates

The surprising discovery of a primitive Middle Eocene rhinoceros (*Hyrachyus*) from a part of eastern present-day Jamaica that was a part of a block of ancient Honduras and Nicaragua must have been in contact with the mainland fauna (MacPhee 2005; Iturralde-Vinent 2006). This land was carried eastward by the Caribbean plate to later unify with what is now Western Jamaica, which was always isolated, even from Cuba. From *Hyrachyus* evolved all modern rhinoceros. This is the only ancient ungulate discovered in the Antilles (Fig. 11.6).

Fig. 11.7 *Solenodon paradoxus* from Hispaniola. (By Seb az86556)

11.2.5 Solenodons

Solenodons (Cuba, Hispaniola) separated from eulipotyplan insectivores (moles, shrews, and hedgehogs) around 76 MYA and represent the oldest West Indian mammals (Roca et al. 2004). They are very rare and endangered. There are two species (Fig. 11.7).

11.3 How Did the Mammals Get There?

Several theories have been put forward to explain the populating of the West Indies with the various mammals that lived there until the arrival of human beings. These theories have been variously classified as vicariant, the landspan hypothesis, and dispersal over water. It is difficult to fit known distributions of sloths and primates to a vicariance scheme, which implies that the West Indian fauna is the result of an earlier fauna that was continuous with that of North and South America. Some geological evidence does suggest a proto-Caribbean Island chain that might have connected the two continents (via islands or solid ground) during the Cretaceous, but this seems too early to explain the later fauna from South America. However it would be consistent with the molecular dates that have been calculated for Caribbean solenodons, which have been shown to be of a very ancient lineage (Roca et al. 2004). Vicariant hypotheses are also difficult to apply given the geological history of the region except within the context of the landspan hypothesis that posits the populating of the Antilles via an uninterrupted peninsula that was then fragmented or split into many islands, especially because the earliest known dates have been found in Cuba, Puerto Rico, and Hispaniola and not in the Lesser Antilles to the south, assumedly toward the source populations of these animals. It is difficult to imagine over water dispersal to so many islands, leaving many isolated subpopulations (Iturralde-Vinent 2006; Iturralde-Vinent and MacPhee 1999).

The landspan or landspan-vicariance hypothesis seems to be a strong contender for West Indian biogeography. There is good evidence for a general uplift of marine topography during the Late Eocene throughout the Caribbean (this was coupled by a lowered sea level due to increased glaciation at the South Pole). As a result a hypothetical peninsula was formed along the fracture line of the Caribbean-Atlantic plates. This reached as far as central Cuba and persisted for about three million years at about 35–33 MYA, permitting migration of some faunal elements (sloths, primates, rodents) along the Aves-Antilles Ridge of the peninsula (Iturralde-Vinent 2006; Iturralde-Vinent and MacPhee 1999). This ridge or the "Garlandia landspan" might have been in the form of many islands connected by very shallow seas. Whichever form it permitted short-term dispersal of several elements of the West Indian fauna (sloths, primates, caviomorph rodents). The uplift included Puerto Rico, which was connected to Hispaniola from the Late Eocene to the Pliocene (Fig. 11.8).

There are surely more extinct species to be found. No efforts to look for evidence have been made for many islands, and in the case of the Greater Antilles, the area is very large and will surely yield more.

But Hedges (2001) presents important points arguing for a dispersal scenario, rather than an overland dispersion, especially because (1) the West Indian fauna is depauperate, that is, it has been filtered such that many freshwater fishes, salamanders, caecilians, marsupials, carnivores, lagomorphs, and most families of frogs, turtles, and snakes are absent and they would have been able to pass over a

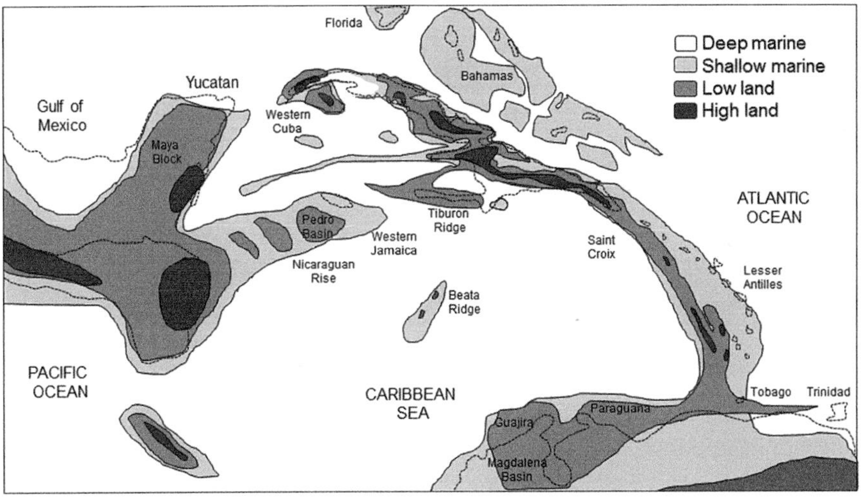

Fig. 11.8 Caribbean latest Eocene–Early Oligocene paleography (35–33 MYA). This scenario provides the possibility of terrestrial route to the Antilles although overwater rafting is also strongly hypothesized. (By Diego Casallas-Pabón, Applied Biodiversity Foundation)

land bridge with the sloths, primates, and caviomorph rodents and (2) molecular clock estimates of divergence times for terrestrial vertebrates should be clustered around the time for the proposed existence of the land bridge, but they are not (Hedges et al. 1992).

References

Allard MD, Baker SD, Emerson GL, Ottenwalder JA, Kilpatrick CW (2001) Characterization of the mitochondrial control region in *Solenodon paradoxus* from Hispañola and implications for the biogeography, systematics and conservation management. In: Woods CA, Sergile FE (eds) Biogeography of the West Indies: patterns and perspectives. CRC Press, Boca Raton, pp 331–334

Anderson A (1984) The extinction of moa in southern New Zealand. In: Martin PS, Klein RG (eds) Quaternary extinctions: a prehistoric revolution. The University of Arizona Press, Tucson, pp 728–740

Anderson RP, Handley CO Jr (2002) Dwarfism in insular sloths: biogeography, selection, and evolutionary rate. Evolution 56(5):1045–1058

Arredondo O, Varona LS (1983) Sobre la validez de *Montaneia anthropomorpha* Ameghino, 1910 (Primates: Cebidae). Poeyana 131:1–12; 533:243–246

Bloch JI, Woodruff ED, Wood AR, Rincon AF, Harrington AR, Morgan GS, Foster DA, Montes C, Jaramillo CA, Jud NA, Jones DS, MacFadden BJ (2016) First North American fossil monkey and early Miocene tropical biotic interchange. Nature 533:243

Brace S, Barnes I, Powell A, Pearson R, Woolaver LG, Thomas MG, Turvey ST (2012) Population history of the Hispaniolan hutia *Plaiodontia aedium* (Rodentia: Capromyidae): testing the model of ancient differentiation on a geotectonically complex Caribbean island. Mol Ecol 21:2239–2253

Burk K (1988) Tectonic evolution of the Caribbean. Annu Rev Earth Planet Sci 16:201–230

Buskirk RE (1985) Zoogeographic patterns and tectonic history of Jamaica and the northern Caribbean. J Biogeogr 12:445–461

Coates AG, Obando JA (1996) The geologic evolution of the Central American isthmus. In: Jackson JBC, Budd AF, Coates AG (eds) Evolution and environment in tropical America. University of Chicago Press, Chicago, pp 21–56

Cooke SB, Rosenberger AL, Turvey S (2011) An extinct monkey from Haiti and the origins of the Greater Antillean primates. Proc Natl Acad Sci U S A 108(7):2699–2704

Dávalos LM (2004) Phylogeny and biogeography of Caribbean mammals. Biol J Linn Soc 81:373–394

Donnelly TW (1985) Mesozoic and Cenozoic plate evolution of the Caribbean region. In: Stehli FG, Webb D (eds) The great American biotic Interchange. Springer-Science, New York, pp 89–121

Fabre P-H, Vilstrup JT, Raghavan M, Sarkissian CD, Willerslev E, Douzery EJP, Orlando L (2014) Rodents of the Caribbean: origin and diversification of hutias unraveled by next-generation museomics. Biol Lett 10:20140266. https://doi.org/10.1098/rsbl.2014.0266

Fleagle JG (2013) Primate adaptation and evolution, 3rd edn. Academic Press, New York

Ford SM (1980) Subfossil platyrrhine tibia (Primates: Callitrichidae) from Hispaniola: A possible further example of Island gigantism. 70(1):47–62

Ford SM (1986) Subfossil platyrrhine tibia (Primates: Callitrichidae) from Hispaniola: A possible further example of Island gigantism. Am J Phys Anthr 70(1):47–62

Ford SM (1988) Platyrrhine evolution in the West Indies. J Hum Evol 19:237–254

Ford SM, Morgan GS (1986) A new ceboid femur from the late Pleistocene of Jamaica. J Vertebr Paleontol 6:281–289

Goodman SM, Jungers WL (2014) Extinct Madagascar. The University of Chicago Press, Chicago

Halenar LB, Cooke SB, Rosenberger AL (2017) New cranium of the endemic Caribbean platyrrhine, *Antillothrix bernensis*, from La Altagracia Province, Dominican Republic. J Hum Evol 106:133–153

Hedges SB (1996) Historical biogeography of west Indian vertebrates. Annu Rev Ecol Syst 27:163–196

Hedges SB (2001) Biogeography of the West Indies: an overview. In: Woods CA, Sergile FE (eds) Biogeography of the West Indies: patterns and perspectives. CRC Press, Boca Raton, pp 15–33

Hedges SB, Hass CA, Maxson LR (1992) Caribbean biogeography: molecular evidence for dispersal in West Indian terrestrial vertebrates. Proc Natl Acad Sci U S A 89:1909–1913

Hershkovitz P (1974) A new genus of late Oligocene monkey (Cebidae, Platyrrhini) with notes on postorbital closure and platyrrhine evolution. Folia Primatol 21:1–35

Horovitz I (1999) A phylogenetic study of living and fossil platyrrhines. Am Mus Novit 3269:1–40

Horovitz I, MacPhee RDE (1998) The quaternary Cuban platyrrhine *Paralouatta varonai* and the origin of Antillean monkeys. J Hum Evol 36:33–658

Iturralde-Vinent MA (2006) Meso-Cenozoic Caribbean paleogeography: implications for the historical biogeography of the region. Int Geol Rev 48(9):791–827

Iturralde-Vinent MA, MacPhee MA (1999) Paleogeography of the Caribbean region: Implications for Cenozoic biogeography. Bull Am Mus Nat Hist 238:95 pp

Kay RF (2013) Biogeography in deep time – What do phylogenetics, geology, and paleoclimate tell us about early platyrrhine evolution? Mol Phylogenet Evol. https://doi.org/10.1016/j.ympev.2013.12.002

Kay RF, Hunt KD, Beeker CD, Conrad GW, Johnson CC, Keller J (2011) Preliminary notes on a newly discovered skull of the exinct monkey *Antillothrix* from Hispaniola and the origin of the Greater Antillean monkeys. J Hum Evol 60:124–128

MacPhee RDE (2005) "First" appearances in the Cenozoic land-mammal record of the Greater Antilles: significance and comparison with South American and Antarctic records. J Biogeogr 32:551–564

MacPhee RDE (2007) Prehistoric sloth extinctions in Cuba: Implications of a new "last" appearance date. Caribb J Sci 43(1):94–98

MacPhee RDE (2009) *Insulae infortunatae*: Establishing a chronology for Late Quaternary mammal extinctions in the West Indies. In: Haynes G (ed) American megafaunal extinctions at the end of the Pleistocene. Springer, New York, pp 169–193

MacPhee RDE, Fleagle JG (1991) Postcranial remains of *Xenothrix mcgregori* (Primates, Xenotrichidae) and other Late Quaternary mammals from Long Mile Cave, Jamaica. Bull Am Mus Nat Hist 206:287–320

MacPhee RDE, Horovitz I (2004) New craniodental remains of the Quatenary Jamaican monkey *Xenothrix mcgregori* (Xenotrichini, Callicebinae, Pitheciidae), with a reconsideration of the *Aotus* hypothesis. Am Mus Novit 3434:1–51

MacPhee RDE, Iturralde-Vinent MA (1995) Earliest monkey from Greater Antilles. J Hum Evol 28:197–200

MacPhee RDE, Meldrum J (2006) Postcranial remains of the extinct monkeys of the Greater Antilles, with evidence for semiterrestriality in *Paralouatta*. Am Mus Novit 3516:1–65

MacPhee RDE, Woods CA (1982) A new fossil Cebine from Hispaniola. Am J Phys Anthropol 58:419–436

MacPhee RDE, Woods CA, Morgan GS (1983) The Pleistocene rodent *Alterodon major* and the mammalian biogeography of Jamaica. Palaeontology 26(4):831–837

MacPhee RDE, Iturralde-Vinent MA, Jiménez Vázquez O (2007) Prehistoric sloth extinctions in Cuba: implications of a new "last" appearance date. Caribb J Sci 43(1):94–98

McFarland DA, Lundberg J (2002) A middle Pleistocene age and biogeography for the extinct rodent *Megalomys curazensis* from Curaçao, Netherlands Antilles. Caribb J Sci 38:278–281

Miller GS (1929) Mammals eaten by Indians, owls, and Spaniards in the coast region of the Dominican Republic (with two plates). Smithson Misc Collect 82(5):1–16

Morgan GS, Woods CA (1986) Extinction and the zoogeography of West Indian land mammals. Biol J Linn Soc 28:167–203
Olson SL, James HF (1984) The role of Polynesians in the extinction of the avifauna of the Hawaiian Islands. In: Martin PS, Klein RG (eds) Quaternary extinctions: a prehistoric revolution. The University of Arizona Press, Tucson, pp 768–784
Paulo Couto C (1979) Tratado de paleomastozoologia. Academia Brasileira de Ciências, Rio de Janeiro. 587 pp
Pregill GK (1981) An appraisal of the vicariance hypothesis of Caribbean biogeography and its application to West Indian terrestrial vertebrates. Syst Biol 30(2):147–155
Rímoli RO (1977) Una nueva especie de monos (Cebidae: Saimirinae: *Saimiri*) de la Hispaniola. Cuadernos de CENDIA 242:1–14
Rivero M, Arredondo O (1991) *Paralouatta varonai*, a new Quaternary platyrrhine from Cuba. J Hum Evol 21:1–11
Roca AL, Bar-Gal GK, Eizirik E, Helgen KM, Maria R, Springer MS, O'Brien SJ, Murphy WJ (2004) Mesozoic origin for West Indian insectivores. Nature 429:649–651
Rosenberger AL (1977) *Xenothrix* and Ceboid phylogeny. J Hum Evol 6:461–481
Rosenberger AL (1978) New species of Hispaniolan monkey: A comment. Anuario Cientifico 3(3):249–251
Rosenberger AL (2002) Platyrrhine paleontology and systematics: the paradigm shifts. In: Hartwig WC (ed) The primate fossil record. Cambridge University Press, New York, pp 151–160
Rosenberger AL, Cooke SB, Rimoli R, Ni X, Cardosa L (2011) First skull of *Antillothrix bernensis*, an extinct relict monkey from the Dominican Republic. Proc R Soc B Biol Sci 278:67–74
Rosenberger AL, Klukkert ZS, Cooke SB, Rímoli R (2013) Rethinking *Antillothrix*: The mandible and its implications. Am J Primatol 75:825–836
Rosenberger AL, Cooker SB, Haenar LB, Tejedor MF, Hartwig WC, Novo NM, Muñoz-Saba Y (2015) Fossil Alouattines and the origins of *Alouatta*: craniodental diversity and interrelationships. In: Howler monkeys, developments in primatology: progress and prospects. Springer Science, New York, pp 21–54
Steadman DW (1995) Prehistoric extinctions of Pacific Island birds: biodiversity meets zooarchaeology. Science 267:1123–1131
Steadman DW (2005) Prehistoric extinctions of Pacific island birds: biodiversity meets zooarchaeology. Science 267(5201):1123–1131
Steadman DW, Vargas Casanova P, Ferrando CC (1994) Stratigraphy, chronology, and cultural context of an early faunal assemblage from Eastern Island. Asian Perspect 33(1):79–96
Steadman DW, Martin POS, MacPhee RDE, Jull AJT, McDonald HG, Woods CA, Iturralde-Vinent M, Hodgins WL (2005) Asynchronous extinction of late Quaternary sloths in continents and islands. Proc Natl Acad Sci U S A 102(33):11763–11768
Trotter MM, MCulloch B (1984) Moas, men and middens. In: Martin PS, Klein RG (eds) Quaternary extinctions: a prehistoric revolution. The University of Arizona Press, Tucson, pp 708–727
Whidden HP, Asher RJ (2001) The origin of the greater Antillean insectivores. In: Woods CA, Sergile FE (eds) Biogeography of the West Indies: patterns and perspectives. CRC Press, Boca Raton, pp 237–252
White JL, MacPhee RD (2001) The sloths of the West Indies: A systematic and phylogenetic review. In: Woods CA, Sergile FE (eds) Biogeography of the West Indies: patterns and perspectives. CRC Press, Boca Raton, pp 201–236
Williams EE, Koopman KF (1952) West Indian fossil monkeys. Am Mus Novit 1546:1–16
Wilson SM (2007) Archaeology of the Caribbean. Cambridge University Press, Cambridge
Woods CA, Ottenwalder JA (1992) The natural history of southern Haiti. Florida Museum of Natural History, Island, Gainesville

Chapter 12
The Genesis of the Modern Amazon River Basin and Andean Uplift and Their Roles in Mammalian Diversification

12.1 Introduction

Because of the volumes of water and the huge area (6,915,000 km^2 or 2,670,000 m^2) that is the Amazon basin, we believe that the Amazon has had a huge influence on the evolution of many mammals. But we know very little about the details of evolution in the Amazon, mainly because of taphonomic difficulties in the formation and preservation of fossils in humid tropical forests in dense vegetation. The acid and humid conditions of tropical forest soils are inimical to fossil formation, and the abundant accumulations of recently deposited sedimentary soils that cover many potential rocky deposits have made the discovery of fossils rather difficult. It has been only in the last 15–20 years that it has been discovered that under certain conditions, it has been possible to recover fossil materials from the rocky outcroppings of many Amazonian rivers. These fossils begin filling in the huge gaps that exist at low latitudes of South America, especially for the later part of the Miocene and the Pliocene. Difficulties in dating these outcroppings and in exploiting strata that are under water during a great part of the year are challenges that are gradually being surmounted.

The thing is, evidence suggests that the actual formation of the Amazon River watershed is not very ancient, and the high to very high precipitations found over the basin (particularly toward the western parts) coupled with the uplift of the Andes mountain chain have resulted in a very active reworking of transported soils that leave sedimentary deposits in some parts that are hundreds of meters deep. That and the essentially flat nature of much of the Amazonian topography have left few sedimentary rocks aboveground.

Much of our understanding of the earlier conditions of the Amazon basin come from a well-known upper Amazon formation found along many affluents in western parts, called the Pebas Formation (also known as the Solimões Formation). Deep perforations of oil companies also have afforded information about even older formations.

T. Defler, *History of Terrestrial Mammals in South America*, Topics in Geobiology 42, https://doi.org/10.1007/978-3-319-98449-0_12

12.2 Evolution of the Tropical Forested Basin and the Proto-Amazonian Forest

During most of the Paleogene, a tropical rainforest existed throughout most of the paleo-Amazon basin, although its modern aspect didn't emerge until the Miocene (Hoorn et al. 2010a). Probably the first tropical moist forest to be studied for Colombia was found in the Cerrejón of the La Guajira Peninsula, outside what is usually defined as Amazonia (which takes its name from the Amazon River, which didn't exist before the late Miocene) (Wing et al. 2009). It is important to realize that during the Paleocene the Cerrejón lay about 15° to the south, due to the drift of the continent to the north for the last 60 million years and this would have placed this Paleocene forest at the equivalent of around 3°S, squarely in the equatorial tropics (Vonhof and Kaandorp 2010).

Though the modern angiosperm forest was just beginning to be defined, the 58 million-year-old forest of the Cerrejón, the earliest known South American rainforest, already had many elements of a modern lowland moist forest (Burnham and Granam 1999). Flowering plants had appeared at least during the early Cretaceous or late Jurassic, although their consolidation into forests took some time (Friis et al. 2006; Frohlich and Chase 2007; Graham 2011). Forests of flowering plants didn't really appear until the early Paleogene (Burnham and Johnson 2004). One study counted 34 extant angiosperm families already in the Paleocene, and the study of the Cerrejón adds 20 new families to that list, but diversity in these studied assemblages had been low compared to Neogene forest floras, probably partly as a result of the Cretaceous extinction event (Morley 2000; Wing et al. 2009; Hoorn et al. 2010b). Early Cenozoic neotropical forests have been placed into two provinces, a northern (Palmae) province dominated by many palms and dominated by megathermal taxa and a southern province or "mixed" province with several families common to Australia and Antarctica (Casuarinaceae, Proteaceae, Fagaceae, Araucariaceae, Podocarpaceae, and *Nothofagus* as well as Sapindaceae, Monimiaceae, and southern relicts such as Restionaceae and Gunneraceae) (Romero 1993; Morley 2000; Jaramillo et al. 2010). The northern province is the progenitor of the modern Amazonian moist tropical forest, and Paleocene families known today were Anacardiaceae, Annonaceae, Araceae, Arecaceae, Combretaceae, Myricaceae, Olacaceae, Rhizophoraceae, Rutaceae, Sapindaceae, and Zamiaceae (Morley 2000; Romero 1993).

Diversity continued to rise during the early Eocene, although there were extinctions as well (Morley 2000; Jaramillo 2002). At the Paleocene-Eocene Thermal Maximum (PETM), subtropical forests reached into North America and to Antarctica, although floral elements were different than would have been found in the proto-Amazon basin. The greatest extent of megathermal forests in the world was during the PETM where forests covered most of South America to the tip of southern Patagonia, reaching King George Island (65°S) off the coast of the Antarctic retreat northward. During this time, pollen diversity also exceeded the values of modern neotropical moist forests but began to drop with the lowering temperatures in the second half of the Eocene (Jaramillo et al. 2010). Noteworthy is the presence of grass pollen in SE Brazil and in Colombia which probably indicated their presence in open swamps, but this was the beginning of what would become the first grassy savannas in the world as temperatures began to drop again and the world cooled off during the Oligocene (Morley 2000).

During the Oligocene moist ever-wet climates persisted in part of equatorial South America. The continuing high diversity of Palmae in South America is evidence of limited extinctions from this family after the ECO (Eocene climatic optimum). The occurrence of wooded grassy savannas increased ahead of other continents, as shown by the great increase in hypsodont dentition among the meridiungulates and the caviomorph rodents, which allowed a diet of tough grasses. Around the ECO of the Eocene, the first known rodents appear in Peruvian tropical forest (Antoine et al. 2011). The oldest known South American caviomorphs (five taxa based on teeth) were very small, like their sister group, the phiomorph rodents from Africa, and had many dental similarities with the Old World sister group. Their presence in pollen-identified tropical moist forest from about 41 Ma demonstrates that the first adaptive radiation of caviomorphs was from tropical forest, not from more xeric habitats as had been posited for the early Oligocene rodents with hypsodont teeth (high-crowned, heavily enameled teeth evolved for eating tough, fibrous, and gritty food like grass) which at first convinced many that most of early rodent evolution took place also in dry, savanna-like niches (Antoine et al. 2011; Wyss et al. 1993; Flynn et al. 2003).

The earliest known South American primates (three taxa) have also been identified from late Eocene (around 36 Ma) tropical forest from Peru (Bond et al. 2015). These were also very small animals and show more resemblance to Eocene African anthropoids, especially of *Perupithecus* to the African Eocene *Tahuapithecus* as sister taxon. Their presence in moist tropical forest also underlines the importance of this ecosystem as being key to the first radiations of primates in South America, well before the Miocene radiations of primates in higher latitudes of the continent and in the Caribbean. Thus, the earliest radiations of both the caviomorphs and the primates are shown to have taken place in tropical moist forest even before the establishment of the Amazon basin, although it cannot be concluded that these first South American primates originated (1) in Africa where they became extinct, (2) they represent Eocene platyrrhines of immediate African extraction, or (3) they represent pre-platyrrhine primates that differentiated in Africa (Bond et al. 2015).

Until the middle Miocene the South American equatorial rainforests were a continuous belt from the Pacific to the Atlantic coast. With the development of the Andean cordillera, the forest fragmented into two parts, one along the northern Pacific coast and the other centered on the proto-Amazonian Basin. With this interruption many sister species of vicariant plants and animals evolved.

12.3 Evolution of the Amazon River

During the late Cretaceous and early Cenozoic, drainage from the Guyana and Brazilian highlands emptied into the Pacific and the proto-Caribbean (Graham 2009). The development of the Amazon River as such begins in the Miocene. During late Oligocene to early Miocene times, continental drainage from the Guyana and Brazilian highlands was toward the northwest, eventually emptying into the Caribbean, but lacustrine and swampy regions increased in a precursor stage for the development of the proto-Amazon. Gradual uplift of Andean mountains accompanied by subsidence of the Andean foreland basins began to create a catchment region for runoff waters both from the incipient Andes and from the southeast drainages. In the northern Andes, the Western and Central Cordillera became established from the Cretaceous and Paleocene much before the Eastern Cordillera (Gregory-Wodzicki 2000). The increasing height of the Andes also increased precipitation to the east (Hoorn et al. 2010). These expanded into immense wetlands that grew toward the east, eventually breaking through to the Atlantic as early as 16 Ma (Hoorn et al. 2010a). In surrounding uplands moist rainforest prevailed. This immense proto-Amazon basin was characterized during the entire Miocene by a subsiding basin that accumulated sediments via avulsive rivers associated with the associated accumulating waters of swamps, lakes, internal deltas, and splays (Latrubesse et al. 2010; Duivenvoorden and Duque 2010). The extent of these mega-wetlands is defined by the presence of the Pebas Phase (~16–11.3 Ma) and Solimões Formations, which at their maximum encompassed about 1.5×10^6 km^2 and attained a thickness of up to 1000 m (Hoorn et al. 2010; Hoorn and Wesselingh 2010). Occasional marine influences occurred during high oceanic levels, but the wetland was predominantly of freshwater.

Sediments from 13.5 to 12.9 Ma indicate that the Central Cordillera drained to the east-southeast into the mega-wetlands, indicating that the Eastern Cordillera was not a barrier (Hoorn et al. 1995; Guerrero 1997). At 11.8 Ma the drainage pattern for La Venta shifted to the north, indicating that the Eastern Cordillera had become a barrier to drainage from the Central Cordillera (Hoorn et al. 1995). During early-late Miocene the Eastern Cordillera became a complete barrier west and, because of uplift, this created a divisoria that affected flow northward into the Orinoquia. This resulted in the ever-increasing wetlands finally cutting a channel to the Atlantic to become the early Amazon River. Earliest evidence of heavy sedimentation at the mouth of the Amazon River dates to 9.4–9 Ma (Hoorn et al. 2017; Figueiredo et al. 2009). The wetlands from ~11.3 to 7 Ma years (Acre Phase) were fluvial-tidal dominated and had a species-rich vertebrate fauna that to date has been poorly studied (Hoorn et al. 2010). Consolidation of the several drainages into the mega-drainage known as the Amazon watershed occurred during the early Pliocene at which point the Amazon began draining water and sediments into the Atlantic Ocean on a very large scale (Hoorn 1994; Latrubesse et al. 2010; Anon 2009).

Previously a trans-Amazonian seaway had been suggested as having connected the proto-Caribbean to inland Argentina, and this would have been an explanation for the presence of dolphins, stingrays, manatees, and other aquatic fauna now found in the freshwater watershed of the Amazon basin (Hoorn 2006). Later evidence supports the hypothesis that the region was converted into an immense freshwater Lake Amazon (Frailey et al. 1988). Best evidence suggests the immense wetlands suggested by some, which was occasionally subject to brackish or saltwater influence as seas rose (Räsänen et al. 1995). This would have provided a highway for the mammals (and fish) that clearly have an ancient oceanic origin (Fig. 12.1).

12.4 Development of Mammalian Diversity in the Neotropical Hylea

Species richness in the western Amazon appears to have a long history of millions of years. Antoine et al. (2017) underline the extraordinary mammalian biodiversity throughout the Cenozoic that is coming to light now that paleontologists have learned how to look for it, and they name this part of the Amazon a "hotspot" of mammalian diversity, although hotspots have usually been defined in terms of conservation, taking into account not only species richness but more importantly endemism and vulnerability (Myers 2000; Mittermeier et al. 1999). Here we are concerned with the origin and development of mammalian species richness in the Amazon basin and not with its conservation, so we find it more correct to discuss species richness and its origin (Carrillo et al. 2015).

Present-day mammalian species richness (measured as the number of sympatric species in a defined area) in the western Amazon reflects what is probably the greatest in the Americas and perhaps in the world (Voss and Emmons 1996; Emmons 1984). Most geographic variation in Amazonian mammal diversity involves lower taxonomic categories. The elevated diversity of western Amazonian nonvolant faunas is therefore attributable to both high generic and congeneric richness. Three mammalian orders—Marsupialia, Primates, and Rodentia—currently account for most diversity differences among nonvolant Amazonian faunas (Emmons 1984). Marsupials and rodents (not primates) dominated past faunas (Antoine et al. 2017), although Miocene tropical species diversity of primates is as yet poorly studied and probably is as rich as marsupials and rodents.

The extraordinary diversity of much Amazonian biota has not been easily explained. Earlier the suggestion that Pleistocene refuges provided isolated forest fragments that acted as a sort of species pump seemed attractive and was much discussed (Haffer 1969, 1997). Posterior molecular analyses, however, have shown that most vertebrate species were formed before the Pleistocene glacial cycles, although this does not eliminate the possible role of Cenozoic forest refuges as a mechanism for driving diversity and (during the Pleistocene) perhaps subspecific diversity, so the original hypothesis of Pleistocene refuges has become modified to the "refuge hypothesis" (Haffer 2008; Antonelli et al. 2010; Pennington and Dick 2010).

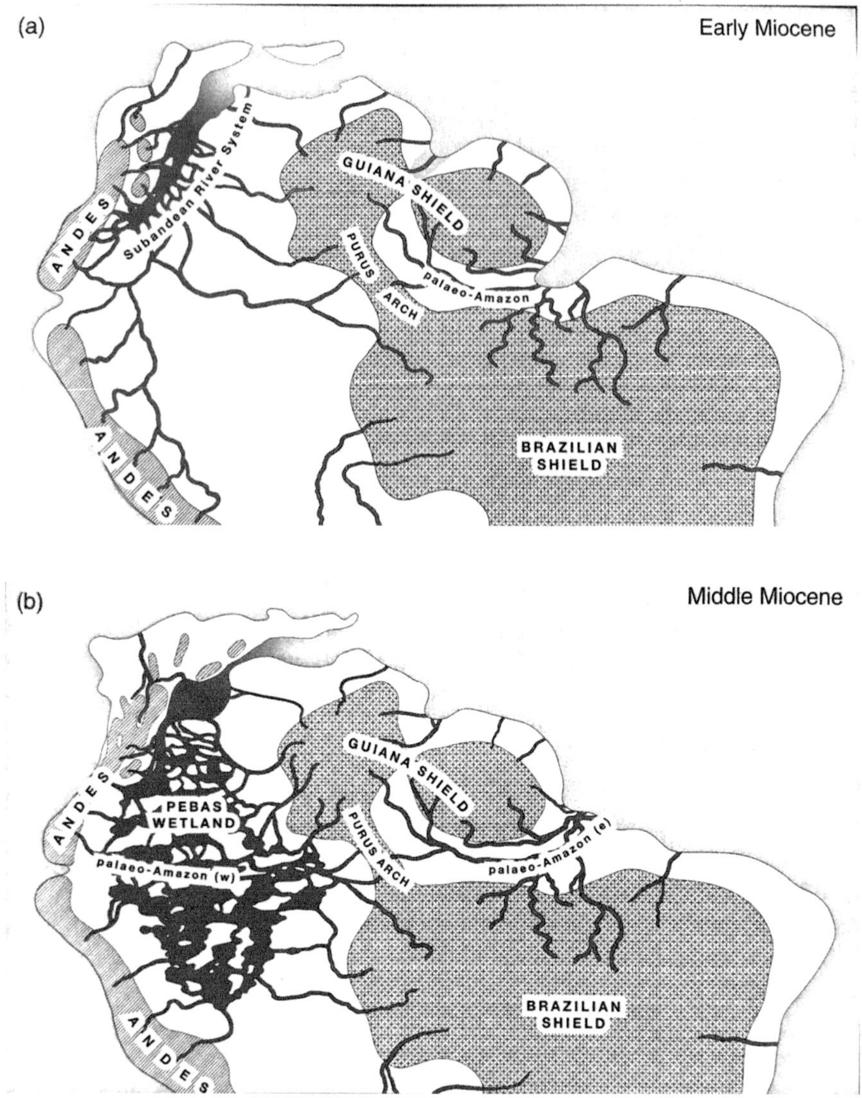

Fig. 12.1 The evolution of the Amazonian wetlands throughout the Miocene: (**a**) latest Oligocene to earliest Miocene with drainage to the north; (**b**) wetlands reaches maximum size due to damning influence of rising Andean Cordillera while damning Purus arch is intact; (**c**) cutting of Purus arch and establishment of Amazon River flowing into the Atlantic Ocean to the east. (From Hoorn et al. 2010; permission Wiley and Sons by Diego Casallas-Pabón, Applied Biodiversity Foundation

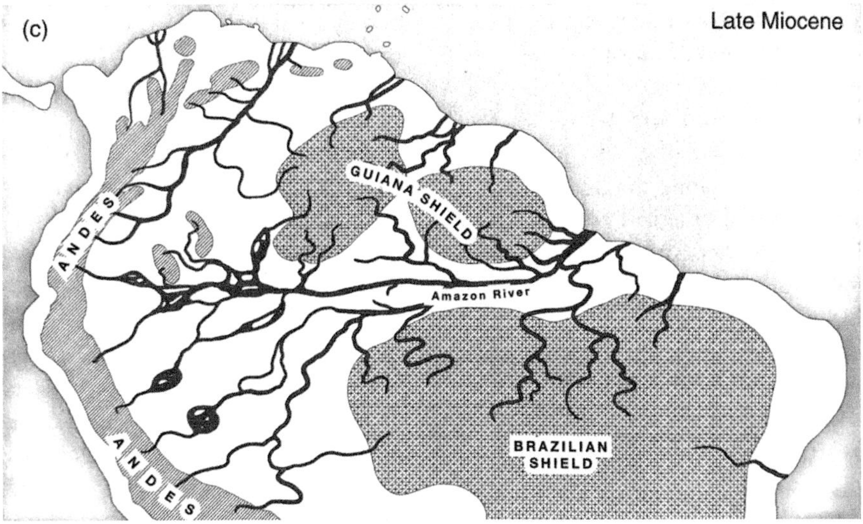

Fig. 12.1 (continued)

Haffer (1997, 2008) sees the generation of diversity due to various mechanisms, having evolved from his support for Pleistocene refuges to a broader view. He lists as follow:

12.4.1 Changes in the Distribution of Land and Sea or in the Landscape Due to Tectonic Movements or Sea-Level Fluctuations (Paleogeography Hypothesis)

The role of ancient geological ridges or arches that are found in various places throughout the basin is still not understood, but because of molecular studies, it is clear that some effect (former hills, different vegetation zones, etc.) somehow separated the ranges of certain mammal species. Patton et al. (2000) describe molecular data that show not only patterns of interrupted gene flow across rivers but also abrupt changes in haplotypes of the same species on the same side of the river. They associated these abrupt changes with a subsurface geological feature called "the Iquitos Arch" which could be related to Andean orogenic events that caused flexure due to tectonic loading and might have influenced landform evolution in terms of changes in fluvial drainages and influenced soil types or the formation of other ecological barriers driving parapatric divergence. Tectonic influences on the history of the Amazon have been little appreciated and are scarcely understood, but the concordance of haplotype shifts with such a landform as the Iquitos Arch suggests an important role for tectonic movements in species diversification. There are a number of such arches that divide the Amazon basin into subbasins, but evidence suggesting their effect on the surface vegetation is slim, and effects on geomorphological features do not appear to be sufficient to form the ecological barriers posited by Patton et al. (2000) (Haffer 2008).

Frailey et al. (1988) find evidence for a huge Pleistocene/Holocene lake in parts of the western basin which would have created a barrier between upland habitats and would have provided a rich habitat of lakeside mosaic of habitats that Emmons (1984) says are especially species-rich compared to upland forest. Hoorn et al. (2010a) describe a mega-wetland rather than one huge lake, and this type of ecosystem would have provided a multitude of opportunities for populations to be isolated from one another. According to multiple sea-level changes, this vast wetland at times became brackish, and such changes would have influenced available dryland and vegetation types. This hypothesis can be applied to the entire Cenozoic (Aleixo 2004; Aleixo and de Fátima Rossetti 2007; Campbell and Frailey 1984).

Hoorn et al. (2013) explain how more new species are found in areas of recent mountain building as a result of related climate changes and the creation of new habitats (Sepulchre et al. 2010; Colinvaux 1998; Colinvaux et al. 1996). Also, particularly rivers reworked the erosional products of the actively growing Andes to form heterogeneous edaphic properties that influence the patterns of Amazonian biota from the Miocene on (Fátima-Rossetti et al. 2005; Irion and Kalliola 2010; Campbell et al. 2006).

Xenarthrans show correlated diversification times with major environmental changes such as the early split between the armadillos and sloths at about 65 Ma and between anteaters and sloths at the Paleocene-Eocene transition at around 55 Ma (Delsuc et al. 2004) suggesting that crucial influences of global changes have influenced the evolutionary history of the xenarthrans and perhaps other mammalian groups. The appearance of hypsodonty in rodents correlates with cooling of the earth and the expansion of grassy habitats (Campbell 2004b).

12.4.2 The Barrier Effect of Amazonian Rivers (River Barrier Hypothesis)

The hypothesis of rivers as barriers to gene flow is the oldest explanation for species diversity in the Amazon. Wallace (1852), on observing different species of primates on each side of the Amazon and Negro rivers, posited the large rivers as barriers to populations on the opposite bank. A test of this hypothesis by Ayres and Clutton-Brock (1992) generally supported this idea with more and more cross-river contact occurring upriver where the rivers are smaller. Hershkovtiz (1977) discussed how a population of mammals might cross via land cut off by river meanders and the eventual connection of that land to the other side of the river. Many observations show that the lower parts of large rivers have separated populations that could have originated from one side of a river and been passively carried by meanders to the other side or even by accidental water dispersal. The upper and narrower parts of the same rivers are less effective as barriers. It has been true for rodents, primates as well as forest birds, reptiles, and amphibians (Patton et al. 2000; Ayres and Clutton-Brock 1992; Peres et al. 1996; Haffer 1985; Hershkovitz 1977). The density of rivers found in lowland forest has also been suggested as affecting biotic diversity (Ayres and Clutton-Brock 1992).

This type of species diversification or stasipatric speciation is further described as a possibility for karyotypic evolution in the *Aotus* or *Callicebus* species complexes. These primates are monogamous and defend a territory. Bush (1975), Bush et al. (1977), Marks (1987), Ridley (1996), and others have developed the idea that high rates of karyotypic evolution may be due to certain social structures and breeding systems (e.g., monogamy and territorial defense), which might give a higher chance of survival of chromosomal changes, so that these changes may become fixed when there is a subdivision of populations into small, inbred demes in the homozygous state (Defler and Bueno 2001). Galbreath (1983: pp. 248–249) explicitly evoked genetic drift in small effective populations as the mechanism most responsible for *Aotus* chromosome variability. Any small population (such as a monogamous family group) crossing a large river could form a small inbred population units, making it possible for a chromosome change to become fixed and to proceed to adaptive change that also might be a barrier to the dominant population on that side of the river (Defler and Bueno 2007).

One scenario is that previously there existed a vast forest without the many wide rivers that crisscross the Amazon now. Then with the development of the Amazon drainage, the new rivers divided previously widespread species into subpopulations that were able to become evolutionarily significant units (Capparella 1988). Haffer (2008) finds it inconceivable that the formation of a great forest and the many wide rivers was not one interrelated process. However, it seems unclear why this should be, especially since many floristic elements of the Amazon date to well before the establishment of the Amazon watershed from around 10 Ma.

There are many problems for the river barrier hypothesis that need to be addressed from the upriver narrower river that allows contact of the two separated species, the passage of animals from one population via a cutoff meander of the river, the considerably reduced flow of the big rivers during the height of glaciation, and the consequent reduced precipitation over the forest and other arguments (Haffer 2008). But like the other mechanisms, a river barrier hypothesis is only one of many hypotheses that could be viable in causing genetic diversification and consequent speciation of populations.

12.4.3 A Combination of the Barrier Effect of Broad Rivers and Vegetation Changes in Northern and Southern Amazonas (River-Refuge Hypothesis or River-Forest Contraction Hypothesis)

There is plenty of evidence for periodic contractions of Amazonian vegetation during dry climatic periods at the height of glaciation periods (Van der Hammen and Absy 1994; Van der Hammen and Hooghiemstra 2000). These vegetation contractions were from the north and south and caused the separation of the upper Amazonian forest from blocks of forest to the east and could have separated animal populations into what Haffer terms "semi-refugia" that in combination with the effects of river barriers would have caused the isolation of animal populations.

Climate variation in the Amazonian basin has also been implicated in studies of the evolution of its biodiversity, especially as it relates to influences on the dynamics of the ITCZ, which impacts rainfall patterns in the basin. Milankovitch cycles, the closure of the Panamanian gap, and the increasing height of the Cordillera have all influenced climate variation by influencing precipitation and its periodicity (Vonhof and Kaandorp 2010). Many of the data for the presence and effect of drier periods and changes in the moist tropical forest cover have been discussed (Haffer and Prance 2001). Climatic forcing of evolution has been proposed, especially driven by Milankovitch cycles, and this has persisted for the entire Cenozoic. It has not been appreciated as an evolutionary force because of the fine details of climate change that are not evident against the backdrop of the larger tendency of geological changes. It is probably because of Pleistocene changes that these cycles have come to our attention. Dry periods have broken up moist forest blocks forming zones of discontinuity separating refuges, especially in peripheral regions and near areas of surface relief (Bennett 1990; Haffer 1997, 2008; Haffer and Prance 2001). Other phenomena also contributed to form moist forest refuges (Aleixo 2004; Aleixo et al. 2007).

12.4.4 The Isolation of Forest Blocks Near Areas of Surface Relief in the Periphery of Amazonia During Dry Climatic Periods of the Tertiary and Quaternary (Refuge Theory or Hypothesis)

Species originated in response to climate and vegetation changes responding to Milankovitch cycles (Pennington et al. 2004). Dry cycles permitted the formation of islands or refugia that were cut off from other members of the species, thus permitting change on a specific or subspecific level (Mayr and O'Hara 1986). When the climate became moist enough again, the fragments joined together and the new taxa could expand if conditions were right, or they did not. Other types of dry forest and savanna existed between these refuges. These cycles were periodic over a 20,000 or a 100,000 cycle throughout a long geological time span. The great diversity of the Amazonian forest is due to the ability of many opportunistic plants to change and survive throughout the Cenozoic (Prance 1973; Morley 1996, 2000; Rancy 1991, 1993; Webb and Rancy 1996). This hypothesis refers to the entire Cenozoic and not only the Pleistocene.

From molecular analyses of neotropical tetrapods as well as from pollen studies, we can now be confident that Amazonian diversity dates especially from the Neogene, associated with the uplift of the Andes, marine incursions and the development of the Amazon basin, as well as climate change (Antonelli et al. 2010; Hoorn 2006). Diversification is likely to have taken place associated with the long history of the forest that dates back to the Eocene, a wet tropical climate and historical perturbations, among which could be counted forest refuges due to climatic oscillations, marine incursions, and geotectonic events (Antonelli et al. 2010). Many of the climatic oscillations throughout the Neogene are apparently due to Milankovitch-driven rainfall variation. Milankovitch cycles are changes in the tilt of the Earth's axis and orbital eccentricity which affect solar insolation patterns and thus the dynamics of the intertropical convergence zone on a scale of 20–100 (ky) (Bennett 1990). Concurrence of these cycles (especially the 100,000 year eccentricity of Earth's orbit and the 41 ky oscillation of the Earth's axial tilt) with a drop in atmospheric CO_2 levels seems to be a driving force for the Quaternary ice age (Bennett 1990).

The role of the late Miocene-Pliocene wetlands or "lake" has been mentioned as a speciation mechanism, suggesting that the accumulation of water isolated many populations on islands and at the periphery. This allowed speciation (Klammer 1984; Marroig and Cerqueira 1997).

12.4.5 *The Canopy-Density Hypothesis*

This hypothesis was suggested by Cowling et al. (2001), who suggest that changes in decreases of atmospheric CO_2 and temperature during glacial and interglacial cycles could result in increases and decreases of the structure (density) of the canopy over large areas and, thus, an interruption of gene flow in many species. This could theoretically result in speciation. This hypothesis is therefore related to the refuge hypothesis.

12.4.6 *The Museum Hypothesis*

The museum hypothesis posits that speciation occurs in localized stable habitat zones in mountainous areas peripheral to the Amazon because of climatic fluctuations. The new species expand into the Amazonian lowlands and accumulate over geological time, preserved as "museum" species (Solomon et al. 2008; Roy et al. 1997; Fjeldsa et al. 1999; Arita and Vasquez-Dominguez 2008).

Molecular phylogenetic relationships were inferred from 21 ant subfamilies to find ancient ages for the modern ants varying from 139 to 158 Ma which also identified 10 periods of significant change. Extremely high ant diversity in the neotropics suggests that the tropics have acted as a museum for this diversity (Moreau and Bell 2013).

12.4.7 Disturbance-Vicariance Hypothesis

Competitive species interactions and local species isolations in peripheral regions of Amazonia occur due to invasion and counterinvasion during cold/warm periods of the Pleistocene (disturbance-vicariance hypothesis). Bush et al. (2011) see temperature variations to be much more important than varied precipitation, at least during the ice age, although they recognize that some precipitations differed.

12.4.8 Parapatric Speciation

Parapatric (the opposite of allopatric speciation, which is the evolution of adjacent populations into distinct species) is speciation across steep environmental gradients without separation of the representative populations (Gradient hypothesis) and may contribute in one degree or another to the evolution of diversity. Clines are often the result of his type of speciation. Ecological divergence may be very important to bimodal hybrid zones and may contribute to speciation (Jiggens and Mallet 2000).

Ecotones have been suggested as being very important in the evolution of closely related species. Because adjacent habitats are often very large, natural selection might drive diversity by acting on subpopulations in these adjacent habitats (Smith et al. 1997). Schneider et al. (1999) found that populations of a widespread leaf-litter skink in closed rainforest and tall open forest were very different in terms of morphological and life history characteristics between the two habitats. Yet their mitochondrial DNA differences were slight. Turner and Mallet (1996) believed that mimicry rings in *Heliconius* were a result of mutation and random drift in parapatry.

During the presence of the mega-wetland of western proto-Amazonia, there existed a huge extent of ecotones between the wetlands and upland forest and forest/savanna that could have driven much diversification. The generation of a cerrado-type ecosystem in close proximity to closed-canopy tropical moist forest and wetland margins would have provide an immense amount of ecotonic habitat.

Edaphic complexity is believed to have been very important in maintaining evolved diversity, and this complexity evolved during the complex geological history of the basin, especially relating to the growth of the Andean Cordillera. These conditions in relation to the generally constant wet and warm climate have maintained a very high biologic diversity and are an argument against any late Miocene seaway having been present. The evidence is for a mostly freshwater environment (Wesselingh et al. 2010; Latrubesse et al. 2007).

Most biologists see mammalian evolution as involving a separation of a population into two by some mechanism, with the posterior development of reproductive barriers that maintain the two populations apart (Mayr 1970; King 1995; Wilkins 2009; Haffer 1997). Mechanisms for reproductive isolation are likely to have been multiple, involving large bodies of water as barriers, including rivers, wetlands, and marine embayments, oscillating climates causing fragmentation of forest types by savanna-like vegetation as well as drier types of forest and complex mosaics of soils. Diversity of invertebrates, plants, and fishes is likely due to other speciation processes such as polyploidy (very important for plants) and for sympatric speciation (Mayr 1942, 1970).

The present consensus on mechanisms of species diversification in the Amazon basin is that it has been driven by multiple processes that have resulted in the remarkable diversity seen. Both general and specific geographical patterns are evident that obligate not only general explications such as habitat heterogeneity in the form of mosaics or edaphic variability and rainfall patterns but other historical developments such as the formation of a proto-Amazon mega-wetland, its many changes, and its extinction. Molecular analyses demonstrate that much of the diversification is pre-Quaternary and not the result of one process but a result of a combination of warm and wet climates, a heterogeneous edaphic substrate as a result of mountain building, changes in a huge lake and swamp and embayments in western Amazonia, and the establishment of the Amazon River and its drainage into the Atlantic Ocean (Antonelli et al. 2010; Behling et al. 2010; Jaramillo et al. 2010; Wesselingh et al. 2010; Pennington and Dick 2010).

12.5 Diversity of Ancient Neotropical Mammalian Fauna

12.5.1 The Contamana Local Fauna

The discovery of the earliest caviomorph rodent fauna *Cachiyacuy contamanensis*, *C. kummeli*, *Canaanimys maquiensis*, cavioid *Eobrasinamys* sp. (similar to *E. romeropittmanae* and *E. riverai* from the Santa Rosa fauna but only half their size), and octodontoid cf. *Eoespina* sp. from 41 Ma Eocene moist forest fauna fossils really did a great deal to point out the importance of forests in the evolution of South American mammals (Pierre-Olivier et al. 2012). From pollen evidence, these animals existed in a tropical rainforest habitat in conjunction with a fauna that included a polydolopimorph marsupial, an astegothiine armadillo, a small archaeohyracid notoungulate, a small pyrothere, and a trigonostylopid astrapothere all of which help to constrain the age of these deposits to 41.6–35.9 Ma. Earlier it was assumed that most rodent evolution had taken place in open habitats (Frailey and Campbell 2004; Wyss et al. 1993; Flynn et al. 2003), but these very early caviomorphs prove that early diversification took place in tropical moist forest. Their very small size is comparable to their phiomorph sister group of Africa that could suggest their short evolutionary history in South America (Antoine et al. 2013). The discovery of these rodents suggests that the first South American rodents began differentiating in the north and then spread south.

12.5.2 The Santa Rosa Local Fauna

The next group representing tropical forest mammals and the first group to be discovered historically as such is the Santa Rosa local fauna of the Yurúa River of eastern Peru and part of the Yahuarango Formation (Campbell 2004a, b). The age is poorly constrained but suggestive of late middle Eocene, considering the stage of evolution and the various taxa of this fauna. Frailey and Campbell (2004) placed the fauna in the early Mustersan SALMA (48–42 Ma), but posterior analysis of notoungulates, litopterns, and cingulates suggests late Eocene and early Oligocene as being more appropriate. Kay et al. (1999) places the fauna in the Casamayoran SALMA using radiometric determinations and magnetic polarity stratigraphy to suggest 36 Ma, but the age is still not certain.

The fauna is made up of 11 species of marsupials, 2 (or more) species of notoungulates, 17 species of rodents, and 3 species of primates. The marsupials include a metatherian sparassodont *Patene campbelli*, which is the largest of the group and was the size of a cat. The rest were very small (20–50 g) and belonged to at least five orders (Goin and Candela 2004). Shockey et al. (2004) list two notoungulates, one (or several) belongs to the Toxodontia while another is a Typotheria. The toxodont specimens probably come from more than one species; one of them would have been the smallest toxodont known. The larger specimens were sheep-sized. The typothere is from the Interatheriidae family and is the only known intheratherid from the Amazon basin. There are 17 species of rodents from this fauna, 6 of them indeterminate and 11 are named. These animals can be placed in the families Erethizontidae, Agoutidae, and Echimyidae. It is worth pointing out that several taxa from the Contamana fauna compare closely to *Eobrasinamys* and *Eoincamys* species from this fauna, except they are half the size of these Santa Rita rodents (Frailey and Campbell (2004).

Primates are, after all, forest animals and are so far best-known from taphonomically ideal sites like La Venta and Patagonia (McFadden 1990). But the recent discovery of the molars of the primate *Perupithecus* and two other unnamed primate taxa from the Santa Rosa local fauna of the Yahuarango Formation in central Peruvian Amazonia represents a tropical forest primate fauna near the beginning of their South American history at about 38–34 Ma. An exactly assigned age has so far proved elusive but has been suggested as being late Eocene (Bond et al. 2015; Shockey et al. 2004). The 41 Ma age for the Yahuarango Formation has been discussed in Supplementary Materials of Bond et al. (2015) with some questions that allow doubt for the firm date and suggest that the Contamana early rodents could possibly be younger and closer to the age of the first primates (Bond et al. 2015). These primates were very small, and their molar characteristics do not suggest a relationship with other South American primates. Rather, they are much more similar to the North African *Talahpithecus*, probably an Oligopithecidae that has been dated from about 37 Ma. Kay (2015) suggests that the small body size suggests insectivory for these primates.

12.5.3 The Laventan Fauna

Neotropical biodiversity really began to increase in the Miocene, probably linked to tectonism and the development of the Andes and climate (Hoorn et al. 2010b). La Venta has been seen erroneously as an Andean fauna when it was in fact during the Middle Miocene an ecosystem peripheral to the mega-wetlands of the proto-Amazon and shared many elements with the Santa Rosa local fauna, especially the Fitzcarrald local mammalian fauna. La Venta is still the most detailed neotropical fauna described and should not be confused with the Andes, even though its location was in a zone which after the 13–14 Ma establishment of the Eastern Cordillera of the Andes became a valley between two cordilleras. During the Laventan SALMA, La Venta's connection with the Santa Rita local fauna seems clear. Tejada-Lara et al. (2015) describe material from 35 localities that can be ascribed to the Fitzcarrald local fauna, while Goillot et al. (2011) describe astrapotheres from the same fauna. At least 30 vertebrate taxa were recovered, including besides mammals, crocodiles, chelonians, snakes, and fishes.

The two faunas, La Venta and the Fitzcarald local fauna share at least ten of the same genera and possibly at least two species (*Greanastrapotherium snorki* and *Prodolichotis pridiana*) based on current material collected at the two sites (Kay et al. 1997; Tejada-Lara et al. 2015). Both sites were characterized by an open vegetation type with savanna mixed with gallery forest. La Venta probably had an annual precipitation of between 1500 and 2000 mm which currently supports a transitional forest between open savanna and closed rainforest (Kay and Madden 1997). Other collections in eastern Peru exist for the middle Miocene and especially around the Laventan age. It seems likely that these two Laventan faunas were distributed throughout a very extensive ecosystem of forest/savanna that stretched like the eastern savannas of Africa from at least La Venta (and probably all of the area that later became the Eastern Cordillera and eastward toward the great proto-Amazon mega-wetlands and extending southward). Judging by the common fauna that has been identified and that are judged to be savanna-forest mammals, this hypothetical forested savanna could have been similar to the cerrado, which is the most biodiverse type of tropical savanna in the world and includes many forest elements.

Croft (2007, 2016, 2018a, b) and others have described the Laventan fauna for southern Bolivia (Quebrada Honda) that includes at this date 7 marsupiales, 8 xenarthrans, 1 sparassodont, 12 rodents, and 7 ungulates or 35 terrestrial mammals. This was a temperate fauna, not tropical and different from the Fitzcarrald fauna. There are no primates known for this site in contrast to the nine primates known for La Venta. There is also very little in common between the two sites. The most abundant fossil from La Venta is the notoungulate *Miocochilius* which is also known for Quebrada Honda. The astrapothere from Quebrada Honda could possibly be *Xenastrapotherium,* but that is uncertain. So far no rodent genus is common to both sites, though these are the most numerous in the collections.

The Quebrada Honda fauna shares almost no genera or species with La Venta of the same age. Also no Argyrolagidae has ever been collected in La Venta, but they have been collected to the south in Argentina. So why is there a big difference between the two sites of the same age? The great mega-wetland might have been a partial barrier between the two.

12.5.4 The Acre Fauna

The latest Paleogene fauna known from tropical lowland Amazon comes from outcrops studied in the Upper Solimões Formation and provisionally dated Huayquerian-Mesopotamian SALMAS (9–6.5 Ma or perhaps to 3.2 Ma) of Acre and Amazonas States in Brazil and eastern Peru (Campbell et al. 2001; Cozzuol 2006; Negri et al. 2010). The Acre, "Mesopotamian" (Argentina and Uruguay), and Urumaco (Venezuela) assemblages are considered faunistically correlated (Cozzuol 2006; Sánchez-Villagra et al. 2010), but although the Urumaco fauna was tropical forest, it is not proto-Amazonian. During the last 25 years, more than 5000 vertebrates have been collected from the Solimões Formation from Acre and Amazonas States from which about 320 mammalian species from 89 families have been described, and the results show a rich terrestrial fauna of caviomorph rodents, marsupials, and xenarthrans and especially ungulates termed the Acre vertebrate fauna (Kay and Cozzuol 2006; Antoine et al. 2017) plus three primates (mentioned above Kay and Cozzuol 2006). This fauna includes the earliest South American gomphothere, as well as the yet unnamed camel, and tapir, and two described tayassuids from Peru (Campbell et al. 2000; Negri et al. 2010). The gomphothere *Amahuacatherium peruvium* can be considered the first precursor of the GABI or Great American Biotic Interchange, or it was second behind the newly discovered *Panamacebus transitus* of Panama that was part of a North American fauna of early Miocene age (Bloch et al. 2016). This Peruvian fauna includes only 2 marsupials, more than 17 xenarthrans, 16 rodents (the newest dinomyid *Ferigolomys pacarana*; Kerber et al. 2017), and 12 ungulates including *Amahuacatherium*, as well as 5 cetaceans, 3 primates, and a couple of bats. The astrapothere *Xenastrapotherium*, present in Laventan faunas, is also present in this fauna. The primate genus *Stirtonia* was found in both the Acre fauna and La Venta and eventually should be found in the Fitzcarrald fauna. Other genera in common with Laventan fauna are megalonychid sloth *Hapalops*, the glyptodonts *Asterostemma* and *Neoglyptastemma*, and the armadillo *Anadasypus*.

The earliest arrival (9.5–9 Ma) in South America of gomphotheres, camelids, tayassuids, and tapirids seems to be closely related to moist forest as well, related to sedimentation in the previous mega-wetlands although the discovery sites might be from broken forest at the edge of the great Amazonian wetland, which seems to have a mixture of swamp vegetation, savanna-like vegetation, and upland forest from the remnants of this wetland (Latrubesse et al. 1997). Close to that Peruvian site are discoveries of a fauna that is similar to La Venta, Colombia, probably proving a similar vegetation type between southern Peru and La Venta, Colombia, which lay west of the great (Pebas) wetland of middle-late Miocene times. Much more evidence of primate evolution should be ultimately found from those forest sites.

The discovery of the taxa *Solimoea acrensis* (a stem atelid), *Acrecebus fraileyi* (a probable stem *Cebus*), and *Stirtonia* sp. (a stem *Alouatta*) from late Miocene moist lowland forests of Acre, Brazil, shows additional tropical forest primates (Kay and Frailey 1993). This is further evidence for lowland moist forest evolution of these mammals and in these cases, we might be seeing stem groups for the larger primates (Kay and Frailey 1993; Kay and Cozzuol 2006).

The known fauna from western Amazonia from the middle Eocene (47.8–38 Ma) is now equal to about 320 mammalian species, and it continues to grow (Antoine et al. 2017). However, these authors do not include La Venta fauna, even though at the time of the middle Miocene the La Venta fauna was connected to all other fauna west of the mega-wetland and has many elements in common with it. La Venta should be analyzed with all western Amazonian middle Eocene fauna, as many elements are common and represent a very extensive cerrado-like ecosystem from La Venta to Santa Rita, Peru. An additional fauna comes from the Paleogene of Santa Rosa, eastern Peru, from the Eocene and records early caviomorph rodents, marsupials, and ungulates of the period (Campbell 2004a, b). This late Eocene fauna would have been at least 8–10° further south from the equator, given that the continent is moving to the north at about 5° per 20 million years.

12.5.5 Invasion of the Northern Cricetid Rodents

It seems likely that the earliest cricetid rodent fossils will also be found in Miocene/Paleocene forest formations, since molecular data indicates an early arrival (Miocene) in South America for these animals, so far without the concurrent fossil evidence. One expects to find the early evidence for the arrival of these rodents in northern South America with the following radiation to the south. But so far, the oldest dated fossils for the cricetid rodents are to be found in Argentina.

References

Aleixo A (2004) Historical diversification of a terra-firme forest bird superspecies: a phylogeography perspective on the role of different hypotheses of Amazonian diversification. Evolution 58:1303–1317

Aleixo A, de Fatima Rossetti D (2007) Avian gene trees, landscape evolution, and geology: towards a modern synthesis of Amazonian historical biogeography. J Ornithol 148(Suppl 2):S443–S453

Anon (2009) How old is the Amazon River? University of Amsterdam Press release. http://www.uva.nl/en/content/news/press-releases/2017/03/amazon-river-no-younger-than-9-million-years-new-study-shows.html?page=1&pageSize=200&origin=zvuyIF7VQlmbnXJokCBg7g

Antoine PO, Maarivaux L, Croft DA, Billet G et al. (2011) Middle Eocene rodents from Peruvian Amazonia reveal the pattern and timing of caviomorph origins and biogeography. Proc R Soc B 279:1319–1326. https://doi.org/10.1098/rspb.2011.1732

Antoine PO, Roddaz M, Brichau S, Tejada-Lara J, Salas-Gismondi R, et al. (2013) Middle Miocene vertebrates from the Amazonian Madre de Dios Subandean Zone, Peru. J S Am Earth Sci 42:91–102

Antoine PO, Salas-ismondi R, Puos F, Ganerød M, Marivaux L (2017) Western Amazonia as a hotspot of mammalian biodviersity throught the Cenozoic. J Mamm Evol 24(1):5–17

Antonelli A, Quijada-Nascareñas A, Crawford AJ, Bates JM, Velazco PM, Wüster W (2010) Molecular studies and phylogeography of Amazonian tetrapods and their relation to geological and climatic models. In: Hoorn C, Wesselingh F (eds) Amazonia: landscape and species evolution: a look into the past. Wiley-Blackwell, New York, pp 386–417

Arita HT, Vásquez-Dominguez E (2008) The tropics; cradle, museum or casino? A dynamic null model for latitudinal gradients of species diversity. Ecol Lett 11:1–12. https://doi.org/10.1111/j.1461-0248.2008.01197.x

Ayres JM, Clutton-Brock TH (1992) River boundaries and species range size in Amazonian primates. Am Nat 140(3):531–537

Behling H, Bush M, Hooghiemstra H (2010) Biotic development of Quaternary Amazonia: a palynological perspective. In: Hoorn C, Wesselingh F (eds) Amazonia: landscape and species evolution: a look into the past. Wiley-Blackwell, New York, pp 335–348

Bennett KD (1990) Milankovitch cycles and their effects on species in ecological and evolutionary time. Paleobiology 16(1):11–21

Bloch JI, Woodruff ED, Wood AR, Rincon AF et al (2016) First North American fossil monkey and early Miocene tropical biotic interchange. Nature 533:213–228. https://doi.org/10.1038/nature17415

Bond M, Tejedor MF, Campbell KE Jr, Chornogubsky L, Novo N, Goin F (2015) Eocene primates of South America and the African origins of New World monkeys. Nature 520:538

Burnham RJ, Granam A (1999) The history of Neotropical vegetation: new developments and status. Ann Mo Bot Gard 86:546–589

Burnham RJ, Johnson KR (2004) South American palaeobotany and the origins of neotropical rainforests. Philos Trans R Soc Lond B 359:1595–1610

Bush GL (1975) Modes of animal speciation. Annu Rev Ecol Syst 6:339–364

Bush GL, Case SM, Wilson AC et al. (1977) Rapid speciation and chromosomal evolution in mammals. Proc Natl Acad Sci USA 74(9):3942–3946

Bush MB, Gosling WD, Colinvaux PA (2011) Climate change in the lowlands of the Amazon basin. In: Bush M, Flenley J, Gosling W (eds) Tropical rainforest responses to climatic change. Springer, New York, pp 61–84

Campbell KE (ed) (2004a) The Paleogene Mammalian Fauna of Santa Rosa, Amazonian Peru. Natural History Museum of Los Angeles County, Los Angeles

Campbell KE (2004b) The Santa Rosa local fauna: a summary. In: Campbell KE (ed) The Paleogene Mammalian Fauna of Santa Rosa, Amazoniana Peru. Natural History Museum of Los Angeles County, Los Angeles, pp 155–163

Campbell KE Jr, Frailey CD (1984) Holocene flooding and species diversity in southwestern Amazonia. Quat Res 21:369–375

Campbell KE Jr, Frailey CD, Romero-Pittman L (2000) The late Miocene gomphothere *Amahuacatherium peruvium* (Proboscidea: Gomphotheriidae) from Amazonian Peru: implications for the great American faunal interchange. Instituto Geológico Minero y Metalúrgico, Boletín, Série D, Estudios Regionales 23:1–152

Campbell KE Jr, Heizler M, Frailey CD, Romero-Pittman L, Prothero DR (2001) Upper Cenozoic chronostratigraphy of the southwestern Amazon Basin. Geology 29(7):595–598. https://doi.org/10.1130/0091-7613(2001)029<0595:UCCOTS>2.0.CO;2

Campbell KE Jr, Frailey CD, Romero-Pittman L (2006) The pan-Amazonian Ucayali peneplain, late Neogene sedimentation in Amazonia, and the birth of the modern Amazon River system. Palaeogeogr Palaeoclimatol Palaeoecol 239:166–219

Capparella A (1988) Genetic variation in Neotropical birds: implications for the speciation process. Cong Int Ornith 19. Acta 2:1658–1664

Carrillo JD, Forasiepi A, Jaramillo C, Sánchez-Villagra MR (2015) Neotropical mammal diversity and the great American biotic interchange: spatial and temporal variation in South America's fossil record. Front Genet 5(451). https://doi.org/10.3389/fgene.2014.00451

Colinvaux P (1998) A new vicariance model for Amazonian endemics. Global Ecol Biogeogr Lett 7(2):95–99

Colinvaux PA, Liu K-B, de Oliveira PE, Bush MB, Miller MC, Steinitz Kannan M (1996) Temperature depression in the lowland tropics in glacial times. Clim Chang 32:19–33

Cowling SA, Maslin MA, Sykes MT (2001) Paleovegetation simulations of lowland Amazonia and implications for neotropical allopatry and speciation. Quat Res (55):140–149

Cozzuol MA (2006) The Acre vertebrate fauna: age, diversity, and geography. J S Am Earth Sci 21:185–203

Croft D (2007) The Middle Miocene (Laventan) Quebrada Honda fauna, southern Bolivia and a description of its notoungulates. Palaeontology 59(1):277–303

Croft D (2016) Horned armadillos and rafting monkeys. Indiana University Press, Bloomington

Croft D (2018a) Blog. Darren A. Croft, PhD: New Mammal Species. https://dcpaleo.org/new-species/

Croft D (2018b) Blog. Darren A. Croft, PhD: Quebrada Honda, Bolivia. https://dcpaleo.org/quebrada-honda-bolivia/

Campbell KE Jr., Heizler M, Frailey CD, Romero-Pittman L, Prothero DR (2001) Upper Cenozoic chronostratigraphy of the southwestern Amazon Basin Geology 29(7):595–598. https://doi.org/10.1130/0091-7613(2001)029>0595:UCCOTS<.0.CO;2

Defler TR, Bueno ML (2001) Aotus diversity and the species problem. Primate Conserv 22(1):55–70

Defler TR and Bueno ML (2007) Aotus diversity and the species problem. Primate Conservation 22(1):55–70

Delsuc F, Vizcaíno SF, Douzery EJP (2004) Influence of Tertiary paleoenvironmental changes on the diversification of South American mammals: a relaxed molecular clock study within xenarthrans. BMC Evol Biol 4:1–13

Duivenvoorden JF, Duque A (2010). Composition and diversity of northwestern Amazonian rainforest in a geological context. In: Hoorn C, Wesselingh F (eds). Amazonia: Landscape and species evolution a look into the past. Oxford: Blackwell Publishing, Oxford, pp 360–372

Emmon LH (1984) Geographic variation in densities and diversities of non-flying mammals in Amazonia. Biotropica 16(3):210–222

Fátima-Rossetti, Toledo P, Goes AM (2005) New geological framework for Western Amazonia (Brazil) and implicatios for biogeography and evolution. Quat Res 63:78–89

Figueiredo J, Hoorn C, van der Ven P, Soares E (2009) Late Miocene onset of the Amazon River and the Amazon deep-sea fan: evidence from the Foz do Amazonas Basin. Geology:619–622

Fjeldsa J, Lambin E, Mertens B (1999) Correlation between endemism and local ecoclimatic stability documented by comparing Andean bird distributions and remotely sensed land surface data. Ecography (22):63–78

Flynn JJ, Wyss AR, Croft DA, Charrier R (2003) The Tinguiririca fauna, Chile: biochronology, paleoecology, biogeography, and a new earliest Oligocene South American Land Mammal "Age". Palaeoecology 195:229–259

Frailey CD, Campbell KE (2004) Paleogene rodents from Amazonian Peru: the Santa Rosa local fauna. In: Campbell KE (ed) The paleogene mammalian fauna of Santa Rosa, Amazonian Peru. Natural History Museum of Los Angeles County, Los Angeles, pp 71–130

Frailey CD, Lavina EL, Rancy A, de Souza Filho JP (1988) A proposed Pleistocene/Holocene lake in the Amazon Basin and its significance to Amazonian geology and biogeography. Acta Amazon 18:119–143

Friis EM, Pedersen KR, Crane PR (2006) Cretaceous angiosperm flowers: innovation and evolution in plant reproduction. Palaeogeogr Palaeoclimatol Palaeoecol 232:251–293

Frohlich MW, Chase MW (2007) After a dozen years of progress the origin of angiosperms is still a great mystery. Nature 450(20/27):1184–1189

Galbreath GJ (1983) Karyotypic evolution in Aotus. Am J Primatol 4(3):245–251

Goillot C, Pierre-Olivier A, Tejada J, Pujos F, Salas Gismondi R (2011) Middle Miocene Uruguaytheriinae (Mammalia, *Astrapotheria*) from Peruvian Amazonia and a review of the astrapotheriid fossil record in northern South America. Geodiversitas 33(2):331–345. https://doi.org/10.5252/g2011n2a8

Goin FJ, Candela AM (2004) New Paleogene marsupials from the Amazon basin of eastern Peru. In: Campbell KE Jr (ed) The Paleogene mammalian fauna of Santa Rosa, Amazonian Peru, Science series, vol 40. Natural History Museum of Los Angeles County, Los Angeles, pp 1–163

Graham A (2009) The Andes: a geological overview from a biological perspective. Ann Mo Bot Gard 96:371–385

Graham A (2011) The age and diversification of terrestrial New World ecosystems through Cretaceous and Cenozoic time. Am J Bot 98(3):336–351

Gregory-Wodzicki K (2000) Uplift history of the Central and Northern Andes: a review. GSA Bull 112(7):1091–1105

Guerrero J (1997) Stratigraphy, sedimentary environments, and the Miocene uplift of the Colombian Andes. In: Kay RF, Madden RH, Cifelli RL, Flynn JJ (eds) Vertebrate paleontology in the neotropics: the Miocene fauna of La Venta, Colombia. Smithsonian Institution Press, Washington, DC, pp 15–43

Haffer J (1969) Speciation in Amazon forest birds. Science 165(3889):131–137

Haffer J (1985) Avian zoogeography of the Neotropical lowlands. Ornithol Monogr 36:113–146

Haffer J (1997) Alternative models of vertebrate speciation in Amazonia: an overview. Biodivers Conserv 6:451–476

Haffer J (2008) Hypotheses to explain the origin of species in Amazonia. Braz J Biol 68(4):917–947

Haffer J, Prance GT (2001) Climatic forcing of evolution in Amazonia during the Cenozoic: on the refuge theory of biotic differentiation. Inst Estudo Avançados da Univers São Paulo

Hershkovitz P (1977) Living new world monkeys (Platyrrhini), vol 1. The University of Chicago Press, Chicago

Hoorn C (1994) An environmental reconstruction of the palaeo-Amazon River system (Middle-Late Miocene, NW Amazonia). Palaeogeogr Palaeoclimatol Palaeoecol 112:187–238

Hoorn C (2006) The birth of the mighty Amazon. Sci Am 294(5):52–59

Hoorn C, Wesselingh F (eds) (2010) Amazonia: landscape and species evolution: a look into the past. Wiley-Blackwell, New York

Hoorn C, Guerrero J, Sarmiento GA, Lorente MA (1995) Andean tectonics as a cause for changing drainage patterns in Miocene northern South America. Geology 23:237–240

Hoorn C, Wesselingh FP, Hovikoski J, Guerroro J (2010a) The development of the Amazonian mega-wetland (Miocene; Brazil, Colombia, Peru, Bolivia). In: Hoorn C, Wesselingh F (eds)

Amazonia: landscape and species evolution: a look into the past. Wiley-Blackwell, New York, pp 123–142

Hoorn C, Wesselingh FP, ter Steege H, Bermudez MA et al (2010b) Amazonia through time: Andean uplift, climate change, landscape evolution, and biodiversity. Science 330:927–931

Hoorn C, Mosbrugger V, Mulch A, Antonerlli A (2013) Biodiversity from mountain building. Nat Geosci 6(154)

Hoorn C, Bogotá-A GR, Romero-Baez M, Lammertsma EI, Flantua SGA, Dantas EL, Dino R, do Carmo DA, Chemale F Jr (2017) The Amazon at sea: onset and stages of the Amazon River from a marine record, with special reference to Neogene plant turnover in the drainage basin. Glob Planet Change. https://doi.org/10.1016/j.gloplacha2017.02.005

Irion G, Kalliola R (2010) Long-term landscape development processes in Amazonia. In: Hoorn C, Wesselingh F (eds) Amazonia: landscape and species evolution: a look into the past. Wiley-Blackwell, New York, pp 185–197

Jaramillo CA (2002) Response of tropical vegetation to Paleogene warming. Paleobiology 28(2):222–243

Jaramillo CA, Hoorn C, Silva SAF, Leite F, Herrera F, Quiroz L, Dino R, Antonioli L (2010) The origin of the modern Amazon rainforest: implications of the palynological and palaeobotanical record. In: Hoorn C, Wesselingh F (eds) Amazonia: landscape and species evolution: a look into the past. Wiley-Blackwell, New York, pp 317–334

Jiggens CD, Mallet J (2000) Bimodal hybrid zones and speciation. Trends Ecol Evol 15(6):250–255

Kay RF (2015) New World monkey origins. Science 347(6226):1068–1069

Kay RF, Cozzuol MA (2006) New platyrrhine monkeys from the Solimões Formation (late Miocene, Acre State, Brazil). J Hum Evol 50:673–686

Kay RF, Frailey CD (1993) Large fossil platyrrhines from the Río Acre fauna, late Miocene, western Amazonia. J Hum Evol 25:319–327

Kay RF, Madden RH (1997) Mammals and rainfall: paleoecology of the middle Miocene at La Venta (Colombia, South America). J Hum Evol 32(2–3):161–199

Kay RF, Madden RH, Vucetich MG, Carlini AA et al (1999) Revised age of the Casamayoran South American land mammal "Age"—climatic and biotic implications. Proc Natl Acad Sci USA 96(23):13235–13240

Kay RF, Madden RH, Cifelli RL, Flynn JJ (1997) Vertebrate paleontology in the Neotropics. Smithsonian Institution Press, Washington, DC

Kerber LO, César Bissario M Jr, Negri FR, Pereira de Souza-Filho J, Guilherme E, Schmaltzz Hsiou A (2017) A new rodent (Caviomorpha: Dinomyidae) from the upper Miocene of southwestern Brazilian Amazonia. Hist Biol. https://doi.org/10.1080/08912963.2017.1237529

King M (1995) Species evolution: the role of chromosome change. Cambridge University Press, Cambridge

Klammer G (1984) The relief of the extra-Andean Amazon basin. In: Sioli H (ed) The Amazon. Limnology and landscape ecology of a mighty tropical river and its basin. Junk Publishers, Dordrecht, pp 47–83

Latrubesse EM, Bocquentin J, Santos JCR, Ramonell CG (1997) Paleonenvironmental model for the Late Cenozoic of southwestern Amazonia: paleontology and geology. Acta Amazon 27(2):103–118

Latrubesse EM, da Silva SAF, Cossuol M, Absy ML (2007) Late Miocene continental sedimentation in southwestern Amazonia and its regional significance: biotic and geological evidence. J S Am Earth Sci 23:61–80

Latrubesse EM, Cossuol M, da Silva-Caminha AF, Rigsby CA, Absy ML, Jaramillo C (2010) The late Miocene paleogeography of the Amazon Basin and the evolution of the Amazon River system. Earth Sci Rev 99:99–124

Marks J (1987) Social and ecological aspects of primate cytogenetics. In: Kinzey WG (ed) Primate models: the evolution of human behavior. State University of New York Press, Albany, pp 139–151

Marroig G, Cerqueira R (1997) Plio-Pleistocene South American history and the Amazon Lagoon hypothesis: a piece in the puzzle of Amazonian diversification. J Comp Biol 2:103–119
Mayr E (1942) Systematics and the origin of species. Columbia University Press, New York
Mayr E (1970) Populations, species, and evolution. The Belknap Press of Harvard Univerisy Press, Cambridge
Mayr E, O'Hara RJ (1986) The biogeographic evidence supporting the Pleistocene forest refuge hypothesis. Evolution 40(1):55–67
McFadden BJ (1990) Chronology of Cenozoic primate localities in South America. J Hum Evol 19:7–21
Moreau CS, Bell CD (2013) Testing the museum versus cradle tropical biological diversity hypothesis: phylogeny, diversification, and ancestral biogeographic range evolution of the ants. Evolution 67-8:2240–2257
Morley RJ (1996) Amazonian diversity: a river doesn't run through it. Special news report. Science 273:1496–1498
Morley RJ (2000) Origin and evolution of tropical rain forests. John Wiley & Sons Ltd., Chichester-New York
Mittermeier RH, Myers N, Mittermeier CG, Robles G (1999) Hotspots: Earth's biologically richest and most endangered terrestrial ecoregions. Cemex, México City.
Myers N (1988) Threatened biotas: "hot spots" in tropical forests. Environmentalist 8(3):187–208
Myers N (1990) The biodiversity challenge: expanded hot-spots analysis. Environmentalist 10(4):243–256
Negri FR, Bocquentin-Villanueva J, Ferigolo J, Antoine P-O (2010) A review of Tertiary mammal faunas and birds from western Amazonia. In: Hoorn C, Wesselingh F (eds) Amazonia: landscape and species evolution: a look into the past. Wiley-Blackwell, New York, pp 245–258
Patton JL, Nazareth da Silva MF, Malcolm JR (2000) Mammals of the Rio Juruá and the evolutionary and ecological diversification of Amazonia. Bull Amer Mus Nat Hist 244:1–306
Pennington RT, Dick CW (2010) Diversification of the Amazonian flora and its relation to key geological and environmental events: a molecular perspective. In: Hoorn C, Wesselingh F (eds) Amazonia: landscape and species evolution: a look into the past. Wiley-Blackwell, New York, pp 373–385
Pennington RT, Lavin M, Prado DE, Pendry CA, Pell SK, Butterworth CA (2004) Historical climate change and speciation: neotropical seasonally dry forest plants show patterns of both Tertiary and Quaternary diversification. Philos Trans R Soc Lond B 359:515–537
Peres CA, Patton JL, da Silva MNF (1996) Riverine barriers and gene flow in Amazonian saddleback tamarins. Folia Primatol 67:113–124
Prance GT (1973) Phytogeographic support for the theory of Pleistocene forest refuges in the Amazon Basin, based on evidence from distribution patterns in Caryocaraceae, Chrysobalanaceae, Dichapetalaceae and Lecythidaceae. Acta Amazon 3(3):5–26
Rancy A (1991) Pleistocene mammals and paleoecology of the western Amazon. PhD dissertation. University of Florida, Gainseville
Rancy A (1993) A paleofauna da Aazonia indica áreas de patagem com pouca cobertura vegetal. Ciência Hoje 16(93):48–51
Räsänen M, Linna AM, Santos JCR, Negri FR (1995) Late Miocene tidal deposits in the Amazoniana foreland basin. Science 269:386–389
Ridley M (1996) Evolution, 2nd edn. Blackwell Science, Cambridge, MA
Romero EJ (1993) South American paleofloras. In: Goldblatt P (ed) Biological relationships between Africa and South America. Yale University Press, New Haven, pp 62–85
Roy MS, Silva JMC, Arctander P, Garcia Moreno J, Fjeldsa J (1997) The speciation of South American and African birds in montane regions. In: Mindell DP (ed) Avian molecular evolution and systematics. Academic Press, New York, pp 325–343
Sánchez-Villagra MR, Aguilera OA, Carlini AA (eds) (2010) Urumaco and Venezuelan Paleontology. Indiana University Press, Bloomington

Schneider CJ, Smith TB, Larison B, Moritz C (1999) A test of alternative models of diversification in tropical rainforests: ecological gradients vs. rainforest refugia. Proc Natl Acad Sci U S A 96:13869–13873

Sepulchre P, Sloan LC, Fluteau F (2010) Modelling the response of Amazonian climate to the uplift of the Andean mountain range. In: Hoorn C, Wesselingh F (eds) Amazonia: landscape and species evolution: a look into the past. Wiley-Blackwell, New York, pp 211–222

Shockey BJ, Hitz R, Bond M (2004) Paleogene notoungulates from the Amazon basin of Peru. In: Campbell KE Jr (ed) The Paleogene mammalian fauna of Santa Rosa, Amazonian Peru, Science series, vol 40. Natural History Museum of Los Angeles County, Los Angeles, pp 61–69

Smith TB, Wayne RK, Girman DJ, Bruford MW (1997) A role for ecotones in generating rainforest biodiversity. Science 276:1855–1857

Solomon SE, Bacci M Jr, Martins J Jr, Gonçalves Vinha G, Mueller UG (2008) Paleodistributions and comparative molecular phylogeography of leafcutter ants (Atta spp.) provide new insight into the origins of Amazonian diversity. PLoS One 3(7):1–15

Tejada-Lara JV, Sala-Gismondi R, Pujos F, Baby P, Benammi M, Brusse S, de Franceschi D, Espurt N, Urbina M, Pierre-Olivier A (2015) Life in proto-Amazonia: Middle Miocene mammals from the Fitzcarado Arch (Peruvian Amazonia). Paléo 58(2):341–378

Turner JRG, Mallet JLB (1996) Did forest islands drive the diversity of warningly coloured butterflies? Biotic drift and the shifting balance. Philos Trans R Soc Lond B 351:835–845

Van der Hammen T, Absy ML (1994) Amazonia during the last glacial. Palaeogeogr Palaeoclimatol Palaeoecol 109:247–261

Van der Hammen T, Hooghiemstra H (2000) Neogene and Quaternary history of vegetation, climate, and plant diversity in Amazonia. Quat Sci Rev 19:725–742

Vonhof HB, Kaandorp RJG (2010) Climate variation in Amazonia during the Neogene and the Quaternary. In: Hoorn C, Wesselingh F (eds) Amazonia: landscape and species evolution: a look into the past. Wiley-Blackwell, New York, pp 202–210

Voss RS, Emmons LH (1996) Mammalian diversity in Neotropical lowland rainforests: a preliminary assessment. Bull Am Mus Nat Hist 230:1–120

Wallace AR (1852) On the monkeys of the Amazon. Proc Zool Soc London 20:107–110

Webb SD, Rancy A (1996) Late Cenozoic evolution of the neotropical mammal fauna. In: Jackson JB, Budd AF, Coates AG (eds) Evolution and environment in tropical America. University of Chicago Press, Chicago, pp 335–358

Wesselingh FP, Hoorn C, Kroonenberg SB, Antonelli A, Lundberg JG, Vonhof HB, Hooghiemstra H (2010) On the origin of Amazonian landscapes and biodiversity: a synthesis. In: Hoorn C, Wesselingh F (eds) Amazonia: landscape and species evolution: a look into the past. Wiley-Blackwell, New York, pp 421–431

Wilkins JS (2009) Species: a history of the idea. University of California Press, Berkeley

Wing SL, Herrera F, Jaramillo CA, Gómez-Navarro C, Wilf P, Labandeira CC (2009) Late Paleocene fossils from the Cerrejón Formation, Colombia, are the earliest record of neotropical rainforest. Proc Natl Acad Sci U S A 106(44):18627–186622

Wyss AR, Flynn JJ, Norell MA, Swisher CC II et al (1993) South America's earliest rodent and recognition of a new interval of mammalian evolution. Nature 365:434–437

Chapter 13
The Great American Biotic (Faunal) Interchange

13.1 Introduction

The connection of North and South America was a biological revolution in the development of the South American mammalian fauna. It was a revolution in the sense that the old was replaced with the new; a balanced fauna of millions upon millions of years was upset, greatly perturbed and changed by the highly competitive fauna that arrived from the north (Webb and Rancy 1996). There was certainly a movement of southern fauna to the north as well, but in the majority of cases, the movement to the north was of tropical South Americans moving into tropical North America and no further. There are no examples of the extinction of northern mammals because of the invasion of southern mammals. The contrary was certainly true.

The principal questions of this story are as follows: (1) When did the connection or land bridge finally become established between the two Americas; (2) what fauna from the north invaded South America; (3) what fauna from the south invaded North America; (4) how were the two faunas affected and changed from the original fauna?

13.1.1 The Establishment of the Isthmus of Panama

The formation of the Isthmus of Panama was a complex process of more than 15 million years involving many processes of tectonic and volcanic forces, sedimentation, and uplift (Coates and Obando 1996; Coates et al. 2004). The isthmus became a complete barrier to the Pacific and Atlantic Oceans between 3.1 and 2.8 Ma. But long before the completion of the land bridge, fauna began to filter north and south, the first elements of which probably passed swimming from island to island (Fig. 13.1).

At the time of the continental connections, world conditions were cooling, and glaciations were beginning, and savannas were opening up in the tropics. The

T. Defler, *History of Terrestrial Mammals in South America*, Topics in Geobiology 42, https://doi.org/10.1007/978-3-319-98449-0_13

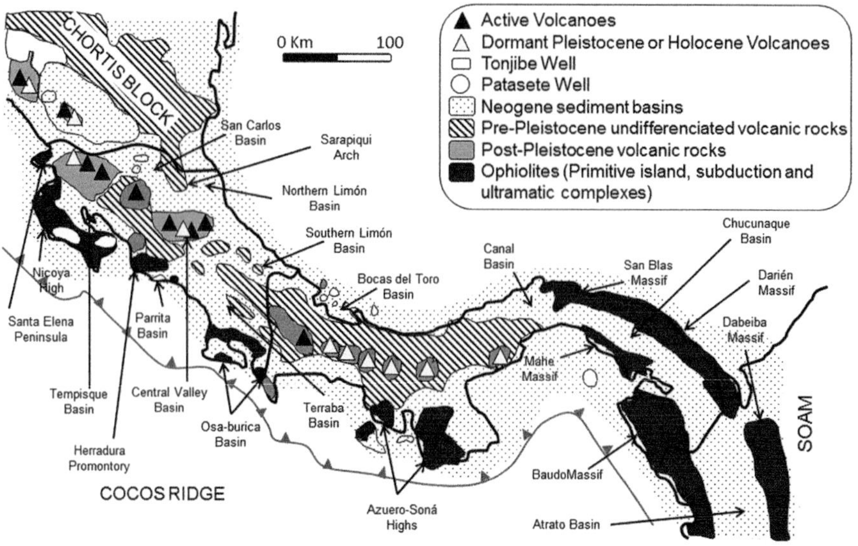

Fig. 13.1 Relationship of Late Neogene sedimentary basins to the volcanic arc of the Chorotega Block and the subduction complexes of the Choco Block

appearance of savannas in the newly formed Central America greatly facilitated the movement of fauna (Graham 1998; Burnham and Graham 1999; Webb 2006; Woodburne 2010). A thick tropical rainforest would have stopped most of the dispersion, as that habitat was inimical to much of the fauna that invaded the south, whereas forest/savanna satisfied many of the needs of the first northern mammals that arrived in South America (By Diego Casallas-Pabón, Applied Biodiversity Foundation, based on Coates and Obando 1996).

It's best to keep in mind that three groups appeared in South America long before the establishment of a solid land connection between the two Americas. Somehow the primitive meridiungulates and marsupials arrived in South America from North America probably during the Late Cretaceous. The caviomorph rodents arrived around 41 Ma or before but from Africa and not from North America. And perhaps not much later than the caviomorphs, the primates arrived around 36 Ma and perhaps as early as 40 Ma, although our fossils don't yet go that far back. Most believe that these groups arrived by waif dispersal, probably by rafting across the Atlantic Ocean and across a proto-Caribbean sea gap (Fig. 13.2).

13.2 The "Invasions" from the North

With time a similar number of mammalian families were exchanged between the continents, depending on what you count and how you classify them. The list includes only land mammals, leaving out vampire bats and manatees that invaded from the south to the north (Table 13.1).

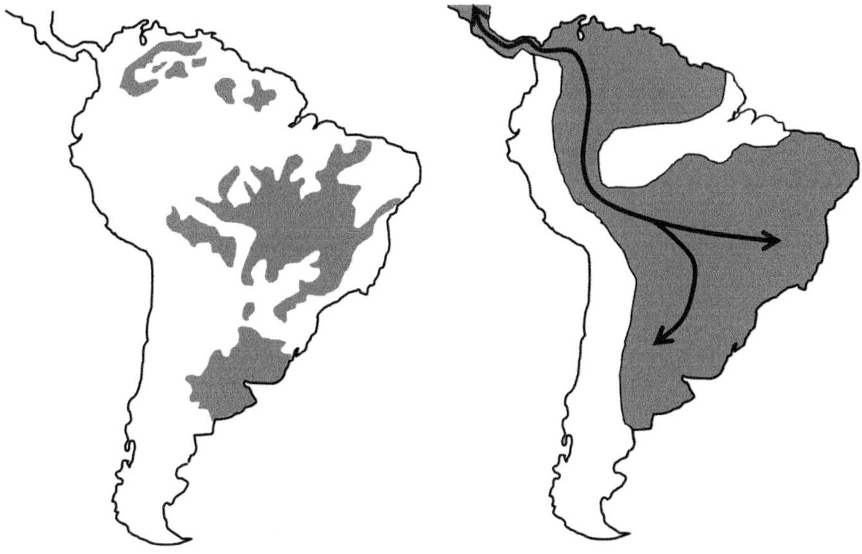

Fig. 13.2 Glacial advances and retreats determined whether many animals could enter South America or not. As glaciers advanced, lowlands tended to become drier, and forest was replaced by grassland savanna (dark grey areas), an advantage for many mammals that entered South America. As glaciers retreated, added humidity allowed forest (white areas) to advance, acting as a barrier to many non-forest mammals (By Diego Casallas-Pabón, Applied Biodiversity Foundation)

Considering the land mammal fauna in South America, this invasion from the north doubled the number of land mammalian families from 18 to 36 families and resulted in a hyper-saturation of mammals that for a time represented the richest mammalian fauna ever known in the world. In this consideration we use the number of families as a proxy for a number of species, which is much more difficult to calculate because of historical considerations as well as different views of what a species is.

13.3 The First Pulse of Mammals to South America

Campbell et al. (2000, 2001, 2009, 2010) have reported a new view of the entrance into South America of northern fauna that challenges the traditional view. They describe the presence of the proboscidean *Amahuacatherium peruvium* from the Late Miocene (around 9.5 Ma) lowlands of Peru, along with fossils of camels, tapirs, and peccaries. These would predate even the arrival of the procyonid *Cyonasua*. Others are cautious about this data, since despite radioscopic and paleomagnetic dates supporting this age, there is an absence of this fauna at any other site until recent ages reported below (Webb 2006; Woodburne 2010; Goin et al. 2012).

Table 13.1 The table lists the families of land mammals that arrived in South America from the north and that arrived in Central and North America from the south. All families from the north prospered and still exist as part of the southern fauna, except for the Gomphotheriidae and the Equidae, which became extinct in the Pleistocene. Of all families that went North from South America, only the families in darkened letters survived (9 of 18) with a new distribution to the North, although some remain restricted to tropical Central America (Cebidae, Callithrichidae, Agoutidae, Dasyproctidae, Myrmecophagidae, Bradypodidae)

Family	Common name
Northern families	**To the South**
Soricidae	Shrews
Leporidae	Rabbits
Heteromyidae	Pocket mice
Geomyidae	Pocket gophers
Sciuridae	Squirrels
Muridae	Field mice
Felidae	Cats
Mustelidae	Otters
Mephistidae	Skunks
Canidae	Dogs
Procyonidae	Raccoons
Ursidae	Bears
Gomphotheriidae	Elephantoids
Tapiridae	Tapirs
Equidae	Horses
Agoutidae	Pacas
Dasyproctidae	Agoutis
Echimyidae	Spiny rats
Tayassuidae	Peccaries
Camelidae	Camels
Cervidae	Deer
Southern families	**To the North**
Dasypodidae	**Armadillos**
Pampatheriidae	Giant armadillos
Glyptodontidae	Glyptodonts
Megalonychidae	Two-toed sloth
Mylodontidae	Ground sloth
Megatheriidae	Ground sloth
Bradypodidae	**Three-toed sloth**
Myrmecophagidae	**Anteater**
Callitrichidae	**Tamarins, marmosets**
Cebidae	**Other primates**
Hydrochoeridae	Capybaras
Caviidae	Guinea pigs
Toxodontidae	Toxodonts
Didelphidae	**Opossums**

But it would appear that this could represent the first pulse of mammals from North America. This is GABI 1, not the pulse that took place 2.6–2.4 Ma (Woodburne 2010), which could be renamed GABI 2.

There is evidence that there was an interruption of the east Pacific with the Caribbean between 10.71 and 9.36 Ma indicated by calcareous nannofossils, surely caused by a drop in the ocean levels that could have allowed a brief small land connection that was again inundated between 8.35 and 3.65 Ma (Kameo and Sato 2000). Geological evidence suggests a regional shallowing from bathyal (depths of 1000–4000 m) to neritic (shallow waters between the low tide mark and the continental shelf) depths from 12.8 to 9.5 Ma and the rise and emergence of the volcanic arc in the Panama Canal basin, such that much of the Darien region was emergent by about 8.6 Ma (Coates et al. 2004).

So a first faunal pulse to South America was possible, especially by large, active mammals that had no trouble swimming water gaps. Elephants are known as strong swimmers, and one Asiatic elephant is known to have swum 162 km in the sea (Johnson 1980). Tayassuids and tapirs commonly swim across very wide rivers in the Amazon (pers. Obs.). Camels are "natural swimmers" and have been known to swim across the Nile river; they are not necessarily desert animals, and many earlier camels might have had forest or woodland adaptations; many of their fossils have been found in association with elephant, horse, hippo, buffalo, bison, and deer (Robinson 1936). But why have we not found evidence of fossils at other sites in South America up until much later (Goin et al. 2012)?

13.3.1 Gomphotheriidae (Proboscidea) South American Elephants

We now know that the first proboscids arrived early in the Late Miocene (Chasicoan), apparently over a transitory connection caused by a drop in the ocean levels and emergence of the volcanic arc in the Panama Canal basin (Kameo and Sato 2000). *Amahuacatherium peruvium* a gomphotherid was found in the southeastern Peruvian lowlands along the Madre de Dios River in deposits of about 9.5 Ma ago and represents the first known North American mammal to arrive in South America. The animal was four-tusked with vestigial lower incisors rather than two-tusked like the later gomphotherids of South America (Campbell et al. 2000, 2001, 2009, 2010).

Mothé et al. (2012, 2013) have combined the several taxa described for Plio-Pleistocene times in South American lowlands into one variable species, *Notiomastodon* (=*Stegomastodon*) *platensis* as had been previously suggested by Madden (1984). This taxon is closely related to the upland *Cuvieronius hyodon*. These two species first appear in the fossil record at the connection of the land bridge that joins North and South America at 3.1–2.8 Ma (Prado et al. 2003), but the discovery of *Amahuacatherium* presents the possibility that the other gompotherids

Fig. 13.3 *Cuvieronius hyodon* was a Neogene proboscid that was found at higher altitudes than the other proboscids in South America (By Roman Uchytel)

could have been present much earlier, although several have insisted that *Amahuacatherium* is synonymous with *Notiomastodon* (=*Stegomastodon*). At any rate, we must date the beginning of the Great American Biotic Interchange (GABI) as beginning much earlier than tradition has presented it at between 10.71 and 9.36 Ma (Mothé et al. 2013). Lucas (2013) suggests that *Amahuacatherium* is a misidentification of *Notiomastodon*, and several analyses claim that there are no diagnostic features that distinguish *Amahuacatherium* from a specimen of *Notiomastodon* (Lucas 2013; Alberdi et al. 2002, 2004; Ferretti 2008) and that more importantly, arguing from biostratigraphy and disavowing the previous magnetostratigraphy dating to the Late Miocene, that this gomphotherid and associated fauna is actually Pleistocene in age (Lucas 2013). Analysis of plant microfossils and enamel microwear was made in one Pleistocene gomphothere assemblage found in Minas Gerais, Brazil, and indicated that this species was a feeding generalist and consumed wood elements, leaves, and C3 grasses (Asevedo et al. 2012).

An analysis of a mass death of a large number of gomphotheres between 60,000 and 55,000 years ago in Minas Gerais, Brazil, suggests that around 40 individuals died in an event related to the Middle Pleniglacial interval characterized by a cold and arid environment that perhaps caused a severe climate and scarcity of resources (dos Santos Avilla et al. 2013). Such severe events could have contributed to the later extinction of this and several other species of megafauna by fragmenting population distributions and lowering population sizes (Figs. 13.3 and 13.4).

Fig. 13.4 *Notiomastodon platensis* includes all the lowland Neogene proboscids so that we recognize two genera and two species in the South American Pleistocene (Mothé et al. 2012) and additionally the Miocene *Amahuacatherium peruvium*

13.3.2 Tayassuidae (the Peccaries)

It is generally agreed that the peccaries originated in North America. There were many forms, and the date when they arrived in South America is controversial. On the one hand, earlier citations mention the first fossils having been found in Argentina about 4–3.3 Ma before the present, and they have been seen as one of the earliest arrivals from North America (Woodburne 2010). But Campbell et al. (2000) and Campbell et al. (2010) list two undetermined (now described by Frailey and Campbell (2012) as *Sylvochoerus woodburnei* and *Waldochoerus bassleri*) forms of peccaries discovered in the Late Miocene Madre de Dios Formation outcrops in southeastern Peru which is well-dated at about 9.5 Ma. There are no fossils known between the two dates of 9.5–4 Ma (or are there? Frailey and Campbell 2012). The family is widely known throughout South America from the Middle Pliocene to recent times, or according to some, in many cases, the known fossils may be erroneously dated due to the reigning belief that the Tayassuidae did not enter South America until Early Pliocene times (Frailey and Campbell 2012; Gasparini 2007, 2013; Gasparini and Ubilla 2011). Nevertheless, the two new genera recently described and dated from upper Miocene of about 9.5–8.9 Ma increase the number of South American tayasuids known to six (Frailey and Campbell 2012). Perhaps *Sylvochoerus* and *Waldochoerus* were forest-adapted forms that fossilize poorly and so were not found until now, while the later *Platygonus* was savanna-adapted and so given to leaving fossils (as opposed to tropical forest forms). The makeup of South

Fig. 13.5 *Platygonus leptorhinus*. (By Charles R. Knight)

American forest mammalian fauna throughout the Cenozoic is still very poorly known, basically because the wealth of fossils found in other southern parts of the continent overshadowed what had been found in the tropics.

Platygonus is the best-known extinct form, and this genus was widely distributed both in North and South America from the Pliocene until about 11,000 years ago. It may not have arrived so early to the southern savannas for which it seems best adapted. Although there were several species of *Platygonus*, they tended to be larger than modern tayassuids at about 1 m in height and 120–160 kg with formidable tusks (Mendoza et al. 2006). Using molecular assessments, *Tayassu* and *Catagonus* diverged in the Late Pliocene, probably in North America (Theimer and Keim 1998) and the newly described Late Miocene species (Fig. 13.5).

13.4 Camelidae

The camelids first appeared in North America as *Protylopus* during the Eocene (50–40 Ma), and there were many species in North America, but it was *Hemiauchenia* (Fig. 13.6) that went to South America, either at the traditional date of the Great American Biotic Interchange 3–5 Ma or much earlier in the Miocene, where it gave rise to the modern camelids, the guanaco, llama (both *Lama*), and the alpaca (*Vicugna*) as well as *Palaeolama* (Fig. 13.7) and *Eulamaops*, both now extinct.

Fig. 13.6 *Hemiauchenia macrocephala*, an extinct form, species known from Brazil, genus that first appeared in North America about 10 Ma, seems to be the ancestor of all south American camelids (by Roman Uchytel)

Fig. 13.7 *Palaeolama mirifica*, an extinct form of camel whose Pleistocene distribution is known from Brazil (By Roman Uchytel)

Nevertheless, *Palaeolama* enjoyed a mild success of diversification into around eight species (Guérin and Faure 1999). It was very common and extended to both Americas from California and Florida to Patagonia (Wheeler 1995). According to studies from eastern Peru, the very earliest South American camels appear 9 Ma years ago in the Late Miocene, along with the earliest Tayassuidae and Tapiridae (Campbell et al. 2000; Campbell et al. 2010; Robinson 1936). Information available does not permit a more exact phylogenetic determination for these Early Miocene camels, which appear to have been adapted to moist tropical forest or to the ecotone beween forest and cerrado.

Evidence suggests that neither the guanaco nor the vicuña was found at high-altitude Andean sites until after the last glacial maximum when these camelids adopted the high-altitude puña as part of their habitat (Wheeler 1995). Thus, habitat changes have been a part of this group's history and might explain the absence of fossils between the earliest discoveries in the Late Miocene and those known in the Plio-Pleistocene.

13.5 Procyonidae

The Procyonidae (raccoon family) are usually associated with the first pulse of the Great American Faunal Interchange, the invasion of the northern fauna, even though this interesting group made it to South America about 7 Ma (end of the Miocene) ago, three million years before the main invasion. How they did it is only a guess, but one figures they were successful island hoppers along the developing Central American isthmus, since by then the short bridge (if it existed) that allowed *Amahuacatherium* to enter the southern continent had been inundated. Perhaps the arrival of the procyonids was two million years before the date usually accepted and thus could have entered South America at the same time as the other "first" northern mammals.

Cyonasua argentina is the first fossil procyonid to be found in South America (Argentina) and dated to about 7.3 Ma. Six other species have been described (Berman 1989). Although the evolution of this family is not clear, *Cyonasua* probably evolved *Chapalmalania*. *Cyonasua* was a small dog-like animal with strong canines. It was probably more of a hunter than modern procyonids.

Chapalmalania looked like a huge raccoon at about 2 m long and 1 m high. Its discoverers at first took it to be a bear. Two species have been described (Soibelzon and Prevosti 2007). It probably had a varied diet and also scavenged (de los Reyes et al. 2013). The genus disappeared about 3.5 Ma BP (Fig. 13.7).

A recent phylogenetic analysis based on morphological characteristics of the group suggested that the main diversification of the family occurred in Central America and that the above clade (i.e., *Cyonasua, Chapalmalania* considered to belong to the Potosini Tribe), associated with the North American *Arctonasua*, can be distinguished from the *Bassariscus* and *Probassariscus* clades (considered as the Procyonini Tribe) which probably evolved the modern *Nasua* (*Bassariscus*) and

Procyon (*Probassariscus*). According to these morphology-based studies, two genera had their origin in North America in the Middle Miocene (Baskin 2004).

However, a molecular-based study did not support the above phylogeny and concluded instead that there are three distinct lineages in the family, a *Nasua-Bassaricyon* clade, a *Bassariscus-Procyon* clade, and a *Potos* clade, and that there may have been two (or several) invasions of South America (Koepfli et al. 2007). The lineage of *Cyonasua*, closely related to the North American *Arctonasua*, entered South America early on and resulted in the evolution of *Chapalmalania* and other procyonids that have become extinct. However divergence times for *Nasua* and *Procyon* suggest that they entered South America about the same time as *Cyonasua*, *Nasua narica*, and *N. nasua* diverged around 7–8 Ma and *Procyon cancrivorus* and *P. lotor* diverged around 5–5.7 Ma years in the Early Pliocene (Koepfli et al. 2007).

13.6 The Leaders of the Flood of New Mammals in the Second Pulse

As the two continents came together, the first of the multitude of mammalian invaders were the skunks *Conepatus* (Mephitidae), the early peccaries *Platygonus* (Tayassuidae), and the early horse *Hippidion* (Equidae), first dated at 2.8–2.5 Ma.

The first *Conepatus* may be associated with Late Pliocene deposits del Ensenada of Argentina and Bolivia (1.2–0.8 Ma) (Reig 1952, cited in Wang and Carranza-Castañeda 2008). After this age, the genus has been commonly found throughout South America (González G et al. 2010; Soibelzon and Prevosti 2007).

Platygonus is known in both North and South America, although the oldest fossils of peccaries (not *Platygonus*) are of uncertain age and have been placed in the Late Miocene (Campbell et al. 2010) before the connection of the two continents. The best old records are from the Middle Pliocene of Argentina (Prevosti et al. 2006; Gasparini 2007), and the genus is widespread in the Late Pliocene. There are various species registered (Gasparini and Ubilla 2011).

13.6.1 Equidae (Horses)

The oldest fossil *Hippidion devillei* was found in Argentina and dated close to the end of the Neogene at 2.588 Ma (Prado et al. 1998). A review of *Hippidion* recognized three neotropical species of this endemic South American genus, *H. principale*, *H. devillei*, and *H. saldiasi* (Alberdi 1993). DNA has clarified that *Hippidion* and *Equus* are separate lineages, in contrast to a previous molecular study identifying *Hippidion* and *Equus* as being closely related (Ludovic et al. 2009; Weinstock et al. 2005) (Fig. 13.10). *H. devillei* was apparently a high-altitude species, while the other two were lowland species (Ludovic et al. 2009). *Equus caballus*, the ancestor

Fig. 13.8 *Chapalmalania* appeared at around 5 Ma, considerably before the main invasion of mammals from the north (By Roman Uchytel)

of the domestic horse did not enter South America until around 0.99 Ma during the Middle Pleistocene Ensenadan SALMA, according to new biostratigraphic evidence (MacFadden 2013) (Fig. 13.8).

13.6.2 Canidae (Canines, Dogs)

The first Canidae appeared in the mid-Eocene of North America and evolved there until Late Miocene times when they began to disperse to all other continents except for Antarctica (Munthe 1998; Perini et al. 2009, 2010). For South America there are 28 species and 10 genera of living and fossil canids known, the most diversity of any continent (Berta 1987), and this diversity arose from 3 to 4 invasions from the north (Perini et al. 2009; Prevosti 2010). The South American canids have been amply shown to be monophyletic (Prevosti 2010).

The first canine to enter South America seems to have been *Dusicyon cultridens*, based on the register in the Late Pliocene (Vorohuean-Sanandresian) (2.5–3 Ma) (Berman 1994; Soibelzon and Prevosti 2007; Soibelzon et al. 2007). This was probably the ancestor of the extinct Falkland Islands fox *Dusicyon australis*, and the genus is phylogenetically related to the extinct *Protocyon* and *Theriodictis* (Austin et al. 2013). These all were related in a subtribe *Cerdocyonina* that gave rise to the South American canids but most diversified in Central America and North America. *Canis*, *Cerdocyon*, *Theriodictis*, and *Chrysocyon* entered South America when the connection was established (Berta 1987; Soibelzon and Prevosti 2007; Wang and

Tedford 2008). These canines explosively diversified, given the great huge offering of prey available. They may have been instrumental in the extinction of the South American borhyaenids. During Lujanian (0.8–0.011 Ma) times, maximum diversity of South American canines was attained, including the wolf-like canids *Theriodictis*, *Protocyon*, and *Canis* are known, although *Canis*, *Theriodictis*, and *Chrysocyon* had appeared earlier during the Ensenadan SALMA (1.2–0.8 Ma) (Berta 1987; Prevosti et al. 2009a, b).

Theriodictis was a genus of large, hypercarnivorous canines with two species known from Bolivia and Argentina. They are known from 1.8 to 0.5 Ma and weighed about 37 kg. Their potential prey was camelids, cervids, equids, and giant rodents (Prevosti and Palmqvist 2001). *Protocyon* lived in open habitats in the southern parts of South America during the Pleistocene, the oldest record being 0.78 Ma BP. Four species are known (Prevosti et al. 2005). *Canis dirus* fossils (only about three fossils are known for South America (Fig. 13.11), so they might not have been very common) were found in lowland pampas and alpine grassland environments of the temperate south, exactly where llamas were also abundant and so might have been a specialist on llamas (Wang and Tedford 2008). *Canis nehringi* and *C. gezi* are also known for the Pleistocene but were smaller than *C. dirus*. *Canis* is only known from the latest Pleistocene (<25,000–27,000 years BP) (Prevosti et al. 2009) (Fig. 13.9).

About two million years ago, a great many other mammals from North America appeared in South America.

Fig. 13.9 Studies of the rostal morphology of hippidiform hourses suggests a very developed, prehensile upper lift that ecologically separates these horses (*Onohippidion and Hippidion*) from *Equus*. The upper lip would have facilitated browsing from trees (from Bernardes C, Sicuro FL, Avilla LS, Pinheiro AEP 2013. Rostral reconstruction of South American hippidiform equids: New anatomical and ecomorphological inferences)

Fig. 13.10 *Hippidion* is the first horse genus to appear in South America about 2.5 Ma and became extinct around 8,500 years ago (by Roman Uchytel)

Fig. 13.11 *Canis dirus* is generally seen as a specialist on large prey and must have effectively hunted large ungulates of South America (By Roman Uchytel)

13.6.3 *Felidae (Felines, Cats)*

The cat tribe began evolving in Asia. The modern South American cats have come from three molecular clades of eight that have been identified: the small spotted cats, the jaguar, and the puma and jaguarundi (O'Brien et al. 2008; Prevosti 2006). These clades are between 10 and 6.4 Ma old for the large panther-like cats, 2.9 Ma old for the small spotted cats, and 4.9 Ma old for the puma group. Within each group these cats are most related (O'Brien and Johnson 2005, 2007).

The ocelot group diverged in North America from the lynx group, and between 3 and 2 Ma, they entered South America and diversified into the seven species of small cats that we know today.

The jaguar evolved in Asia from an unidentified clouded leopard-like (*Neofelis nebulosa*) ancestor of other large panther-like cats. It dispersed across the Bering Strait during the Late Villafranchian (3–2 Ma) at the end of the Pliocene or beginning of the Pleistocene). As a subspecies *Panthera onca augusta* was 15–20% larger and longer-legged than the jaguar of today. Although the modern jaguar is quite variable in size, an "average" size seems quite misleading, because of the occasional very large males that have been weighed. They continued expanding into South America (Fig. 13.12).

The American lion (*Panthera leo atrox*) might have been present in northern South America in the Pleistocene as well, although the Peruvian fossil specimens

Fig. 13.12 The last of the big cats to arrive in South America (1.8 Ma–11,000 years ago) was the jaguar *Panthera onca mesembrina*, which, on arrival in North America from Asia, weighed 50% more than the current races. The subspecies *P. onca mesembrina* was a Pleistocene subspecies native to both North and South America. Two specimens have been estimated at 46.3 and 129.1 kg (By Roman Uchytel)

Fig. 13.13 An American lion confronts a saber-toothed cat over prey with *Megatherium* in the background. Either of these cats could be backed-up by members of its own pride (By Roman Uchytel)

known, according to some, could have been exceptionally large jaguars. However, the species was very widely present, not only in Africa but also in Europe, Asia, and North America, and it seems probable that it was able to arrive in South America as well (Fig. 13.13).

About the same time that the small spotted cats entered South America, the large saber-toothed cats *Smilodon populator* (endemic to South America) and *Smilodon fatalis* arrived from North America. These cats (Felidae) were a sister group (different subfamily, Machairodontinae, tribe Smilodontini) to all modern cats, and they had their own radiation in Asia. The South American species include the largest known saber-toothed (*S. populator*) and probably evolved from *S. fatalis* to prey on the many large ungulates, rodents, and perhaps sloths available in South America during most of the Pleistocene. It weighed in the region of 200–360 kg (and probably could exceed 400 kg) and probably was a social animal that hunted in family groups, just as is suspected for the species of western North America, *Smilodon fatalis,* which probably weighed in the vicinity of 160–280 kg (size of a Siberian tiger) (Christiansen and Harris 2005) (Fig. 13.14).

Recently three additional species of saber-toothed cat have been tentatively identified from Uruguay and Venezuela (Mones and Rinderknecht 2004; Rincón et al. 2011). The scimitar-toothed cat *Xenosmilus hodsonae* was first described from Pleistocene Irvingtonian age (1.8–0.3 Ma) deposits in Florida (Martin et al. 2000), and a lower jaw agreeing in all respects with the fossils found in Florida was discovered in Pleistocene (Ensenaden age) deposits in Uruguay. This is the first record of

Fig. 13.14 *Smilodon populator* is perhaps the largest feline known, sometimes exceeding 400 kg and was an important Pleistocene predator in South America (By Roman Uchytel)

the tribe Homotheriini for South America (Mones and Rinderknecht 2004). These cats were large and muscular, second in size only to *S. populator*. Their bite was apparently more generalized and less precise with smaller canines than *Smilodon*; skulls had both an elongated sagittal crest and enlarged temporalis muscle, arguing for a very strong bite (*Smilodon* did not bite in this way, it probably severed an artery so that the prey bled out) (Martin et al. 2000). This cat had a relatively small head and short, muscular legs, and overall it was more massive than a *Smilodon* of the same size. The fossil cat was found in a cave deposit with many skeletons of *Platygonus*, suggesting a preferred prey (Fig. 13.15).

Another recent report describes the discovery of a new species of Machairodontinae *Homotherium venezuelensis* and the first record of *Smilodon gracilis* for South America, both from Venezuela (Rincón et al. 2011).The *Homotherium* skull was found in an ancient tar seep deposit from Early to Middle Pleistocene age (1.0–0.5 Ma). The fossil possesses hypertrophied canines with course serrations as well as serrations on incisors with the ramus of the mandible being much more massive than in *Smilodon*. The scarcity of *Homotherium* and *Xenosmilus* fossils in South America probably correlates with very low densities of these cats (Fig. 13.16).

Until the recent Venezuelan find, *Smilodon gracilis* was only known from North America. This was the smallest species of the three known and probably weighed around 50–100 kg, comparable in size to leopards, pumas, and jaguars (Christiansen and Harris 2005) (Fig. 13.17).

The origins of the puma or couger (*Panthera puma*) and the jaguarundi (*Puma jaguarondi*) are in North America. The common ancestor of the puma lineages

Fig. 13.15 *Xenosmilus hodsonae (By Roman Uchytel)*

Fig. 13.16 *Homotherium venezuelensis (By Roman Uchytel)*

migrated across the Bering Land Bridge into the Americas 8.0–8.5 million years ago (Ma). The puma and the jaguarondi then evolved in North America and invaded South America at the closing of the gap between the two Americas. Molecular studies have suggested that the cougar and jaguarundi are most closely related to the modern cheetah of Africa and Western Asia (Johnson et al. 2006).

Fig. 13.17 *Smilodon gracilis* is well-known as part of North American fauna but only recently has been discovered in South America (By Roman Uchytel)

13.6.4 Ursidae (Bears)

South American bears belong to the group Tremarctini which had its origin in North America with the appearance of *Plionarctus* in the Late Miocene. *Plionarctus* was ancestral to the bears *Arctodus*, *Arctotherium*, and the modern spectacled bear *Tremarctos ornatus*. *Tremarctos floridanus Artodus*, known from North America, gave rise to several species in North America, but the genus *Arctotherium* evolved in South America. The earliest bear remains for South America are *Arctotherium angustidens* in early Middle Pleistocene of 1.2 Ma, and one individual of this species might have weighed in at 1588–2042 kg, depending on the means of calculation (Soibelzon and Schubert 2011). This bear is not only the oldest bear species found in South America and might have been the largest bear ever found and is certainly one of the largest carnivorous land mammals known. The species is a conspicuous element of Ensenadan fauna (Early–Middle Pleistocene, 1.76–0.78 Ma) (Soibelzon et al. 2005). Ecologically such a large bear would need a reliable source of animal protein to feed its bulk. During the Middle Pleistocene, the South American savannas had one of the highest mammalian diversities of megafauna that has even been known, so the great variety of ungulates, glyptodonts, and ground sloths were typical for the time.

During the Bonaerian SALMA, diversity increased to three species of *Arctotherium*, and during the Lujanian SALMA, three more species have been recorded, all of which were smaller than *A. angustidens* and probably less carnivorous, as more and more carnivores arrived from the north (Soibelzon et al. 2005). The youngest *Arctotherium* fossils, corresponding to *A. tarijense*, have been dated

Fig. 13.18 *Arctotherium angustidens* may be the largest bear ever found at somewhere between 1000 and 2000 kg. In contrast a large polar bear is around 800 kg and sometimes reach 1000 kg

to 11,210 and 11,600 years BP from southern Argentina (Prevosti et al. 2003; Prevosti et al. 2013; Soibelzon 2002 cited in Soibelzon et al. 2005). The last living bear of the Tremarctini, *Tremarctos ornatus*, is not known from fossils, although a related species, *Tremarctos floridanus*, is known from Florida, North America (Soibelzon et al. 2005). The only *Arctotherium* known in northern South America are two specimens of *A. wingei*, found in Venezuela (Fig. 13.18).

13.6.5 Mustelidae (Weasels, Otters)

There were probably five invasions of this family into South America comprising *Lutra*, *Pteronura*, *Galictis*, *Lyncodon* (an extinct sister group of *Galictis*), *Mustela*, and *Eira* (Eizirik 2012; Harding and Smith 2009) (Fig. 15.33).

The oldest appearance of a mustelid fossil dates to Late Miocene around the time of the first procyonids at around 5.7–5.8 Ma and is found in central Argentina, accompanied by the first rodents in the family Cricetidae as well (Verzi and Montalvo 2008). This was a mandible fragment of an indeterminate mustelid, although the fragment showed some similarity to living Lutrinae (Verzi and Montalvo 2008). Molecular data suggest that *Eira* and *Pteronura* probably arrived around 7 Ma (Eizirik 2012) (Verzi and Montalvo 2008 Soibelzon and Prevosti 2007). *Pteronura*, which is considered the most basal lineage of the Lutrinae, seems to have diverged from the other otters around 7.4 Ma BP, implying that either this evolution occurred outside of South America or that the genus entered South America around the same time as did the procyonid *Cyanosura* (Koepfli et al. 2008; Eizirik 2012; Eizirik et al. 2010).

The tribe Galictini includes a number of fossil ancestors, two genera of which appeared in South America: *Galictis* and *Lyncodon*. The South American *Galictis* evolved after the invasion of a North American *Galictis* around 2.5 Ma. The first *Lyncodon* (*L. bosei*) is known only from the Middle-lower Pleistocene and probably evolved from the sister genus *Galictis* (Pascual 1958; Eizirik 2012).

Lutra in South America evolved from the North American genus and didn't arrive in South America until <2.5 Ma (Koepfli et al. 2008; Eizirik 2012).

Mustela seems to have evolved from a common ancestor of the North American mink (*Neovison*) around 3.2 Ma and probably entered South America around 2.4 Ma BP in a single colonization when they diversified into the known South American species (Koepfli et al. 2008; Eizirik 2012). *Eira*, the basal-most living member of the martens, had its last common ancestor with the martens around 6.8 Ma and may have invaded South America around the time of *Pteronura* (Koepfli et al. 2008; Eizirik 2012).

13.6.6 Mephitidae (the Skunks)

North American skunks (*Conepatus*, *Mephitis*, *and Spilogale*) share a common ancestor. *Conepatus* appears widely in South America at least 2 Ma BP during the Lujanian age and presently reflects four extinct species as compared to three extant living species (Hunt 1996; Wang and Carranza-Castañeda 2008; González G et al. 2010). *Conepatus altiramus* is the oldest species to be found in South America and is associated with the Chapadmalalan SALMA at about 4.5 Ma years (Wang and Carranza-Castañeda 2008) and is among the first fauna of the second pulse (Fig. 15.34).

13.6.7 Tapiridae (the Tapirs)

The tapirs are an ancient lineage of Perissodactyla (odd-toed mammals) that first appear in the North American Eocene where they spread throughout Asia and Europe. The genus *Tapirus* is also ancient, dating to the Miocene and also found in both the New and Old World (Hulbert 2005; Wilson and Mittermeier 2011). It would appear that tapirs arrived in South America by about 9.5 Ma, along with gomphotherids, tayassuids, and camelids (Campbell et al. 2000). There are three living species recognized for the neotropics (*Tapirus pinchaque*, *T. terrestris*, *T. bairdii*, and *T. kabomani* [Cozzuol et al. 2013]), and their evolution is not at all clear, although phylogenetic analysis (based on only three of the known species) suggests three distinct mitochondrial lineages based on one South American invasion and *T. terrestris* is most closely related to *T. pinchaque* (Ashley et al. 1996). A much more extensive molecular analysis (201 individuals) based on the known three (not *T. kabomani which has been discounted as a species*) species of *Tapirus* suggest that

T. bairdii split from *T. indicus* 17 Ma, *T. bairdii* then split from *T. pinchaque* and *T. terrestris* 10.9 Ma, and the last two species split from each other only 3.9 Ma (Ruiz-García et al. 2012). These neotropical *Tapirus* present recognizable molecular lineages as well, and in this study diversification of haplotypes began only about 2.5 Ma BP (Ruiz-García et al. 2016), suggesting that the earlier Late Miocene tapirs could have been a previous invasion that became extinct (Fig. 15.36).

Another analysis using morphometric characteristics of fossils suggests that *Tapirus webbi*, known from the Late Miocene of Florida, dispersed to South America where it evolved a variety of species (seven described so far, besides the four modern neotropical species recognized). Much work remains in order to understand the evolution of these animals (Holanda and Ferrero 2013).

13.6.8 Cervidae (the Deer)

Of the more or less 40 deer species in the world, 6 genera and 16 species have colonized the neotropics (Grubb 2005). These animals all belong to one tribe, the Odocoileini of the Capreolinae subfamily, and their ancestors did not arrive in North America until late in the Miocene at around 5 Ma BP (Gilbert et al. 2006). Many fossils (at least four genera) suddenly appear in South America at the Plio-Pleistocene boundary that have traditionally been used to explain the common ancestors of the South American Cervidae. However, due to the wide range of morphological and ecological adaptations found and molecular evidence, an early Pliocene invasion and a later invasion at 3–3.5 Ma has been suggested. Molecular studies show that the genus *Mazama* is polyphyletic, since both *Mazama gouazoubira* and *M. nemorivaga* (the gray brocket deer) group molecularly with (marsh deer) *Blastocerus* (huemul and Taruka), *Hippocamelus*, and *Pudu*, while the red brocket deer *Mazama americana*, *M. bororo*, *M. nana*, and *M. temama* group with *Odocoileus* (Gilbert et al. 2006). There is evidence as well that *M. americana* is actually a complex of cryptic species with varied chromosome numbers, and great chromosomal differences are typical of this group (Barbanti Duarte et al. 2008) (Fig. 15.39).

13.6.9 Sigmodontine Rodents (Cricetidae)

According to some, the first northern (sigmodontine cricetid) rodents arrived in South America by 6.7 Ma (between 7 and 5 Ma) (Marshall et al. 1979; Goin et al. 2012), although these sources do not cite evidence. A later fauna has documented the so far earliest Cricetidae at the Late Miocene age of about 6 Ma, based on the evolutionary stage of some accompanying caviomorph rodents as compared to a more clearly established fauna of similar rodents (Verzi and Montalvo 2008). This date has been criticized for the lack of isotopic or paleomagnetic evidence that could more solidly define the age of this material (Prevosti and Pardiñas 2009). Paleontological evidence suggests that the sigmodontine rodents were not present in South America before 3.5 Ma (Reig 1978; Baskin 1986), and these fossils could be 0.8 million years younger, due to problems of dating. Finally, Reig (1972) suggested that the sigmodontine rodents could have arrived as early as the Early Miocene, based on the diversification of the group and that the absence in the fossil record could be due to sampling error. From these conflicting views, it is not at all clear when the sigmodontine rodents finally arrived in South America (Webb 1999) (Fig. 15.43).

Historically, the first arrivals or rather the earliest fossils of many mammalian groups have been discovered in Central Argentina, mostly because that part of the continent has offered the great majority of Late Miocene, Pliocene, and Pleistocene deposits so that nearly 80% of "first" appearances of families have been found there (Verzi and Montalvo 2008). It should not be a great surprise that concentrated work at Peruvian Amazonian sites around Santa Rita have begun revealing a much earlier fauna than further south and that the Peruvian sites have been diverse and educational (Campbell 2004; Campbell et al. 2000, 2001, 2009, 2010; Frailey and Campbell 2004). Dispersal of any mammalian group would have always had to contend with very different ecozones, and certainly for many recent immigrants, adaptations to tropical zones might partially explain why they were not detected in Argentina until much later, especially since the continent was in the grip of a cooling off. It seems probable that further work in the Amazon, where possible, will offer more surprises.

The populating of South America has proved to be much less concentrated in Late Pliocene times; after all, the emergence of the Isthmus of Panamá developed through stages of ever-accumulating dry land, but before a complete land bridge was available, islands and increasingly shallow water barriers must have made dispersal north and south more and more possible to mammals, beginning with large sloths (to the north) and gomphotheres, smaller tayassuids, tapirs (to the south), and apparently camels, all of which in part had to resort to swimming the last oceanic barriers around 9.5–9 million years ago to be followed by procyonids at around 7.3 Ma. Geological evidence doesn't leave much doubt that dry land was available and that it led south.

13.7 Southern Mammals that Went North

The xenarthran sloths were by far the most successful invasion into the north. They are first recorded from about 9 to 9.5 Ma ago, and they managed to widely populate North America up to Alaska (Marshall et al. 1979). They apparently immigrated north at about the same time as the first gompotherids, tayassuids, and tapirids moved south, probably by partially swimming the gaps that still existed between the emerging land.

Fig. 13.19 *Glyptotherium asper.* (By Arent derived from image by Wolfman, http://de.wikipedia.org/wiki/Bild:Glyptodon-1.jpg from the Naturhistorisches Museum Wien)

Other xenarthrans to go north were the giant anteater (*Myrmecophaga tridactyla*) which has been found as far north as northwestern Sonora, Mexico (Shaw and McDonald 1987). pampatheres that made it to Florida and the glyptodonts (*Glyptotherium*) of Texas and Florida. This xenarthra still ranges to eastern Guatemala and southern Belize. glyptodonts (Fig. 13.19).

References

Alberdi MT (1993) Review of the genus *Hippidion* Owen, 1869 (Mammalia: Perissodactyla) from the Pleistocene of South America. Zool J Linnean Soc 108:1–22

Alberdi MT, Prado JL, Cartell C (2002) El registro de *Stegomastodon* (Mammalia, Gomphotheriidae) en el Pleistoceno superior de Brasil. Rev Esp Paleontol 17:217–235

Alberdi MT, Prado JL, Salas R (2004) The Pleistocene Gomphotheriidae (Proboscidea) from Peru. Neues Jahrb Geol P-M 231:423–452

Asevedo L, Winck GR, Mothé D, Avilla LS (2012) Ancient diet of the Pleistocene gomphothere *Notiomastodon platensis* (Mammalia, Proboscidea, Gomphotheriidae) from lowland mid-latitudes of South America: Stereomicrowear and tooth calculus analyses combined. Quat Int 255:42–52

Ashley MV, Norman JE, Stross L (1996) Phylogenetic analysis of the perissodactylan family Tapiridae using mitochondrial cytochrome c oxidase (COII) sequences. J Mamm Evol 3(4):315–326

Austin JJ, Soubrier J, Prevosti FJ, Prates L, Trejo V, Mena F, Cooper A (2013) The origins of the enigmatic Falkland Islands wolf. Nat Commun. https://doi.org/10.1038/ncomms2570

Barbanti Duarte JM, González S, Maldonado JE (2008) The surprising evolutionary history of South American deer. Mol Phylogenet Evol 49:17–22

Baskin JA (1986) The late Miocene radiation of Neotropical sigmodontine rodents in North America. Contrib Geol, Special Paper 3:287–303

Baskin JA (2004) *Bassariscus* and *Probassariscus* (Mammalia, Carnivora, Procyonidae) from the early Barstovian (Middle Miocene). J Vertebr Paleontol 24(3):709–720

Berman WD (1989) Notas sobre la sistemática y paleobiogeografía del grupo *Cyonasua* (Carnivora, Procyonidae) de la Argentina: Actas de las VI Jornadas Argentinas de Paleontología Vertebrados 17–18

Berman, Walter Daniel (1994) Los carnívoros continentales (Mammalia: Carnivora) del Cenozoico en la provincia de Buenos Aires. PhD Thesis, Universidad Nacional de La Plata, Buenos Aires

Berta A (1987) Origin, diversification, and zoogeography of the South American Canidae. In: Patterson BD, Timm RM (eds) Studies in neotropical mammalogy: essays in honor of Philip Hershkovitz, Fieldiana Zoology new series, vol 6. Field Museum of Natural History, Chicago, pp 455–472

Burnham RJ, Graham A (1999) The history of neotropical vegetation: new developments and status. Ann Mo Bot Gard 86:546–589

Campbell KE Jr (2004) The Paleogene mammalian fauna of Santa Rosa, Amazonian Peru, Science series, vol 40. Natural History Museum of Los Angeles County, Los Angeles, p 163

Campbell KE, Frailey DC, Romero Pittman L (2000) The Late Miocene Gomphothere *Amahuacatherium peruvium* (Proboscidea: Gomphotheriidae) from Amazonian Peru: Implications for the great american faunal interchange. INGEMMET. Boletín, Serie D: Estudios Regionales; n° 23, 152 p

Campbell KE Jr, Heizler M, Frailey CD, Romero-Pittman L, Prothero DR (2001) Upper Cenozoic chronostratigraphy of the southwestern Amazon basin. Geology 29(7):595–598

Campbell KE Jr, Frailey DC, Romero-Pittman L (2009) In defense of *Amahuacatherium* (Proboscidea: Gomphotheriidae). Neues Jahrb Geol P-A 252:113–128

Campbell KE Jr, Prothero DR, Romero-Pittman L, Hertel F, Rivera N (2010) Amazonian magnetostratigraphy: dating the first pulse of the great American faunal interchange. J S Am Earth Sci 26:619–626

Christiansen P, Harris JM (2005) Body size of *Smilodon* (Mammalia: Felidae). J Morphol 266:369–384

Coates AG, Obando JA (1996) The geologic evolution of the Central American Isthmus. In: Jackson JBC, Budd AF, Coates AG (eds) Evolution and environment in tropical America. The University of Chicago Press, Chicago, pp 21–56

Coates AG, Collins LS, Aubry M-P, Berggren WA (2004) The geology of the Darien, Panama, and the late Miocene-Pliocene collision of the Panama arc with northwestern South America. GSA Bull 116(11/12):1327–1344

Cozzuol MA, Clozato CL, Holanda EC, Rodrigues FVHG, Nienow S, De Thoisy B, Redondo RAF, Santos FCR (2013) A new species of tapir from the Amazon. J Mammal 94(6):1331. https://doi.org/10.1644/12-MAMM-A-169

dos Santos Avilla L, Figueiredo AMG, Kinoshita A, Bertoni-Machado C, Mothé D, Asevedo L, Baffa O, Dominato VH (2013) Extinction of a gomphothere population from Southeastern Brazil: Taphonomic, paleoecological and chronological remarks. Quat Int 305:85–90

De los Reyes M, Poiré D, Soibelzon L, Zurita AE, Arrouy MJ (2013) First evidence of scavenging of a glyptodont (Mammalia, Glyptodontidae) from the Pliocene of the Pampean región (Argentina): taphonomic and paleoecological remarks. Palaeontol Electron 16(2):15A, p 13

Eizirik E (2012) A molecular view on the evolutionary history and biogeography of neotropical carnivores (Mammalia, Carnivora). In: Patterson BD, Costa LP (eds) Bones, clones, and biomes. University of Chicago Press, Chicago, pp 123–142

Eizirik E, Murphy WJ, Koepfli K-P, Johnson WE, Dragoo JW, Wayne RK, O'Brien SJ (2010) Pattern and timing of diversification of the Mammalian order Carnivora inferred from multiple nuclear gene sequences. Mol Phylogenet Evol 56:49–63

Ferretti MP (2008) A review of South American proboscideans. N M Mus Nat Hist Sci Bull 44:381–391

Frailey CD, Campbell KE Jr (2012) Two new genera of peccaries (Mammalia, Artiodactyla, Tayassuidae) from upper Miocene deposits of the Amazon basin. J Paleontol 86(5):852–877

Frailey CD, Campbell KE Jr (2004) Paleogene rodents from Amazonian Peru: the Santa Rosa local fauna. Sci Ser 40:71–130

Gasparini GM (2007) Sistemática, biogeografía, ecología y bioestratigrafía de los Tayassuidae (Mammalia, Artiodactyla) fósiles y actuales de América del Sur, con especial énfasis en las especies fósiles de la provincia de Buenos Aires. PhD dissertation, Universidad Nacional de La Plata, Argentina

Gasparini GM (2013) Records and stratigraphical ranges of South American Tayassuidae (Mammalia, Artiodactyla). J Mamm Evol 20:57–68

Gasparini GM, Ubilla M (2011) *Platygonus* sp. (Mammalia: Tayassuidae) in Uruguay (Raigón? Formation; Pliocene-early Pleistocene), comments about its distribution and palaeoenvironmental significance in South America. J Nat Hist 45(45–46):2855–2870

Gilbert C, Rpóqiet A, Hassainin A (2006) Mitochondrial and nuclear phylogenies of Cervidae (Mammalia, Ruminantia): systematics, morphology, and biogeography. Mol Phylogenet Evol 40:101–117

Goin FJ, Gelfo JV, Chornogubsky L, Woodburne MO, Martin T (2012) Origins, radiations, and distribution of South American Mammals. In: Patterson BD, Costa LP (eds) Bones, clones, and biomes. The University of Chicago Press, Chicago, pp 20–50

González G E, Prevostt FJ, Pino Q M (2010) Primer registro de Mephitidae (Carnivora: Mammalia) para el Pleistoceno de Chile. Magallania (Chile) 38(2):238–248

Graham A (1998) Studies in Neotropical paleobotany. XI. Late Tertiary vegetation and environments of southeastern Guatemala: palynofloras from the Mio-Pliocene Padre Miguel Group and the Pliocene Herrería Formation. Am J Bot 85(10):1409–1425

Grubb P (2005) Order Artiodactyla. In: Wilson DE, Reeder DAM (eds) Mammal Species of the World. The Johns Hopkins University Press, Baltimore, pp 637–722

Guérin C, Faure M (1999) *Palaeolama (Hemiauchenia) niedae* nov. sp., nouveau Camelidae du Nordeste brésilien et sa place parmi les Lamini d'Amérique de Sud. Geobios 32(4):629–659

Harding LE, Smith FA (2009) *Mustela* or *Vison*? Evidence for the taxonomic status of the American mink and a distinct biogeographic radiation of American weasels. Mol Phylogenet Evol 52:632–642

Holanda EC, Ferrero BS (2013) Reappraisal of the genus *Tapirus* (Perissodactyla, Tapiridae): systematics and phylogenetic affinities of the South American tapirs. J Mamm Evol 20:33–44

Hulbert RC Jr (2005) Late Miocene *Tapirus* (Mammalia, Perissodactyla) from Florida with a review of Blancan tapirs from the state. Bull Florid Mus Nat Hist 49(3):67–126

Hunt RM (1996) Biogeography of the order Carnivora. In: Gittleman JL (ed) Carnivore behavior, ecology, and evolution. Cornell University Press, New York City, pp 485–541

Johnson DL (1980) Problems in the land vertebrate zoogeography of certain islands and the swimming powers of elephants. J Biogeogr 7:383–398

Johnson WE, Eizirik E, Pecon-Slattery J, Antunes A, Teeling E, Murphy WJ, O'Brien SJ (2006) The late Miocene radiation of modern Felidae: A genetic assessment. Science 311:73–77

Kameo K, Sato T (2000) Biogeography of Neogene calcareous nannofossils in the Caribbean and the Eastern equatorial Pacific – floral response to the emergence of the Isthmus of Panama. Mar Micropaleontol 39:201–218

Koepfli K-P, Gompper ME, Eizirik E, Ho C-D, Linden L, Maldonado JE, Wayne RK (2007) Phylogeny of the Procyonidae (Mammalia: Carnivora): molecules, morphology and the Great American Interchange. Mol Phylogenet Evol 43:1076–1094

Koepfli K-P, Deere KA, Slater GJ, Begg C, Begg K, Grassman L, Lucherini M, Veron G, Wayne RK (2008) Multigene phylogeny of the Mustelidae: Resolving relationships, tempo and biogeographic history of a mammalian adaptive radiations. BMC Biol 6(10):10. https://doi.org/10.1186/1741-7007-6-10

Lucas SG (2013) The paleobiogeography of South American gomphotheres. J Palaeogeogr 2(1):19–40

Ludovic O, Metcalf JL, Alberdi MT, Telles-Antunes M, Bonjean D, Otte M, Martin F, Eisenmann V, Mashkour M, Morello F, Prado JL, Salas-Gismondi R, Shockey BJ, Wrinn PF, Vasil'ev

SK, Ovodov ND, Ml C, Hopwood B, Male D, Austin JJ, Hänni C, Cooper A (2009) Revising the recent evolutionary history of equids using ancient DNA. Proc Natl Acad Sci U S A 106(51):21754–21759

MacFadden BJ (2013) Dispersal of Pleistocene *Equus* (Family Equidae) into South America and Calibration of GABI 3 based on evidence from Tarija, Bolivia. PLoS One 8(3):e59277. https://doi.org/10.1371/journal.pone.0059277

Madden CT (1984) New and primitive species of progressive gomphothere proboscidean *Stegomastodon* from Pliocene of New Mexico. Geological Society of America Abstracts with Programs 16:274

Marshall LG, Butler RF, Drake RE, Curtis GH, Tedford RH (1979) Calibration of the great American interchange. Science 204:272–279. https://doi.org/10.1126/science.204.4390.272

Martin LD, Babiarz JP, Naples VL, Hearst J (2000) Three ways to be a saber-toothed cat. Naturwissenschaften 87:41–44

Mendoza M, Janis CM, Palmqvist P (2006) Estimating the body mass of extinct ungulates: a study on the use of multiple regression. J Zool 270:90–101

Mones A, Rinderknecht A (2004) The first South American Homotherinii (Mammalia: Carnivora: Felidae). Com Paleont Mus Nac Hist Nat Antrop 2(35):201–212

Mothé D, Avilla LS, Cozzuol M, Winck GR (2012) Taxonomic revision of the Quaternary gomphotheres (Mammalia: Proboscidea: Gomphotheriidae) from the South American lowlands. Quat Int 276–277:2–7

Mothé D, Avilla LS, Cozzuol MA (2013) The South American Gomphotheres (Mammalia, Proboscidea, Gomphotheriidae): taxonomy, phylogeny, and biogeography. J Mamm Evol 20:23–32

Munthe K (1998) Canidae. In: Janis CM, Scott KM, Jacobs L (eds) Evolution of tertiary mammals of North America. Cambridge University Press, Cambridge, pp 124–143

O'Brien SJ, Johnson WE (2005) Big cat genomics. Annu Rev Genomics Hum Genet 6:407–429

O'Brien SJ, Johnson WE (2007) The evolution of cats. Sci Am 297(1):68–75

O'Brien SJ, Johnson W, Driscol C, Pontius J, Pecon-Slattery J, Menotti-Raymond M (2008) State of cat genomics. Trends Genet 24(6):268–279

Pascual R (1958) "*Lyncodon bosei*" nueva especie del Ensenadense. Un antecessor del Huroncito Patagónico, Serie Paleontología. Rev Mus La Plata 4:1–34

Perini FA, Guedes PG, Moraes Neto CR et al (2009) Carnivores (Mammalia, Carnivora) from the quaternary of Serra da Bodoquena, Mato Grosso do Sul, Brazil. Arquivos do Museu Nacional, Rio de Janeiro 67(1-2):119–128

Perini FA, Russo CAM, Schrago CG (2010) The evolution of South American endemic canids: a history of rapid diversification and morphological parallelism. J Evol Biol 23:311–322

Prado JL, Alberdi MT, Reguero MA (1998) El registro más antiguo de *Hippidion* Owen, 1869 (Mammalia, Perissodactyla) en América del Sur. Estud Geol 54:85–91

Prado JL, Alberdi MT, Sánchez B, Azanza B (2003) Diversity of the Pleistocene gomphotheres (Gomphotheriidae, Proboscidea) from South America. In: Reumer JWF, De Vos J, Mol D (eds) Advances in mammoth research. Proceedings of the Second International Mammoth Conference, Rotterdam, 16–20 May 1999. Deinsea 9:347–363

Prevosti FJ (2006) New material of Pleistocene cats (Carnivora, Felidae) from southern South America, with comments on biogeography and the fossil record. Geobios 39:679–694

Prevosti FJ (2010) Phylogeny of the large extinct South American Canids (Mammalia, Carnivora, Canidae) using a "total evidence" approach. Cladistics 26:456–481

Prevosti J, Palmqvist P (2001) Análisis ecomorfológico del cánido hipercarnívoro *Theriodictis platensis* Mercerat (Mammalia, Carnivora), basado en un nuevo ejemplar del Pleistoceno de Argentina. Ameghiniana 38(4):375–384

Prevosti FJ, Pardiñas FJ (2009) Comment on "The oldest South American Cricetidae (Rodentia) and Mustelidae (Carnivora): Late Miocene faunal turnover in central Argentina and the Great American Biotic Interchange". Palaeogeogr Palaeoclimatol Palaeoecol 280:543–547

Prevosti FJ, Soibelzon LH, Prieto A, San Roman M, Morello F (2003) The southernmost bear: *Paractotherium* (Carnivora, Ursidae, Tremarctinae) in the latest Pleistocene of southern Patagonia, Chile. J Vertebr Paleontol 23(3):709–712

Prevosti FJ, Zurita AE, Carlini AA (2005) Biostratigraphy, systematics, and paleoecology of *Protocyon* Giebel 1855 (Carnivora, Canidae) in South America. J S Am Earth Sci 20:5–12

Prevosti FJ, Gasparini GM, Bond M (2006) On the systematic position of a specimen previously assigned to Carnivora from the Pliocene of Argentina and its implications for the Great American Biotic Interchange. Neues Jahrbuch für Geologie und Paläontologie - Abhandlungen 242(1):133–144

Prevosti FJ, Ubilla M, Perea D (2009a) Large extinct canids from the Pleistocene of Uruguay: systematic, biogeographic and paleoecological remarks. Hist Biol 21(1–2):79–89

Prevosti FJ, Tonni EP, Bidegain JC (2009b) Stratigraphic range of the large canid (Carnivora, Canidae) in South America, and its relevance to quaternary biostratigraphy. Quat Int 210:76–81

Reig O (1952) Sobre la presencia de mustélidos mefitinos en la Formación Chapadmalal. Revista de Museo Municipal de Ciencias Naturales de Mar del Plata, vol 1, pp 45–51

Reig O (1972) The evolutionary history of the South American cricetid rodents. PhD dissertation, The University of London, London

Reig O (1978) Roedores cricétidos del Plioceno superior de la Provincia de Buenos Aires (Argentina): Publicaciones Museo Municipal de Ciencias Naturales de Mar del Plata 2:164–190

Rincón A, Prevosti FJ, Parra GE (2011) New saber-toothed cat records (Felidae: Machairodontinae) for the Pleistocene of Venezuela, and the great American biotic interchange. J Vertebr Paleontol 31(2):468–478

Robinson AE (1936) The camel in antiquity. Sudan Notes Rec 19(1):47–69

Ruiz-García M, CVásquez C, Pinedo-Castro M, Sandoval S et al (2012) Phylogeography of the Mountain Tapir (*Tapirus pinchaque*) and the Central American Tapir (*Tapirus bairdii*) and the origins of the three Latin-American tapirs by means of mtCyt-B sequences. In: Anamthawat-Jónsson K (ed) Current topics in phylogenetics and phylogeography of terrestrial and aquatic systems. IntechOpen. https://doi.org/10.5772/35361

Ruiz-García M, Castellanos Luz A, Bernal A, Pinedo-Castroa M et al (2016) Mitogenomics of the mountain tapir (Tapirus pinchaque, Tapiridae, Perissodactyla, Mammalia) in Colombia and Ecuador: Phylogeography and insights into the origin and systematics of the South American tapirs. Mamm Biol - Zeitsch Säugetier 81(2):163–175. https://doi.org/10.1016/j.mambio.2015.11.001

Shaw CA, McDonald HG (1987) First record of giant anteater (Xenarthra, Myrmecophagidae) in North America. Science 236:186–188

Soibelzon L (2002) Los Ursidae (Parnivora, Fissipedia) fósiles de la República Argentina. aspectos sistemáticos y paleoecológicos. PhD dissertation. Universidad Nacional de La Plata

Soibelzon L, Prevosti FJ (2007) Los carnívoros (Carnivora, Mammalia) terrestres del Cuaternario de América del Sur. In: Pons GX, Vicens D (eds) Geomorfologia litoral i quaternari. Homenage a Joan Cuerda Barceló. Museo de La Plata 14:49–68

Soibelzon LH, Schubert BW (2011) The largest known bear, *Arctotherium angustidens*, from the early Pleistocene Pampean region of Argentina: with a discussion of size and diet trends in bears. J Paleontol 85(1):69–75

Soibelzon LH, Tonni EP, Bond M (2005) The fossil record of South American short-faced bears (Ursidae, Tremarctinae). J S Am Earth Sci 20:105–113

Soibelzon LH, Cenizo MM, Prevosti FJ, Soibelzon E, Tartarini VB (2007) Dos nuevos registros de *Dusicyon* Hamilton-Smith, 1839 (Canidae, Mammalia) en el Plioceno y Pleistoceno de la región Pampeana (Argentina). aspectos sistemáticos, tafonomicos y bioestratigraficos. V Congreso Uruguayo de Geología

Theimer TC, Keim P (1998) Phylogenetic relationships of peccaries based on mitochondrial cytochrome B DNA sequences. J Mammal 79(2):566–572

Verzi D, Montalvo CI (2008) The oldest South American Cricetidae (Rodentia) and Mustelidae (Carnivora): late Miocene faunal turnover in central Argentina and the great American biotic interchange. Palaeogeogr Palaeoclimatol Palaeoecol 267:284–291

Wang X, Carranza-Castañeda O (2008) Earliest hog-nosed skunk, *Conepatus* (Mephitidae, Carnivora), from the early Pliocene of Guanajuato, Mexico and origin of South American skunks. Zool J Linn Soc Lond 154:386–407

Wang X, Tedford RH (2008) Dogs: their fossil relatives and evolutionary history. Columbia University Press, New York

Webb SD (1999) Isolation and interchange: A deep history of South American mammals. In: Eisenberg JF, Redfored KH (eds) Mammals of the Neotropicas: the central Neotropics: Volume 3, Ecuador, Peru, Bolivia, Brazil. The University of Chicago Press, Chicago, pp 13–19

Webb SD (2006) The great American biotic interchange: patterns and processes. Ann Mo Bot Gard 93:245–257

Webb SD, Rancy A (1996) Late Cenozoic evolution of the Neotropical mammal fauna. In: Jackson JBC, Budd AF, Coates AG (eds) Evolution and environment in tropical America. University of Chicago Press, Chicago, pp 335–358

Weinstock J, Willerslev E, Sher A, Tong W et al (2005) Evolution, Systematics, and Phylogeography of Pleistocene Horses in the New World: A Molecular Perspective. PloS Biol 3(8):e241. https://doi.org/10.1371/journal.pbio.0030241

Wheeler JC (1995) Evolution and present situation of the South American Camelidae. Biol J Linn Soc 54:271–295

Wilson DE, Mittermeier RA (eds) (2011) Handbook of the mammals of the world. Vol. 2. Hoofed mammals. Lynx Ediciones, Barcelona, pp 182–197

Woodburne MO (2010) The great American biotic interchange: dispersals, tectonics, climate, sea level and holding pens. J Mamm Evol 17:245–264

Chapter 14
Pleistocene Mammal Communities and Their Extinction

14.1 Introduction

As North and South America became more and more connected, the diversity of land mammals increased in South America due to the increase of mammals arriving from the north. But considering ecological theory, there are only a limited number of niches in any region, island, or continent that define the number of species that can exist in that physical location. There is a relation between species richness and area. Species richness increases with an increase in area (MacArthur and Wilson 1967). This is the oldest ecological relationship known.

Considering the land mammal fauna in South America, this invasion from the north doubled the number of land mammalian families from 18 to 36 families and resulted in a hyper-saturation of mammals that for a time represented the richest mammalian fauna ever known in the world. In this consideration, we use the number of families as a proxy for numbers of species, which is much more difficult to calculate because of historical considerations as well as different views of what a species may be.

The Theory of Island Biogeography suggests that the richness of species increases as the area increases and that there is a limit to species richness that is mediated by an equilibrium between the extinction rate and the immigration rate of species (MacArthur and Wilson 1967). It has been accepted that this theory be used in continent-wide spaces as well as more conventional island spaces, and the theory predicts that species richness, if drastically increased, will eventually lose species and fall back to the former species equilibrium. The arrival of 18 new families of mammals caused a hyperdiversity in South America, never before seen on any continent, and one supposes that this hyperdiversity was not ecologically stable.

T. Defler, *History of Terrestrial Mammals in South America*, Topics in Geobiology 42, https://doi.org/10.1007/978-3-319-98449-0_14

14.2 Mammalian Families Invading from the North to the South

A measure of the success of the invading mammals is the amount of generic diversification that occurred in each family of mammals up to recent times. Some families became very successful, while other families are only represented by one genus since the family arrived. Table 14.1 might not reflect all the genera that appeared, since some probably have not been discovered as yet, either as fossils or as new taxa.

14.3 Mammalian Families Invading from the South to the North

In contrast to the new families in South America, the fate of mammalian families going north was much more restricted with only slight success for the tropical mammals of South America which became established in tropical parts of Central America and southern Mexico. The most successful families going north were, of course, the terrestrial sloths, but their success was interrupted by total Pleistocene extinction (Table 14.2).

Table 14.1 Diversification of northern fauna in South America

Family	No. of genera	No. of living species in SA[a]
Soricidae	1	11
Heteromyidae	1	3
Geomyidae	1	1
Leporidae	1	2
Mephitidae	1	4
Tapiridae	1	4
Ursidae	1	1
Sciuridae	3	20
Gomphotheriidae	3–4	0
Equidae	3	0
Tayassuidae	4	3
Felidae	5	13
Camelidae	5	3
Procyonidae	6	13
Mustelidae	8	11
Cervidae	9	16
Canidae	10	13
Muridae Cricetidae	74	352

Patton et al. 2015

Table 14.2 Diversification of southern fauna in North and Central America

Family	No. of genera	No. living species in NA or CA
Pampatheriidae	1	0
Toxodontidae	1[a]	0
Bradypodidae	1[a]	1
Callitrichidae	1[a]	1
Glyptodontidae	1	0
Agoutidae	1[a]	1
Dasypodidae	2	2
Dasyproctidae	2[a]	2
Myrmecophagidae	2[a]	2
Hydrochoeridae	2	0
Megatheriidae	2	0
Erethizontidae	2	2
Didelphidae	3	3
Echimyidae	3[a]	3
Megalonychidae	3	0
Mylodontidae	3	0
Cebidae	5[a]	5

[a]Central America, includes *Panamacebus*, now extinct Chap. 8)

14.4 Patterns of Extinctions

Since these faunal interactions occurred over the course of at least 9.5 million years, stability in South American diversity took a long time to be reached. Few extinctions occurred during most of the Pleistocene over the last two million years, and finally a cascade of extinctions began during the early history of human beings on the southern continent. Forty-five genera and 83 species of large (over 44 kg) and megamammals (over 1000 kg) are known (Table 14.3) to have become extinct at the end of the Pleistocene (the largest loss on any continent and about half the genera lost worldwide for this period), almost all of which were adapted to open environments during the period 20,000–10,000 BP (Martin 1986; Cione et al. 2009; Barnosky et al. 2004) A statistical analysis of the patterns of this extinction implicated size as the highly significant factor for extinction with the concomitant factor of low reproduction, while trophic niche and origin were poor predictors (Lessa and Fariña 1996; Lessa et al. 1997; Lyons et al. 2004). Other taxonomic views would lower this number somewhat, for example, recognizing only one *Cuvieronius* species (Ficcarelli et al. 1995) and perhaps only two (or three) horse species (Alberdi and Prado 1993; Prado and Alberdi 1996) although actually there is no consensus in the literature on the number of species of horses or proboscids that lived in South America (Weinstock et al. 2005).

Table 14.3 Mammalian taxa that became extinct from about 125,000 years to the Holocene (Lujanian SALMA). (Cione et al. 2009)

Megamammals >1000 kg	Large mammals >45 kg
Cuvieronius humboldtii = C. tarijensis	*Antifer niemeyeri*
Cuvieronius hyodon = Haplomastodon chimborazi	*Arctotherium bonariense*
Doedicurus clavicaudatus	*Arctotherium brasiliense*
Eremotherium carolinense	*Arctotherium tarijense*
Eremotherium laurillardi	*Brasiliochoerus stenocephalus*
Eremotherium mirabile	*Equus (A. merhippus) andium*
Eremotherium rusconii	*Equus (A.) insularis*
Glossotherium (Oreomylodon)	*Equus (A.) lasallei wagneri*
Glossotherium lettsomi	*Equus (A.) neogeus*
Glossotherium (Pseudolestodon)	*Equus (A.) santa-elenae myloides*
Glossotherium robustum	*Eulamaops paralellus*
Glossotherium tropicorum	*Eutatus punctatus*
Glyptodon clavipes	*Glyptotherium* sp.
Glyptodon perforates	*Hippidion principale*
Glyptodon reticulatus	*Holmesina occidentalis*
Glyptodon cf. *cylindricum*	*Holmesina paulacoutoi*
Hemiauchenia paradoxa	*Hoplophorus euphractus*
Lestodon armatus	*Lama gracilis*
Lestodon trigonidens	*Morenelaphus lujanensis*
Macrauchenia patachonica	*Mylodopsis ibseni*
Magalonyx sp.	*Neochoerus aesopi*
Megatherium americanum	*Neochoerus sirasakae*
Megatherium medinae	*Neosclerocalyptus paskoensis*
Mixotoxodon larensis	*Neuryurus n.* sp.
Mylodon darwinii	*Nothropus priscus*
Mylodon listai	*Nothrotherium roverei*
Neothoracophorus depressus	*Ocnopus gracilis*
Panochthus frenzelianus	*Ocnohippidion saldiasi*
Panochthus morenoi	*Palaeolama niedae*
Panochthus tuberculatus	*Palaeolama weddelli*
Plaxhaplous canaliculatus	*Pampatherium humboldti*
Stegomastodon platensis	*Pampatherium typum*
Stegomastodon guayasensis	*Paraceros fragilis*
Stegomastodon waringi	*Parapanochthus jaguaribensis*
Toxodon burmeisteri	*Propraopus grandis*
Toxodon platensis	*Propraopus humboldtii*
Xenorhinotherium bahiense	*Propraopus magnus*
Medium-sized mammals	*Scelidon cuvieri*

(continued)

Table 14.3 (continued)

Megamammals >1000 kg	Large mammals >45 kg
	Scelidodon chiliensis
	Scelidodon reyesi
	Scelidotherium leptocephalus
	Smilodon populator
	Tapirus cristatellus
	Trigonodops lopesi

Adapted from Cione et al. (2009), Ficcarelli et al. (1995)

14.4.1 When Did Humans Arrive in South America and How Did They Impact the Fauna?

But what part did humans play in these extinction events? It is strange that around the arrival of humans so many species did become extinct and the association suggests strongly that humans had a part in these extinctions, despite much controversy about the human role (Diamond 1984; Miotti and Solemme 2003).

The arrival of humans has been implicated in the extinction of many species of mammals in Australia (Murray 1984), North America (Martin 1984), New Zealand (Cassels 1984), Hawaii (Olson and James 1984), Madagascar (Dewar 1984), and, of course, South America. But what other influences could explain such extinctions?

Climate change has been the most often mentioned factor that might have caused Pleistocene extinctions in South America (Barnosky and Lindsey 2010; Pascual and Ortiz Jaureguizar 1990). Yet studies of glaciations and glacial interstadial have not identified anything different from the present warm period that began some 12,000 years ago in comparison with previous warm periods during the entire Pleistocene (Cione et al. 2009). Pleistocene glaciations and interglacials are purported by many to have had strong effects on vegetation types, and the last great glaciation caused extensive open savanna-like vegetation throughout the continent, including throughout the Amazon region, as detected by local palynological, geomorphological, and fossil studies (Ab'Sáber 1982; Salgado-Labouriau 1982; van der Hammen 1982; Rancy 1991, 1999; Absy and van der Hammen 2001; Carneiro-Filho et al. 2002; Bonaccorso et al. 2006). There are extensive collections of Pleistocene savanna mammals even from parts of west Amazon previously thought to have maintained forest throughout the glaciations (Rancy 1991, 1999; Webb and Rancy 1996). A complication for interpretation of the cyclic changes on open and closed vegetation comes from the discovery that precipitation regimens in the Amazon were different and complex during the Neogene with a climate in the north different from the south that were driven more by Milankovitch cyclic events than by glaciation (Vonhof and Kaandorp 2010).

Analysis of the effect of the Holocene climatic optimum suggests that there was a decrease of open vegetation areas (savanna and related vegetation types) that certainly would have decreased populations of megamammals (Van der Hammen et al. 1992; Iriondo 1997, 1999; Iriondo and Garcia 1993; Van der Hammen and Absy 1994; Behling et al. 1999; Van der Hammen and Hooghiemstra 2000; de Vivo and Carmignotto 2004; Rabassa et al. 2005; Cione et al. 2003, 2009), particularly if they were open savanna grazers as generally thought. However, stable carbon isotope data for a series of Late Quaternary toxodonts (mostly *Toxodon* but a few *Mixotoxodon* from Honduras) show that despite hypsodonty (generally interpreted as grazing), there is some evidence for a mixed diet from tropical rainforest habitat for some populations of these ungulates (MacFadden 2005). This was probably a result of adaptation to the ecotone between savanna and forest that formed much of the cerrado vegetation that was a common occurrence during the comparatively dry glacial periods.

Populations of mammals had increased and decreased during millennia as glaciation increased the area of savannas during cool, dry climatic phases. Then open areas decreased with warming and the increasing dominance of other types of vegetation to which the mammals were not so well-adapted. During the low end of the population scale, many species might have been fragmented and close to minimum level of population viability. Even some populations that were in isolated fragments might have been driven to extinction, but during the Pleistocene the populations always recovered due to new glaciation and increase of open habitats. This characterization is known as the zigzag hypothesis, a characterization of the modifications in distribution and biomass of mammalian fauna in reaction to modifications of South American landscapes (Cione et al. 2003, 2009) (Fig. 14.1).

14.4.2 Ecological Factors Impacting Mammalian Extinctions

An enormous expansion of cerrado vegetation characterizes the equatorial last glacial maximum (LGM) landscape in South America. This biodiverse ecosystem included park savanna, gramineous-woody savanna, savanna wetlands, and gallery forests and is considered currently to be the most diverse tropical savanna vegetation in the world (Eiten 1972; Pires and Prance 1985; Da Fonseca et al. 1999), and its incidence would have provided ample pressures to adapt to mixed feeding.

A dissenting view of glacial savanna vegetation in Amazonia, based on pollen profiles of various Amazonian lakes and on the Amazon River marine fan at the mouth of the river, asserts that the Amazon was stable and un-fragmented throughout the Pleistocene (Colinvaux 1996; Colinvaux et al. 2000; Haberle and Maslin 1999; Kastner and Goñi 2003; Hoorn 1997; Colinvaux and De Oliveira 2001). It seems probable that the pollen profiles used for these conclusions were formed from extensive gallery forests and forests surrounding the lakes, preventing grass pollen from reaching the water. Other information leading to conclude that the Amazon became fragmented during glaciated times, including the Rancy fossil collection from the Napo River, has been dismissed as a fauna living in the middle of forest, though it is amply accepted that the species reported in this research were savanna

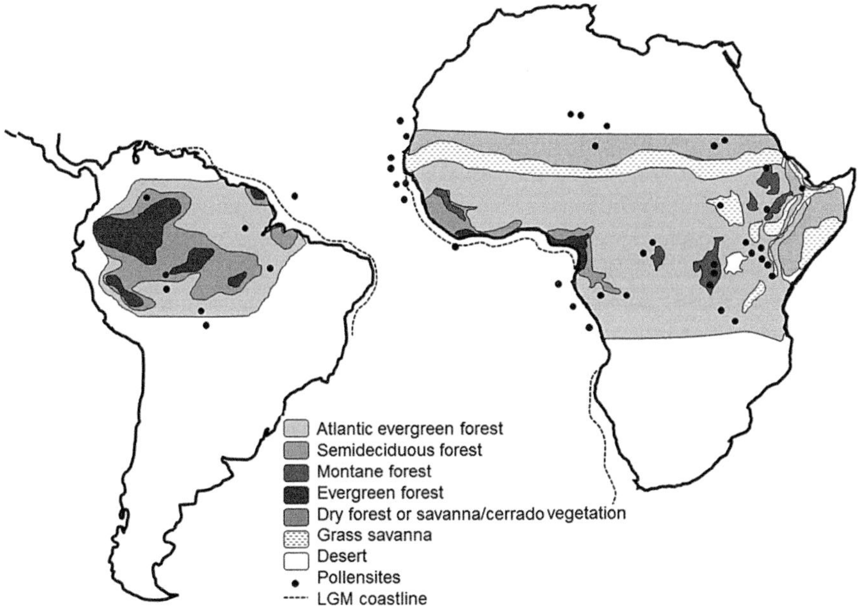

Fig. 14.1 Equatorial vegetation at time of last maximum glaciation (18,000 BP) in South America and Africa. (Modified from Anhuf et al. 2006). Tropical moist forest in South America shrunk to about 54% of its present-day expansion. Previous forest minima and savanna maxima created habitats allowing continent-wide expansion of open-area megamammals, many of which were not forest animals. Pollen sites have given data on previous precipitation. (By Diego Casallas-Pabón, Applied Biodiversity Foundation)

species (Van der Hammen et al. 1992; Anhuf et al. 2006; Webb and Rancy 1996; Rancy 1999).

The largest mammals most likely had low densities and very low reproductive rate (k-selected) and invested much effort of various years to each young, so that even a very low harvesting of individuals could eventually spell extinction for that population (Kiltie 1984; Johnson 2002; Cione et al. 2003; Cardillo 2003). If the population were fragmented into subunits, extinction of that subpopulation would be much more likely. These are key characteristics causing a mammal to be easily endangered by disruptions in their ecology (Pim 1984). Fragmentation of habitat and the concomitant habitat loss due to changed landscapes and of populations of megamammals during the postglacial age would have drastically reduced some isolated populations and caused others to go extinct (Fahrig 1997, 2003). Earlier vegetation cycles, without human hunters, allowed habitats and species populations from proximate glaciations to recover, but with human hunters, smaller and isolated populations of large mammals could easily have been driven to extinction (Whittington and Dyke 1984).

The extinction of herbivores would have sounded the death knell for large predators such as dire wolves, saber-toothed cats, and American lions, while the other, smaller predators adapted to smaller prey persisted. Cascading effects of the disappearance of large prey to which these three predators were adapted explain their disappearance from the scene (Thébault et al. 2006, 2007). The largest predators alive today in South America (the jaguar and the puma) hunt and survive very successfully nocturnally in closed-canopy forest, a habitat much less successfully hunted and disrupted by early humans, those cats freely hunt in open areas as well.

Well-established carbon dates for human presence in South America begin around 15,000 BP, although several sites suggest dates from before (Barnosky and Lindsey 2010) and there were probably low densities of humans in South America much earlier (**30,000 years**?). Differences between ages of contact with humans and last known appearance can be at least 1000–3000 years for many species, and some taxa persisted for >6000 years (the glyptodont *Doedicurus*, the giant sloth *Megatherium*), though future data may lengthen human history in South America (Barnosky and Lindsey 2010). This indicates that mammalian extinction may have often been spread over long time periods (Borrero 2009). There were, however, probably low densities of humans in South America much earlier than we can prove.

Overall data on human/mammal interaction are not very numerous, and most of the data suggests opportunistic, generalist exploitation rather than selective hunting techniques (Miotti and Salemme 1999; Gutiérrez and Martínez 2008; Borrero 2009). Hunting large mammals on the savanna precluded carrying the prey to a camping site because of its size and left carcasses that attracted scavengers and thus left no fossil record, while anything butchered in a semipermanent camp might be protected enough for the bones to become buried in soil or human refuse, after the dogs picked over what was left unguarded. Only in the case of mastodonts is there some evidence of organized hunting substantiated (Bryan et al. 1978; Lyons et al. 2004). At best the evidence proves coexistence of humans and many megamammals though there is little evidence of actual hunting, which must have occurred with some evidence of preparation at camp sites (Borrero 2009). There is no evidence for human interaction with mammals in rainforest or even much information about which mammals inhabited the rainforest. Evidence of hunting today in rainforest quickly disappears in an environment of decay and scavengers with only some few skulls and bones to be found, but most often in the kitchen middens of ancient settlements.

Lack of evidence of mammalian presence in tropical forest is a basic weakness of the entire process of interpreting the history of Pleistocene and earlier mammals in South America, especially since comparatively new evidence suggests that many northern mammals had arrived in South America and were present in tropical forests as long ago as 9.5 Ma (Campbell et al. 2000, 2001, 2009, 2010; Frailey and Campbell 2012). Now that we know that such evidence can be found, one expects more such data to turn up in the future.

However, adding the pressure of diminishing open areas and human hunting in competition with the many new predators that had become active in South America could have just been enough to tip these many mammalian species into oblivion. It must not be forgotten that the hyperdense populations of mammals meant an unbalanced fauna that was vulnerable to competition for the limited number of niches available. Many tropical savannas could have had very low carrying capacity just as they do today, thus populations outside of the rich pastures of Patagonia probably sustained very low biomass.

Barnosky et al. (2016) attempted to identify major "ecological state shifts" in both North and South America due to defaunation of the two continents during the Pleistocene. They were seeking to compare this Pleistocene defaunation with contemporary loss of species richness, which they identified as "changes in vegetation structure and species composition, reductions in environmental heterogeneity, species richness, evenness, seed dispersal, nutrient cycling and distribution, and ecosystem services, coextinction of dependent species and increases in disease-transmitting organisms and fire frequency and/or intensity." Their hypothesis was that (1) vegetation should change noticeably, consistent with ecological release from browsing, grazing, and trampling, and this should be apparent in fossil-pollen time series, and (2) surviving mammal communities should demonstrate changes in species composition, richness, and evenness. But in South America, there is no site that adequately tracks surviving mammal communities, and evidence suggests that vegetation change from an increasingly moist climate occurred well before the megamammal extinctions. In North America it was possible to confirm vegetation changes as well as (in northern California) small mammal community changes (Blois et al. 2010).

However, end-Pleistocene climate change and the hyper-species richness and thus ecological imbalance that existed during the Pleistocene need to be considered in the case of a major ecological shift. In fact the hyper-richness that obtained during the Pleistocene as the invasions from the north supplemented the southern natives must have, in themselves, caused a massive impact on the Pleistocene ecosystems, especially the savannas and their vegetation (Gingerich 1984; Graham and Lundelius 1984). Though we know little enough about the tropical forest species richness during this time, that somehow must be discovered. For example, if the great selva was diminished 50% during the height of the last glaciation, then great extensions of savanna were available to the many savanna species of megamammals. At the end of the Pleistocene, warming and increased moisture regimes would fill in huge extensions of land, diminishing savannas and in many cases fragmenting and isolating populations of savanna mammals.

Those herbivorous mammals that went extinct were diurnal and large, and many were probably vulnerable to the hunting tactics of intelligent and coordinated human beings that had become efficient at hunting this particular fauna (Gruhn and Bryan 1984). For various reasons, but particularly because of litter size, large animals are more vulnerable to extinction (Cardillo 2003). And how many sloths and glyptodonts were slow and unable to avoid such a predator?

The largest land mammal that now exists in South America, the tapir (up to about 300 kg), is largely nocturnal and mostly uses the forest rather than the savanna. No

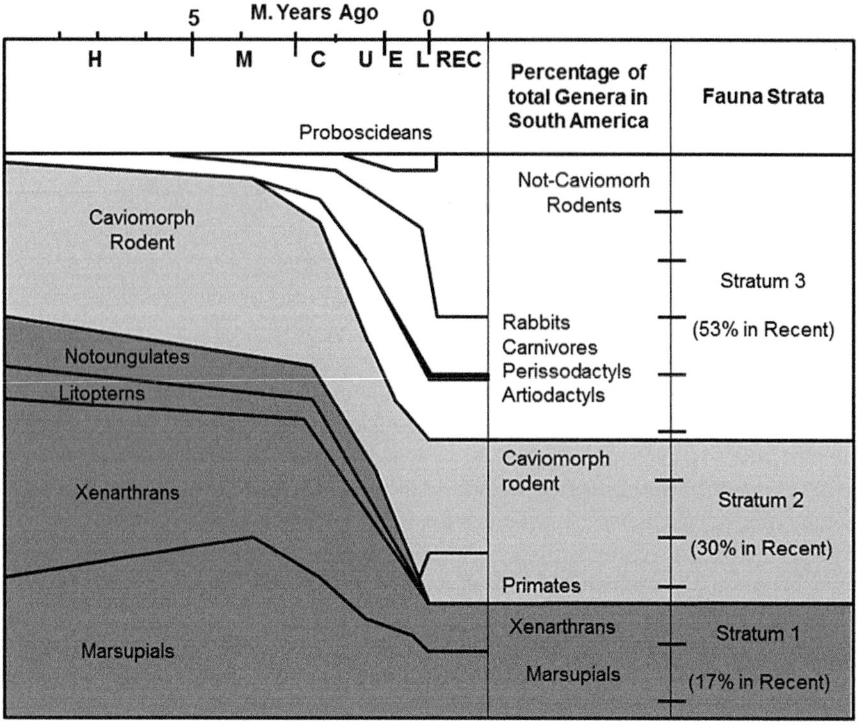

Fig. 14.2 Changes in South American land mammal fauna by percentages of total genera during the last nine million years. Stratum 1 genera declined from 70% to less than 20%, while stratum 3 increased from zero to more than 50%. (By Diego Casallas-Pabón, Applied Biodiversity Foundation)

megamammals survived the Early Holocene. The only other large mammals that survived are the giant armadillos (*Priodontes maximus*) and the caviomorphs *Hydrochoerus hydrochaeris* because one was nocturnal and lived in the forest and the other lived near bodies of water and was very aquatic and could escape hunters and had a high reproductive rate (Moreira et al. 2013) (Fig. 14.2).

References

Ab'Sáber AN (1982) The paleoclimate and paleoecology of Brazilian Amazonia. In: Prance GT (ed) Biological diversification in the tropics. Columbia University Press, New York, pp 41–59

Absy ML, van der Hammen T (2001) Some palaeoecological data from Rondonia, southern part of the Amazon Basin. Acta Amazon 6(3):293–299

Alberdi MT, Prado JL (1993) Review of the genus Hippidion Owen, 1869 (Mammalia: Perissodactyla) from the Pleistocene of South America. Zool J Linn Soc Lond 108:1–22

Anhuf D, Da Cruz FW Jr, Cordeiro RC, Van der Hammen T, Karmann I, Marengo JA, De Oliveira PE, Pessenda L, Siffedine A, Albuquerque AL, Da Silva Dias PI, Ledru MP, Behling H (2006) Paleo-environmental change in Amazonian and African rainforest during the LGM. Palaeogeogr Palaeoclimatol Palaeoecol 239:510–527

Barnosky AD, Lindsey EL (2010) Timing of Quaternary megafaunal extinction in South America in relation to human arrival and climate change. Quat Int 217:10–29

Barnosky AD, Koch PL, Reranec RS, Wing SL, Shabel AB (2004) Assessing the causes of late Pleistocene extinctions on the continents. Science 306:70–75

Barnosky AD, Lindsey EL, Villavicencio NA, Bostelmann E, Hadly EA, Wanket J, Marshall CR (2016) Variable impact of late-Quaternary megafaunal extinction in causing ecological state shifts in North and South America. Proc Natl Acad Sci U S A 113(4):856–861

Behling H, Berrio JC, Hooghiemstra H (1999) Late Quaternary pollen records from the middle Caquetá river basin in central Colombian Amazon. Palaeogeogr Palaeoclimatol Palaeoecol 145:193–213

Blois JL, McGuire JL, Hadly EA (2010) Small mammal diversity loss in response to late-Pleistocene climatic change. Nature 465(7299):771–774

Bonaccorso E, Koch I, Townsend Peterson A (2006) Pleistocene fragmentation of Amazon species' ranges. Divers Distrib 12:157–164

Borrero LA (2009) The elusive evidence: the archeological record of the South American megafauna. In: Haynes G (ed) American megafaunal extinctions at the end of the Pleistocene. Springer, New York, pp 145–168

Bryan A, Casamiquela J, Cruxent J, Gruhn R, Ochsenius C (1978) An El Jobo Mastodon Kill at Taima-Taima, Venezuela. Science 200:1275–1277. https://doi.org/10.1126/science.200.4347.1275. PMid:17738722

Campbell KE Jr, Frailey DC, Romero-Pittman L (2000) The late Miocene gomphothere *Amahuacatherium peruviam* (Proboscidea: Gomphotheriidae) from Amazonian Peru: implications for the great American faunal interchange. Boletin-Instit Geol Minero y Metalúrgico Serie D: Estudios Regionales 23:1–152

Campbell KE Jr, Frailey DC, Romero-Pittman L (2009) In defense of *Amahuacatherium* (Proboscidae: Gompohotheriidae). Neues Jahrb Geol P-A 252:113–128

Campbell KE Jr, Prothero DR, Romero-Pittman L, Hertel F, Rivera N (2010) Amazonian magnetostratigraphy: dating the first pulse of the great American faunal interchange. J S Am Earth Sci 29(3:619–626

Campbell KE Jr, Heizler M, Frailey CD, Romero-Pittman L, Prothero DR (2001) Upper Cenozoic chronostratigraphy of the southwestern Amazon Basin. Geology 29(7):595–598

Cardillo M (2003) Biological determinants of extinction risk: why are smaller species less vulnerable. Anim Conserv 6:63–69

Carneiro Filho A, Schwartz D, Tatumi SH, Rosique T (2002) Amazonian paleodunes provide evidence for drier climate phases during the late Pleistocene-Holocene. Quat Res 58:2005–2009

Cassels R (1984) Faunal extinction and prehistoric man in New Zealand and the Pacific islands. In: Martin PS, Klein RG (eds) Quaternary extinctions: a prehistoric revolution. The University of Arizona Press, Tucson, pp 741–767

Cione AL, Tonni EP, Soibelzon L (2003) The broken zig-zag: Late Cenozoic large mammal and tortoise extinction in South America. Rev Mus Argentino Cienc Nat 5(1):1–19

Cione AL, Tonni EP, Soibelzon L (2009) Did humans cause the late Pleistocene-Early Holocene mammalian extinctions in South America in a context of shrinking open areas? In: Haynes G (ed) American megafaunal extinctions at the end of the Pleistocene. Springer, New York, pp 125–144

Colinvaux PA (1996) Quaternary environmental history and forest diversity in the neotropics. In: Jackson JBC, Budd AF, Coates AG (eds) Evolution and environment in tropical America. University of Chicago Press, Chicago, pp 359–405

Colinvaux PA, De Oliveira PE (2001) Amazon plant diversity and climate through the Cenozoic. Palaeogeogr Palaeoclimatol Palaeoecol 166:51–63

Colinvaux PA, De Oliveira PE, Bush MB (2000) Amazonian and neotropical plant communities on glacial time-scales: the failure of the aridity and refuge hypotheses. Quat Sci Rev 19:141–169
de Vivo M, Carmignotto AP (2004) Holocene vegetation change and the mammal faunas of South America and Africa. J Biogeogr 31:943–957
Da Fonseca GAB, Mittermeier RA, Cavalcanti RB, Mittermeier CG (1999) The Brazilian cerrado. In: Mittermeier RA, Myers N, Robles Gil P, Goettsch Mittermeier C (eds) Hotspots: Earth's biologically richest and most endangered terrestrial ecoregions. Cemex, México
Dewar RE (1984) Extinctions in Madagascar: the loss. In: Martin PS, Klein RG (eds) Quaternary extinctions: a prehistoric revolution. The University of Arizona Press, Tucson, pp 574–593
Diamond J (1984) Historic extinction: a Rosetta Stone for understanding prehistoric extinctions. In: Martin PS, Klein RG (eds) Quaternary extinctions: a prehistoric revolution. The University of Arizona Press, Tucson, pp 824–862
Eiten G (1972) The cerrado vegetation of Brazil. Bot Rev 38(2):201–341
Fahrig L (1997) Relative effects of habitat loss and fragmentation on population extinction. J Wildl Manag 61(3):603–610
Fahrig L (2003) Effect of habitat fragmentation on biodiversity. Annu Rev Ecol Evol Syst 34:487–515
Ficcarelli G, Borselli V, Herrera G, Moreno Espinosa M, Torre D (1995) Taxonomic remarks on the South America mastodons referred to *Haplomastodon* and *Cuvieronius*. Geobios 28(6):745–756
Frailey CD, Campbell KE Jr (2012) Two new genera of peccaries (Mammalia, Artiodactyla, Tayassuidae) from upper Miocene deposits of the Amazon basin. J Paleontol 86(5):852–877
Gingerich PD (1984) Pleistocene extinctions in the context of origination-extinction equilibria in Cenozoic mammals. In: Martin PS, Klein RG (eds) Quaternary extinctions: a prehistoric revolution. The University of Arizona Press, Tucson, pp 211–222
Graham RW, Lundelius EL Jr (1984) Coevolutionary disequilibrium and Pleistocene extinctions. In: Martin PS, Klein RG (eds) Quaternary extinctions: a prehistoric revolution. The University of Arizona Press, Tucson, pp 223–249
Gruhn R, Bryan AL (1984) The record of Pleistocene megafaunal extinctions at Taima-taima, northern Venezuela. In: Martin PS, Klein RG (eds) Quaternary extinctions: a prehistoric revolution. The University of Arizona Press, Tucson, pp 128–137
Gutiérrez MA, Martinez GA (2008) Trends in the faunal human exploitation during the late Pleistocene and early Holocene in the Pampean region (Argentina). Quat Int 191:53–68
Haberle SG, Maslin MA (1999) Late Quaternary vegetation and climate change in the Amazon basin based on a 50,000 year pollen record from the Amazon fan, ODP Site 932. Quat Res 51:27–38
Hoorn C (1997) Palynology of the Pleistocene glacial/interglacial cycles of the Amazon fan (Holes 940A, 944A, and 946A). In: Flood RD, DJW P, Klaus A, Peterson LC (eds) Proceedings of the ocean drilling program, scientific results, vol 155. Ocean Drilling Program, College Station
Iriondo M (1997) Models of deposition of loess and loessoids in the upper Quaternary of South America. J S Am Earth Sci 10(1):71–79
Iriondo M (1999) Climatic changes in the South American plains: records of a continental-scale oscillation. Quat Int 57/58:93–112
Iriondo M, Garcia NO (1993) Climatic variations in the Argentine plains during the last 18,000 years. Palaeogeogr Palaeoclimatol Palaeoecol 101:209–220
Johnson CN (2002) Determinants of loss of mammal species during the Late Quaternary "megafauna" extinctions: life history and ecology, but not body size. Proc R Soc Lond B 269:2221–2130
Kastner TP, Goñi MA (2003) Constancy in the vegetation of the Amazon Basin during the late Pleistocene: evidence from the organic matter composition of Amazon deep sea fan sediments. Geology 31(4):291–294
Kiltie RA (1984) Seasonality, gestation time and large mammal extinctions. In: Martin PS, Klein RG (eds) Quaternary extinctions: a prehistoric revolution. The University of Arizona Press, Tucson, pp 299–314

Lessa EP, Fariña RA (1996) Reassessment of extinction patterns among the late Pleistocene mammals of South America. Palaeontology 39:651–662

Lessa EP, Van Valkenburgh B, Fariña RA (1997) Testing hypotheses of differential mammalian extinctions subsequent to the great American biotic interchange. Palaeogeogr Palaeoclimatol Palaeoecol 135:157–162

Lyons SK, Smith FA, Brown JH (2004) Of mice, mastodons and men: human-mediated extinctions on four continents. Evol Ecol Res 6:339–358

MacArthur RH, Wilson EO (1967) The theory of island biogeography. Princeton University Press, Princeton

MacFadden BJ (2005) Diet and habitat of toxodont megaherbivores (Mammalia, Notoungulata) from the late Quaternary of South and Central America. Quat Res 64:113–124

Martin PS (1984) Prehistoric overkill: the global model. In: Martin PS, Klein RG (eds) Quaternary extinctions: a prehistoric revolution. The University of Arizona Press, Tucson, pp 354–403

Martin PS (1986) Refuting late Pleistocene extinction models. In: Elliot DK (ed) Dynamics of extinction. Wiley, New York, pp 197–109

Miotti L, Salemme M (1999) Biodiversity, taxonomic richness and specialists-generalists during late Pleistocene/early Holocene times in Pampa and Patagonia (Argentina, Southern South America). Quat Int 53/54:53–68

Miotti L, Salemme MC (2003) When Patagonia was colonized: people mobility at high latitudes during Pleistocene/Holocen transition. Quat Int 109–110:95–111

Moreira JR, Herrera EA, Ferraz KMPMB, Macdonald DW (2013) Capybara: biology, use and conservation of an exceptional neotropical species. Springer, New York

Murray P (1984) Extinctions down-under: a bestiary of extinct Australian late Pleistocene monotremes and marsupials. In: Martin PS, Klein RG (eds) Quaternary extinctions: a prehistoric revolution. The University of Arizona Press, Tucson, pp 768–780

Olson SL, James HF (1984) The role of Polynesians in the extinction of the avifauna of the Hawaiian islands. In: Martin PS, Klein RG (eds) Quaternary extinctions: a prehistoric revolution. The University of Arizona Press, Tucson, pp 768–780

Pascual R, Ortiz Jaureguizar E (1990) Evolving climates and mammal faunas in Cenozoic South America. J Hum Evol 19(1–2):23–60

Patton JL, Pardiñas UFJ, D'Elía G (2015) Mammals of South America, volume 2: rodents. University of Chicago Press, Chicago.

Pim SL (1984) The complexity and stability of ecosystems. Nature 307:321–326

Pires JM, Prance GT (1985) The vegetation types of the Brazilian Amazon. In: Prance GT, Lovejoy TE (eds) Key environments: Amazonia. Pergamon Press, New York, pp 109–145

Prado JL, Alberdi MT (1996) A cladistics analysis of the horses of the tribe Equini. Palaeontology 39(3):663–680

Rabassa J, Coronato AM, Salemme M (2005) Chronology of the Late Cenozoic Patagonian glaciation and their correlation with biostratigraphic units of the Pampean region (Argentina). J S Am Earth Sci 20:81–103

Rancy A (1991) Pleistocene mammals and paleoecology of the western Amazon. PhD dissertation, University of Florida, Gainseville

Rancy A (1999) Fossil mammals of the Amazon as a portrait of a Pleistocene environment. In: Eisenberg JF, Redfored KH (eds) Mammals of the neotropics: the central neotropics: Volume 3, Ecuador, Peru, Bolivia, Brazil. The University of Chicago Press, Chicago, pp 20–26

Salgado-Labouriau ML (1982) Climate change at the Pleistocene-Holocene boundary. In: Prance GT (ed) Biological diversification in the tropics. Columbia University Press, New York, pp 74–77

Thébault E, Huber V, Loreau M (2006) Cascading extinctions and ecosystem functioning: contrasting effects of diversity depending on food web structure. Oikos 116(1):163–173

Thébault E, Huber V, Loreau M (2007) Cascading extinctions and ecosystem functioning: contrasting effects of diversity depending on food web structure. Oikos 116:163–173

Van der Hammen T (1982) Paleoecology of tropical South America. In: Prance GT (ed) Biological diversification in the tropics. Columbia University Press, New York, pp 60–66

Van der Hammen T, Absy ML (1994) Amazonia during the last glacial. Palaeogeogr Palaeoclimatol Palaeoecol 109:247–261

Van der Hammen T, Hooghiemstra H (2000) Neogene and Quaternary history of vegetation, climate, and plant diversity in Amazonia. Quat Sci Rev 19:725–742

Van der Hammen T, Duivenvoorden JF, Lips JM, Urrego LE, Espejo N (1992) Late Quaternary of the middle Caquetá River area (Colombian Amazonia). J Quat Sci 7(1):45–55

Vonhof HB, Kaandorp RJG (2010) Climate variation in Amazonia during the Neogene and the Quaternary. In: Hoorn C, Wesselingh F (eds) Amazonia: landscape and species evolution: a look into the past. Wiley-Blackwell, New York, pp 201–210

Webb SD, Rancy A (1996) Late Cenozoic evolution of the neotropical mammal fauna. In: Jackson JBC, Budd AF, Coates AG (eds) Evolution and environment in tropical America. University of Chicago Press, Chicago, pp 335–358

Weinstock J, Willerslev E, Sher A, Tong W, Ho SYW et al (2005) Evolution, systematics, and phylogeography of Pleistocene horses in the New World: a molecular perspective. PLoS Biol 3(8):1373–1379

Whittington SL, Dyke B (1984) Simulating overkill: experiments with the Mosimann and Martin Model. In: Martin PS, Klein RG (eds) Quaternary extinctions: a prehistoric revolution. The University of Arizona Press, Tucson, pp 451–465

Wilson DE, Reeder DM (2005) Mammal species of the world: a taxonomic and geographic reference, 3rd edn. The John Hopkins University Press, Baltimore

Chapter 15
The Modern Terrestrial Mammals of South America

15.1 Introduction

At this point in the history of South American mammals, we can count about 1330 living species, one of the most diverse mammalian faunas on the planet. Many of the old, autochthonous mammals have passed away, to the everlasting sadness of many of us. We will never see a giant ground sloth, a glyptodont, a toxodont, a saber-toothed cat, a gomphothere, or a macrauchenid, so we can only strive to learn more about them through the fossil record. Perhaps we shall find some ancient DNA and be able to study the genetics of some of these wonderful animals. Ancient DNA has been reconstituted from much more ancient mammals, and the entire genome of *Homo neanderthalensis* has been reconstituted from a 130,000-year-old toe bone from a Siberian cave (Prüfer et al. 2014)! These new techniques for ancient DNA recovery make it increasingly likely that in the future we shall be able to study the genetics at least of late Pleistocene fauna from South America, giving us an increased depth of understanding of not only their phylogeny but perhaps of population characteristics that would permit us to better understand their extinction.

In this chapter I shall simply list the species numbers and families of the South American land mammal taxa, using as bases the latest Gardner (2007) *Mammals of South America. Vol 1, Marsupials, Xenarthrans, Shrews and Bats* (Patton et al. 2015); *Mammals of South America. Vol 2, Rodents* (Wilson and Mittermeier 2009); *Handbook of the Mammals of the World. Vol. 1, Carnivores* (Wilson and Mittermeier 2011); *Handbook of the Mammals of the World. Vol. 2, Hoofed mammals* (Mittermeier et al. 2013); and *Handbook of the Mammals of the World. Vol 3, Primates*. For a primate count, I also used the latest literature on revisions of various taxa. I exclude the Sirenia (2 spp.), the Chiroptera (272 spp.), and the Cetacea (56 spp.), as I have done throughout the book. Three species of primates (*Alouatta pigra*, *Ateles geoffroyi*, and *Saimiri oerstedii*), although there is a possibility that

T. Defler, *History of Terrestrial Mammals in South America*, Topics in Geobiology 42, https://doi.org/10.1007/978-3-319-98449-0_15

Ateles geoffroyi enters Colombia from Panama, are limited to outside of South America, and so they are not counted. This gives a count of 991 species of nonflying and nonmarine mammals as well, excluding the two species of manatee and five species of dolphins.

The marsupials (more correctly the metatheres) are divided by Gardner (2005) into three South American orders.

15.2 Order: Didelphimorphia (Gardner 2005a, 2007) 86 Species

Didelphidae (*Caluromys*, 3 spp.; *Caluromysiops*, 1 sp.; *Glironia*, 1 sp.; *Chironectes*, 1 sp.; *Didelphis*, 6 app.; *Gracilinanus*, 9 spp.; *Hyladelphys*, 1 sp.; *Lestodelphys*, 1 sp.; *Lutreolina*, 1 sp.; *Marmosa*, 9 spp.; *Marmosops*, 14 spp.; *Metachirus*, 1 sp.; *Micoureus*, 6 spp.; *Monodelphis,* 18 spp.; *Philander*, 4 spp.; *Thylamys*, 9 spp.; *Tlacuatzin*, 1 sp. **86 species** (Figs. 15.1, 15.2, 15.3, 15.4, 15.5, and 15.6) (Cerqueira and Tribe 2007; Creighton and Gardner 2007a, b, c; Gardner and Creighton 2007a, b; Gardner and Dagosto 2007; Patton and da Silva 2007; Pearson 2007; Pine and Handley 2007; Stein and Patton 2007a, b; Voss et al. 2007a, b)

Fig. 15.1 Bare-tailed woolly opossum—*Caluromys philander.* (By Alex Popovkin)

Fig. 15.2 Common opossum—*Didelphis marsupialis*. (By Juan Tello)

Fig. 15.3 *Marmosa* (*Micoureus*) sp. (By Ramon Campos)

Fig. 15.4 Gray slender opossum—*Marmosops incanus*. (By Ramon Campos)

Fig. 15.5 *Metachirus nudicaudatus*. (By Haplochromis)

Fig. 15.6 *Monodelphis domestica*. (By Paul Samollow)

15.3 Order: Paucituberculata (Gardner 2005b, 2007) 6 Species

Caenolestidae (*Caenolestes*, 4 spp.; *Lestoros*, 1 sp.; *Rhyncholestes*, 1 sp.) (Myers and Patton 2007; Patterson 2007a, b; Timm and Patterson 2007) **6 species**

15.4 Order: Microbiotheria (Gardner 2005c, 2007) 1 Species

Microbiotheriidae (*Dromiciops*—1 sp.) **1 species** (Fig. 15.7) (Patterson and Rogers 2007)

Fig. 15.7 *Dromiciops gliroides*. (By José Luis Bartheld)

15.5 Order: Cingulata (Gardner 2005e, 2007) 19 Species

Dasypodidae (*Dasypus*, 6 spp.; *Calyptophractus*, 1 sp.; *Chlamyphorus*, 1 sp.; *Chaetophractus*, 3 spp.; *Euphractus*, 1 sp.; *Zaedyus*, 1 sp.; *Cabassous*, 3 spp.; *Priodontes*, 1 sp.; *Tolypeutes*, 2 spp.) **19 species** (Figs. 15.8 and 15.9) (Wetzel et al. 2007)

Fig. 15.8 *Dasypus novemcinctus*. (By Tom Friedel)

Fig. 15.9 *Chaetophractus vellerosus*. (By Arnaud Boucher)

15.6 Order: Pilosa (Gardner 2005, 2008) 10 Species

Bradypodidae (*Bradypus*—4 spp.) (Fig. 15.10)

Fig. 15.10 Three-toed Sloth—*Bradypus variegatus*. (By Stefan Laube)

Megalonychidae (*Choloepus*—2 spp.) (Fig. 15.11) (Gardner and Naples 2007)

Fig. 15.11 Two-toed sloth—*Choloepus hoffmanni*. (By Masteraah)

Cyclopedidae (*Cyclopes*—1 sp.) **1 species** (Fig. 15.12)

Fig. 15.12 Pygmy anteater—*Cyclopes didactylus*. (By Quinten Questel)

Myrmecophagidae (*Myrmecophaga*, 1 sp.; *Tamandua*, 2 spp.) **2 species** (Fig. 15.13)

Fig. 15.13 Giant anteater—*Myrmecophaga tridactyla*. (By Marcelo Calzans)

15.7 Order: Primates

Suborder: Platyrrhini (Groves 2005; Mittermeier et al. 2013) **152 species**

Analyses have increased this number by 22 species since 2005 from 121 species to 152 Neotropical species or (subtracting the three species endemic Central America) about **149 species** of primates in South America (Rylands et al. 2012). The increases are due to several factors but especially the description of four new species and molecular analyses showing species-level molecular differences, especially in the *Cebus albifrons* complex, as well as a revision of *Pithecia* (without molecular biology). The total number varies among different authors and different analyses.

Callitrichidae (*Cebuella*, 1 sp.; *Callibella*, 1 sp.; *Callimico*, 1 sp.; *Callithrix*, 6 spp.; *Leontopithecus*, 4 spp.; Mico, 14; *Saguinus*, 20 spp. [adding *Saguinus fuscus*]) = **47 spp.** (Figs. 15.14, 15.15, 15.16, 15.17, and 15.18) (Rylands and Mittereier 2013a)

Fig. 15.14 Pygmy marmoset—*Callithrix pygmaea.* (By Malene Thyssen)

Fig. 15.15 White-headed marmoset—*Callithrix geoffroyi.* (By Mosztics Attila)

Fig. 15.16 Golden lion tamarin—*Leontopithecus rosalia.* (By Trisha Shears)

Fig. 15.17 Emperor tamarin—*Saguinus imperador.* (By Brocken Inaglory)

Fig. 15.18 Cotton-top tamarin—*Saguinus oedipus*. (By Raimond Spekking)

Cebidae *Cebus*, 14 spp. [*Cebus albifrons* consists of more than 1 species, judging by molecular analysis; Ruiz-García et al. 2014]; *Sapajus*, 8 spp.; *Saimiri*, 7 spp. (Mittermeier et al. 2013; Ruiz-García et al. 2010; Rylands and Mittermeier 2013b) **= 29 spp.** (Figs. 15.19 and 15.20)

Fig. 15.19 White-fronted capuchin—*Cebus albifrons*. (By Whaldener)

Fig. 15.20 Squirrel monkey—*Saimiri sciureus*. (By Adrian Pingstone)

Aotidae (*Aotus*—11 spp.) = **11 species** (Fig. 15.21) (Fernandez-Duque et al. 2013)

Fig. 15.21 Panamanian night monkey—*Aotus zonalis*. (By Tom Friedel)

Pitheciidae (*Callicebus*, 5 spp. [Byrne et al. 2016; Ferrari 2013]; *Cheracebus*, 6 spp. [Byrne et al. 2016]; *Plecturocebus*, 17 spp. [Byrne et al. 2016]; *Cacajao*, 3 spp. [Boubli et al. 2008; Ferrari et al. 2013]; *Chiropotes*, 5 spp.; *Pithecia*, 16 spp. [Marsh 2014; Ferrari et al. 2014; Marsh and Ferrari 2013; Pinto 2013; Pinto and Ferrari 2013; Rylands and Mittermeier 2013c]) = **52 species** (Figs. 15.22 and 15.23)

Fig. 15.22 Caquetá tití—*Plecturocebus caquetensis*. (By Javier Garcia)

Fig. 15.23 Hairy saki—*Pithecia hirsuta*. (By Thomas Defler)

Atelidae (*Alouatta*, 12 spp. (Glander 2013; Glander and Pinto 2013) *Ateles*, 7 spp. (Konstant and Rylands 2013); *Brachyteles*, 2 spp. (Talebi 2013); *Lagothrix*, 2 spp. [*Oreonax* synonym for *Lagothrix or Lagothrix flavicauda* (Di Fiore et al. 2014; Ruiz-García et al. 2014); one wide-ranging species (*Lagothrix lagothricha*) with 4 subspecies; (Defler and Stevenson 2014; Botero et al. 2010; Ruiz-Garcia and Alvarez 2003, Ruiz-García and Pinedo-Castro 2009; Ruiz-García et al. 2014; Cornejo 2013; Defler 2013, 2014; Defler and Stevenson 2014; Rylands and Mittermeier 2013d) = **23 spp.** (Figs. 15.24 and 15.25)

Fig. 15.24 Red howler monkey—*Alouatta seniculus.* (By Alessandro Catenazzi)

Fig. 15.25 Humboldt's woolly monkey—*Lagothrix lagothricha.* (By Evgenia Kononova)

15.8 Order: Lagomorpha (Hoffmann and Smith 2005) 2 Species

Leporidae (*Sylvilagus*—2 spp.) **2 species** (Fig. 15.26)

Fig. 15.26 Cottontail rabbit—*Sylvilagus floridanus*. (By Harvey Henkelmann)

15.9 Order: Soricomorpha (Hutterer 2005) 11 Species

Soricidae (*Cryptotis*—10 spp.) **11 species** (Woodman and Péfaur 2015) (Fig. 15.27)

Fig. 15.27 Venezuelan shrew—*Cryptotis venezuelensis*. (By Marcial Quiroga-Carmona)

15.10 Order: Carnivora (Wozencraft 2005) 47 Species

Felidae (*Leopardus*, 9 spp.; *Puma*, 2 spp.; *Panthera*, 1 sp.) **12 species** (Figs. 15.28 and 15.29) (Sunquist and Sunquist 2009)

Fig. 15.28 Margay—*Felis wiedii.* (By Trisha Shears)

Fig. 15.29 Jaguar—*Panthera onca.* (By Hafiz Issadeen)

Canidae (*Atelocynus*, 1 sp.; *Cerdocyon*, 1 sp.; *Chrysocyon*, 1 sp.; *Lycalopex*, 6 spp.; *Speothos*, 1 sp.; *Urocyon*, 1 sp.) **11 species** (Figs. 15.30 and 15.31) (Sillero-Zubiri 2009)

Fig. 15.30 Bush dog—*Speothos venaticus*. (By Markus Bonnevier)

Fig. 15.31 Maned wolf—*Chrysocyon brachyurus*. (By Sage Ross)

Ursidae (*Tremarctos*—1 sp.) **1 species** (Fig. 15.32) (Garshelis 2009)

Fig. 15.32 Spectacled bear—*Tremarctos ornatus*. (By Silvio Tanaka)

Mustelidae (Lontra, 3 spp.; *Pteronura*, 1 sp.; *Eira*, 1 sp.; *Galictis*, 2 spp.; *Lyncodon*, 1 spp.; *Mustela*, 3 spp.) **11 species** (Fig. 15.33) (Larivière and Jennings 2009)

Fig. 15.33 Giant otter—*Pteronura brasiliensis*. (By Renaud d'Avout d'Auerstaedt)

Mephitidae (*Conepatus*—3 spp.) **3 species** (Fig. 15.34)

Fig. 15.34 Humboldt's hog-nosed skunk—*Conepatus humboldtii*. (By Payayita)

Procyonidae (*Bassaricyon*, 4 spp.; *Nasua*, 2 spp.; *Nasuella*, 1 sp.; *Potos*, 1 sp.; *Procyon*, 2 spp.) **10 species** (The number of species found in *Bassaricyon* has recently increased due to the recognition of *Bassaricyon neblina* (Helgen et al. 2013)) (Fig. 15.35).

Fig. 15.35 Crab-eating raccoon—*Procyon cancrivorus*. (By Szop; rakojad.jpg)

15.11 Order: Perissodactyla (Grubb 2005) 3 Species

Tapiridae (*Tapirus*—4 spp.) **3 species** (Medici 2011)

The supposed species *Tapirus kabomani* is falsified with various molecular and morphological data (Cozzuol et al. 2013; Medici 2011; Voss et al. 2014; Ruiz-García et al. 2014) (Fig. 15.36).

Fig. 15.36 Lowland tapir—*Tapirus terrestris*. (By Jean-Marc Rosier; http://www.rosier.pro)

15.12 Order: Artiodactyla (Grubb 2005) 23 Species

Tayassuidae—*Catagonus*, 1 sp.; *Pecari*, 1 sp.; *Tayassu*, 1 sp. = **3 species** (Taber et al. 2011) (Fig. 15.37)

Fig. 15.37 Chacoan peccary—*Catagonus wagneri*. (By Dave Pape)

Camelidae (*Lama*, 2 sp.; *Vicugna*, 2 sp.) = **4 species** (Franklin 2011) (Fig. 15.38)

Fig. 15.38 Vicuña—*Vicugna vicugna*. (By Alexandre Buisse)

Cervidae—*Blastocerus*, 1 sp.; *Hippocamelus*, 2 spp.; *Mazama*, 9 spp.; *Odocoileus*, 1 spp. [However, Molina and Molinari (1999) recognize three species in Venezuela, while Solari et al. (2013) recognize two additional species in Colombia]; *Ozotoceros*, 1 sp.; *Pudu*, 2 spp. = **16 species** (Matioli 2011) (Figs. 15.39 and 15.40)

Fig. 15.39 Marsh deer—*Blastocerus dichotomus*. (By Phillip Capper)

Fig. 15.40 Pudu—*Pudu puda*. (By Jaime E. Jimenez)

15.13 Order: Rodentia 621 Species (Patton et al. 2015)

Because of the use of new techniques of molecular biology, cytology, and field biology and a great expanse of interest in South American mammalogy, accompanied by an increase in upper-level educational opportunities and increased support by Latin American governments and private entities, growing appreciation for Neotropical biodiversity has helped expand knowledge of rodent diversity. A count in Patton et al. (2015) *Mammals of South America. Vol. 2: Rodents* yields 621 species. New genera have also proliferated due to a deeper understanding of rodent phylogeny and taxonomy.

Sciuridae (*Sciurillus*, 1 sp.; *Microsciurus*, 4 spp.; *Sciurus*, 13 spp.) **18 species** (Thorington and Hoffmann 2005; de Vivo and Carmignotto 2015) (Fig. 15.41)

Fig. 15.41 *Guerlinguetus aestuans*. (By Dick Culbert)

Heteromyidae (*Heteromys*—5 spp.) **5 species** (De Vivo and Carmignotti 2005; Hafner 2015b)

Geomyidae (*Orthogeomys*, 2 spp. [Solari et al. 2013; Hafner 2015a]) **2 species**

Cricetidae (*Isthmomys*, 1 sp.; *Reithrodontomys*, 1 sp.; *Abrawayaomys*, 1 sp.; *Abrothrix*, 9 spp.; *Aepeomys*, 2 spp.; *Akodon*, 41 spp.; *Amphinectomys*, 1 sp.; *Andalgalomys*, 3 spp.; *Andinomys*, 1 sp.; *Anotomys*, 1 sp.; *Auliscomys*, 3 sp.; *Bibimys*, 3 sp.; *Blarinomys*, 1 sp.; *Brucepattersonius*, 8 spp.; *Calomys*, 12 spp.; *Chelemys*, 3 spp.; *Chibchanomys*, 2 spp.; *Chilomys*, 1 sp.; *Chinchillula*, 1 sp.; *Delomys*, 3 spp.; *Deltamys*, 1 sp.; *Eligmodontia*, 4 spp.; *Euneomys*, 4 spp.; *Galenomys*, 1 sp.; *Geoxus*, 1 sp.; *Graomys*, 4 spp.; *Handleyomys*, 2 spp.; *Holochilus*, 3 spp.; *Ichthyomys*, 4 spp.; *Irenomys*, 1 sp.; *Juliomys*, 2 spp.; *Juscelinomys*, 3 spp.; *Kunsia*, 2 spp.; *Lenoxus*, 1 sp.; *Loxodontomys*, 2 spp.; Lundomys, 1 sp.; *Megaoryzomys*, 1 sp.; *Melanomys*, 3 spp.; *Microakodontomys*, 1 sp.; *Microyzomys*, 2 spp.; *Necromys*, 16 spp.; *Nectomys*, 5 spp.; *Neotomys*, 1 sp.; *Nesoryzomys*, 4 spp.; *Neusticomys*, 5 spp.; *Noronhomys*, 1 sp.; *Notiomys*, 1 sp.; *Oecomys*, 15 spp.; *Oligoryzomys*, 16 spp.; *Oryzomys*, 34 spp.; *Oxymycterus*, 16 spp.; *Paralomys*, 1 sp.; *Pearsonomys*, 1 sp.; *Phaenomys*, 1 sp.; *Phyllotis*, 13 spp.; *Podoxymys*, 1 sp.; *Pseudoryzomys*, 1 sp.; *Ponomys*, 2 spp.; *Reithrodon*, 2 spp.; *Rhagomys*, 2 spp.; *Rhipidomys*, 17 spp.; *Salinomys*, 1 sp.; *Scaptoromys*, 2 spp.; *Scolomys*, 2 spp.; *Sigmodon*, 4 spp.; *Sigmodontomys*, 2 spp.; *Tapecomys*, 1 sp.; *Thalpomys*, 2 spp.; *Thaptomys*, 1 sp.; *Thomasomys*, 36 spp.; Wiedomys, 1 sp.; *Wilfredomys*, 1 sp.; *Zygodontomys*, 2 spp.; *Tylomys*, 1 sp.) **352 species** (Musser and Carleton 2005; Alvarado-Serrano and D'Elía 2015; Álvarez-Castañeda 2015; Arellano 2015; Bezerra 2015; Bonvicino 2015; Bonvicino and Weksler 2015; Braun 2015; Braun and Pardiñas 2015; Braun and Patton 2015; Carleton 2015a, b; Carleton and Musser 2015; D'Elía 2015; D'Elía and UFJ 2015; de Olivwie and Gonçalves 2015; de Olivwie and Paradiñas 2015; Dowler 2015; Emmons 2015; Gomez-Laverde et al. 2015; Gonçales et al. 2015; González et al. 2015a, b; Lanzone and Braun 2015; Lanzone et al. 2015; Luna 2015; Ortiz and Jayat 2015; Pacheco 2015a, b, c; Pardiñas and Bezerra (2015); Pardiñas and Patton 2015; Pardiñas and Teta 2015a, b, c; Pardiñas et al. 2015a, b, c, d, e; Paresque and Hanson 2015; Patterson et al. 2015; Patton 2015b, c, d, e; Percequillo 2015a, b, c, d, e, f, g, h, i, j, k; Percequillo and Weksler 2015; Salazar-Bravo 2015a, b, c, d, e, f; Salazar-Bravo and Jayat 2015; Steppen and Ramirez 2015; Teta and Paradiñas 2015a, b, c; Teta et al. 2015a, b, c, d, e, f, g, h; Tribe 2015; Vilela et al. 2015; Voss 2015a, b, c, d; Weksler 2015a, b, c; Weksler and Bonvicino 2015a, b; Weksler and Lóss 2015; Weksler and Valqui 2015) (Figs. 15.42 and 15.43)

Fig. 15.42 Sanborn's grass mouse—*Abrothrix sanborni.* (By J. Cardenas)

Fig. 15.43 Atlantic Forest climbing mouse—*Rhipidomys mastacalis.* (By Leonardo Merçon)

Erethizontidae (*Chaetomys*, 1 sp.; *Coendou*, 4 spp.; *Echinoprocta*, 1 sp.; *Erethizon*, 1 sp.; *Sphiggurus*, 9 spp.) **16 species** (Woods and Kilpatrick 2005) (Fig. 15.44)

Fig. 15.44 Rothschild's porcupine—*Coendou rothschildi*. (By Beatrice Murch)

Chinchillidae (*Chinchilla*, 1 sp.; *Lagidium*, 3 spp.; *Lagostomus*, 1 sp.) **5 species** (Woods and Kilpatrick 2005; Spotorno and Patton 2015 (Fig. 15.45)

Fig. 15.45 Long-tailed chinchilla—*Chinchilla lanigera*. (By Trurl66)

Dinomyidae (*Dinomys*—1 sp.) **1 species** (Woods and Kilpatrick 2005; Patton 2015f) (Fig. 15.46)

Fig. 15.46 *Dinomys branickii.* (By Benjamin Frable)

Cavidae (*Cavia*, 6 spp.; *Galea*, 3p.; *Microcavia*, 3 spp.; *Dolichotis*, 2 spp.; *Hydrochoerus*, 2 spp.; *Kerodon*, 2 spp.) **18 species** (Woods and Kilpatrick 2005; Dunuum 2015) (Figs. 15.47 and 15.48)

Fig. 15.47 Mara—*Dolichotis patagonum.* (By Vassil)

Fig. 15.48 Capybara—*Hydrochoerus hydrochaeris*. (By Charles J Sharp)

Dasyproctidae (*Dasyprocta*, 11 spp.; *Myoprocta*, 2 spp.) **13 species** (Woods and Kilpatrick 2005; Patton and Emmons 2015a) (Fig. 15.49)

Fig. 15.49 Red-rumped agouti—*Dasyprocta leporina*. (By Mistvan)

Cuniculidae (*Cuniculus*—2 spp.) **2 species** (Woods and Kilpatrick 2005; Patton 2015g) (Fig. 15.50)

Fig. 15.50 Lowland paca—*Cuniculus paca.* (By Hans Hillewaert)

Ctenomyidae (*Ctenomys*—60 spp.) **60 species** (Woods and Kilpatrick 2005; Bidau 2015) (Fig. 15.51)

Fig. 15.51 Flamarion's tuco-tuco—*Ctenomys flamarioni.* (By Cláudio Dias Timm)

Octodontidae (*Aconaemys*, 3 spp.; *Octodon*, 4 spp.; *Octodontomys*, 1 spp.; *Octomys*, 1 sp.; *Pipanacoctomys*, 1 sp.; *Salinoctomys*, 1 sp.; *Spalacopus*, sp.; *Tympanoctomys*, 1 sp.) **12 species** (Woods and Kilpatrick 2005; Díaz et al. 2015a, b; Verzi et al. 2015a, b, c, d; Honeycutt et al. 2003) (Fig. 15.52)

Fig. 15.52 Common degu—*Octodon degus*. (By Algirdas)

Abrocomidae (*Abrocoma*, 8 spp.; *Cuscomys*, 2 spp.) **10 species** (Woods and Kilpatrick 2005)

Echimyidae (*Dactylomys*, 3 spp.; *Kannabateomys*, 1 sp.; *Olallamys*, 2 spp.; *Callistomys*, 1 sp.; *Diplomys*, 3 spp.; *Echimys*, 3 spp.; *Isothrix*, 4 spp.; *Makalata*, 6 spp.; *Phyllomys*, 12 spp.; *Carterodon*, 1 sp.; *Clyomys*, 2 spp.; *Euryzygomatomys*, 1 spp.; *Hoplomys*, 1 sp.; *Lonchothrix*, 1 sp.; *Mesomys*, 4 spp.; *Proechimys*, 25 spp.; *Thrichomys*, 3 spp.; *Trinomys*, 11 spp.; *Boromys*, 2 spp.; *Brotomys*, 2 spp.; *Heteropsomys*, 2 spp.) **90 species** (Woods and Kilpatrick 2005; Bezerra and Bonvicino 2015a, b; Bonvicino and Bezerra 2015; Emmons and Leite 2015; Emmons and Patton 2015a, b, c, d; Emmons et al. 2015a, b, c; Patton and Emmons 2015c, d; Patton and Leite 2015; Pessôa et al. 2015a, b; Leite and Loss 2015)

Myocastoridae (*Myoscastor*—1 sp.) **1 species** (Woods and Kilpatrick 2005) (Fig. 15.53)

Fig. 15.53 Coupu—*Myocastor coypus*. (By Philippe Amelant)

Above is a graph of all orders and species of South American land mammals (Wilson and Reeder 2005). It is clear that the rodents represent more than half of the South American terrestrial mammals (Fig. 15.54 and Table 15.1).

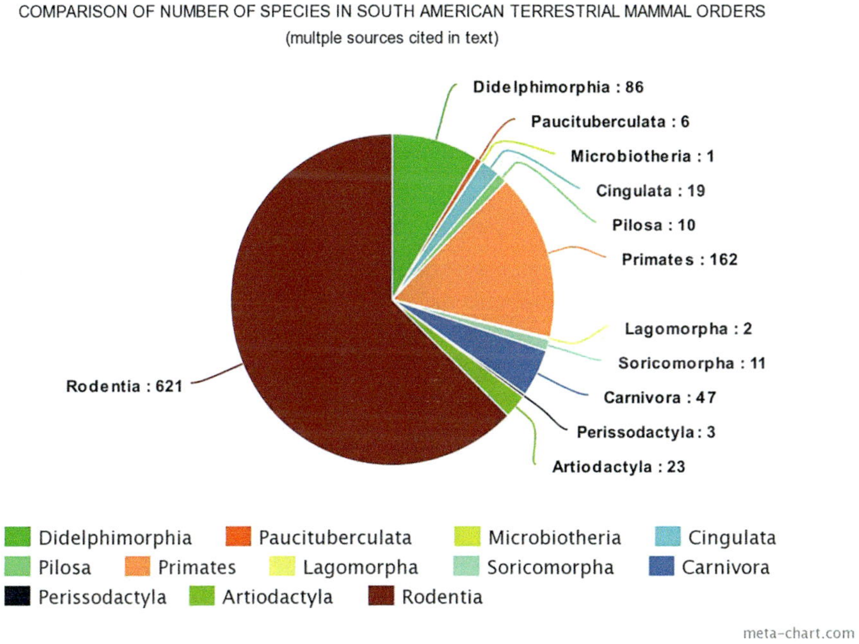

Fig. 15.54 Comparison of species in each mammalian order in South America, according to references cited throughout the chapter and based on 991 species

Table 15.1 Percentage of total terrestrial mammal fauna

Order	Percentage of total terrestrial mammal fauna
Rodentia	62.7
Primates	16.3
Didelphimorphia	8.7
Carnivora	4.7
Artiodactyla	2.3
Cingulata	1.9
Soricomorpha	1.1
Pilosa	1.0
Paucituberculata	0.6
Perissodactyla	0.3
Lagomorpha	0.2
Microbiotheria	0.1

This rich mammalian fauna is under threat as never before, especially because of habitat loss, though many other pressures impact mammals. The greatest concern of those of us who care about South American mammals is that like the many Pleistocene extinctions, the modern mammalian fauna might also become more impoverished through extinction. The take-home message is that these negative processes are because of human activities and that only human solutions can solve the threats that are growing toward the survival of this and other biota. Much has been written about the pressures threatening mammals and other organisms with extinction, and it is vital that we familiarize ourselves with the negative pressures and the possible solutions for preserving this wonderful fauna for the future (Brooks et al. 2006; Ceballos et al. 2005; Estrada et al. 2017; Pires Costa et al. 2005; Mittermeier et al. 2013; Schipper et al. 2008; Solari et al. 2013; Tognelli 2004; Wilson and Mittermeier 2009, 2011 for a small sample of the literature).

References

Alvarado-Serrano DF, D'Elía G (2015) Cricetidae. *Neomicroxus*. In: Patton JL, Pardiñas UFJ, D'Elía G (eds) Mammals of South America. Vol. 2. Rodents. University of Chicago Press, Chicago, pp 96–99

Álvarez-Castañeda ST (2015) Cricetidae. *Tylomys*. In: Patton JL, Pardiñas UFJ, D'Elía G (eds) Mammals of South America. Vol. 2. Rodents. University of Chicago Press, Chicago, pp 686–688

Arellano E (2015) Cricetidae. *Reithrodontomys*. In: Patton JL, Pardiñas UFJ, D'Elía G (eds) Mammals of South America. Vol. 2. Rodents. University of Chicago Press, Chicago, pp 61–83

Barkley LJ (2007) Didelphidae. *Glironia*. In: Gardner AL (ed) Mammals of South America. Volume 1. Marsupials, xenarthrans, shrews, and bats. University of Chicago Press, Chicago, pp 12–14

Bezerra AMB (2015) Cricetidae. *Kunsia*. In: Patton JL, Pardiñas UFJ, D'Elía G (eds) Mammals of South America. Vol. 2. Rodents. University of Chicago Press, Chicago, pp 228–230

Bezerra AMB, Bonvicino CR (2015a) Echimyidae. *Carterodon*. In: Patton JL, Pardiñas UFJ, D'Elía G (eds) Mammals of South America. Vol. 2. Rodents. University of Chicago Press, Chicago, pp 933–935

Bezerra AMB, Bonvicino CR (2015b) Echimyidae. *Clyomys*. In: Patton JL, Pardiñas UFJ, D'Elía G (eds) Mammals of South America. Vol. 2. Rodents. University of Chicago Press, Chicago, pp 935–936

Bidau CI (2015) Ctenomyidae. *Ctenomys*. In: Patton JL, Pardiñas UFJ, D'Elía G (eds) Mammals of South America. Vol. 2. Rodents. University of Chicago Press, Chicago, pp 818–877

Bonvicino CR (2015) Cricetidae. *Wiedomys*. In: Patton JL, Pardiñas UFJ, D'Elía G (eds) Mammals of South America. Vol. 2. Rodents. University of Chicago Press, Chicago, pp 683–685

Bonvicino CR, Bezerra MR (2015) Echimyidae. *Euryzgomatomys*. In: Patton JL, Pardiñas UFJ, D'Elía G (eds) Mammals of South America. Vol. 2. Rodents. University of Chicago Press, Chicago, pp 937–940

Bonvicino CR, Weksler M (2015) Cricetidae. *Nectomys*. In: Patton JL, Pardiñas UFJ, D'Elía G (eds) Mammals of South America. Vol. 2. Rodents. University of Chicago Press, Chicago, pp 359–377

Botero S, Rengifo IY, Bueno MI et al (2010) How many species of woolly monkeys inhabit Colombian forests? Am J Primatol 72(12):1131–1140

Boubli JP, Nazareth F, da Silva M, Amado MV, Hrbek T, Boavista Pontual F, Farias IP (2008) A taxonomic reassessment of *Cacajao melanocephalus* Humboldt (1811), with the description of two new species. Int J Primatol 29:723–741

Braun JK (2015) Cricetidae. *Andalgalomys*. In: Patton JL, Pardiñas UFJ, D'Elía G (eds) Mammals of South America. Vol. 2. Rodents. University of Chicago Press, Chicago, pp 469–472

Braun JK, Pardiñas UFJ (2015) Cricetidae. *Euneomys*. In: Patton JL, Pardiñas UFJ, D'Elía G (eds) Mammals of South America. Vol. 2. Rodents. University of Chicago Press, Chicago, pp 83–89

Braun JK, Patton JL (2015) Cricetidae. *Groomys*. In: Patton JL, Pardiñas UFJ, D'Elía G (eds) Mammals of South America. Vol. 2. Rodents. University of Chicago Press, Chicago, pp 523–531

Brooks TM, Mittermeier RA, da Fonseca GAB, Gerlach J, Hoffmann M, Lamoreux JF et al (2006) Global biodiversity conservation priorities. Science 313:58–61

Byrne H, Rylands AB, Carneiro JC, Lynch Alfaro JW, Bertuol F, da Silva MNF, Messias M, Groves CP, Mittermeier RA, Farias I, Hrbek T, Schneider H, Sampaio I, Boubli JP (2016) Phylogenetic relationships of the New World titi monkeys (*Callicebus*): first appraisal of taxonomy based on molecular evidence. Front Zool 13(10). https://doi.org/10.1186/s12983-016-0142-4

Carleton MC (2015a) Cricetidae. *Microryzomys*. In: Patton JL, Pardiñas UFJ, D'Elía G (eds) Mammals of South America. Vol. 2. Rodents. University of Chicago Press, Chicago, pp 355–359

Carleton MC (2015b) Cricetidae. *Transandinomys*. In: Patton JL, Pardiñas UFJ, D'Elía G (eds) Mammals of South America. Vol. 2. Rodents. University of Chicago Press, Chicago, pp 456–460

Carleton MC, Musser GG (2015) Cricetidae. *Oecomys*. In: Patton JL, Pardiñas UFJ, D'Elía G (eds) Mammals of South America. Vol. 2. Rodents. University of Chicago Press, Chicago, pp 393–417

Ceballos G, Ehrlich PR, Soberón J, Salazar I, Fay JP (2005) Global mammal conservation: what must we manage? Science 309:603–606

Cerqueira R, Tribe CJ (2007) Didelphidae. *Didelphis*. In: Gardner AL (ed) Mammals of South America. Volume 1. Marsupials, xenarthrans, shrews, and bats. University of Chicago Press, Chicago, pp 17–25

Cornejo FM (2013) Atelidae. *Oreonax*. In: Mittermeier RA, Rylands AB, Wilson DE (eds) Handbook of the mammals of the world 3. primates. Platyrrhini. Lynx Editions, Barcelona, p 547

Cozzuol MA, Clozato CL, Holanda EC, Rodrigues FHG, Nienow S, De Thoisy B, Redondo RAF, Santos FR (2013) A new species of tapir from the Amazon. J Mammal 94(6):1331–1345

Creighton GK, Gardner AL (2007a) Didelphidae. *Gracilinanus*. In: Gardner AL (ed) Mammals of South America. Volume 1. Marsupials, xenarthrans, shrews, and bats. University of Chicago Press, Chicago, pp 43–50

Creighton GK, Gardner AL (2007b) Didelphidae. *Marmosa*. In: Gardner AL (ed) Mammals of South America. Volume 1. Marsupials, xenarthrans, shrews, and bats. University of Chicago Press, Chicago, pp 51–61

Creighton GK, Gardner AL (2007c) Didelphidae. *Thylamys*. In: Gardner AL (ed) Mammals of South America. Volume 1. Marsupials, xenarthrans, shrews, and bats. University of Chicago Press, Chicago, pp 107–117

D'Elía G (2015) Cricetidae. *Pearsonomys*. In: Patton JL, Pardiñas UFJ, D'Elía G (eds) Mammals of South America. Vol. 2. Rodents. University of Chicago Press, Chicago, pp 138–140

D'Elía G, UFJ P (2015) Cricetidae. *Scapteromys*. In: Patton JL, Pardiñas UFJ, D'Elía G (eds) Mammals of South America. Vol. 2. Rodents. University of Chicago Press, Chicago, pp 269–274

de Olivwie JA, Gonçalves PR (2015) Cricetidae. *Oxymycterus*. In: Patton JL, Pardiñas UFJ, D'Elía G (eds) Mammals of South America. Vol. 2. Rodents. University of Chicago Press, Chicago, pp 247–268

de Olivwie JA, Paradiñas UFJ (2015) Cridetidae. *Wilfredomys*. In: Patton JL, Pardiñas UFJ, D'Elía G (eds) Mammals of South America. Vol. 2. Rodents. University of Chicago Press, Chicago, pp 105–109

Defler TR (2013) Species Accounts. Family Atelidae. *Lagothrix*. In: Mittermeier RA, Rylands AB, Wilson DE (eds) Handbook of the mammals of the world 3. Primates. Platyrrhini. Lynx Editions, Barcelona, pp 544–547

Defler TR (2014) Colombian *Lagothrix*: analysis of their phenotype and taxonomy. In: Defler TR, Stevenson PR (eds) The woolly monkey: behavior, ecology, systematics and captive studies. Springer, New York, pp 33–58

Defler TR, Stevenson PR (eds) (2014) The woolly monkey: behavior, ecology, systematics and captive studies. Springer, New York

De Vivo, Carmignotti (2015) Family Sciuridae. In: Patton JL, Pardiñas UFJ, D'Elía G (eds) Mammals of South America. Vol. 2 Rodents. The University of Chicago Press, Chicago, pp 1–48

Di Fiore A, Chaves PB, Cornejo FM et al (2014) The rise and fall of a genus: complete mtDNA genomes shed light on the phylogenetic position of yellow-tailed woolly monkeys, *Lagothrix flavicauda*, and on the evolutionary history of the family Atelidae (Primates: Platyrrhini). Mol Phylogenet Evol. https://doi.org/10.1016/j.ympev.2014.03.028

Díaz MM, Barquez RM, Verzi DH (2015a) Octodontidae. *Octodon*. In: Patton JL, Pardiñas UFJ, D'Elía G (eds) Mammals of South America. Vol. 2. Rodents. University of Chicago Press, Chicago, pp 1029–1035

Díaz MM, Barquez RM, Verzi DH (2015b) Octodontidae. *Tympanoctomys*. In: Patton JL, Pardiñas UFJ, D'Elía G (eds) Mammals of South America. Vol. 2. Rodents. University of Chicago Press, Chicago, pp 1043–1048

Dowler RC (2015) Cricetidae. *Nesoryzomys*. In: Patton JL, Pardiñas UFJ, D'Elía G (eds) Mammals of South America. Vol. 2. Rodents. University of Chicago Press, Chicago, pp 390–393

Dragoo JW (2009) Mephitidae. In: Wilson DE, Mittermeier RA (eds) Handbook of the mammals the world. Vol. 1. Carnivores. Lynx Editions, Barcelona, pp 532–562

Dunuum JL (2015) Caviidae. *Cavia, Galea, Microcavia, Dolichotis, Hydrochoerus, Kerodon*. In: Patton JL, Pardiñas UFJ, D'Elía G (eds) Mammals of South America. Vol. 2. Rodents. University of Chicago Press, Chicago, pp 690–726

Emmons LH (2007) Family Didelphidae. *Caluromysiops*. In: Gardner AL (ed) Mammals of South America. Volume 1. Marsupials, xenarthrans, shrews, and bats. University of Chicago Press, Chicago, pp 11–12

Emmons LH (2015) Cricetidae. *Jusxwlinomya*. In: Patton JL, Pardiñas UFJ, D'Elía G (eds) Mammals of South America. Vol. 2. Rodents. University of Chicago Press, Chicago, pp 225–228

Emmons LH, Leite YLR (2015) Echimyidae. *Callistomys*. In: Patton JL, Pardiñas UFJ, D'Elía G (eds) Mammals of South America. Vol. 2. Rodents. University of Chicago Press, Chicago, pp 889–890

Emmons LH, Patton JL (2015a) Echimyidae. Diplomys. In: Patton JL, Pardiñas UFJ, D'Elía G (eds) Mammals of South America. Vol. 2. Rodents. University of Chicago Press, Chicago, pp 890–893

Emmons LH, Patton JL (2015b) Echimyidae. *Isothrix, Makalata, Pattonomys*. In: Patton JL, Pardiñas UFJ, D'Elía G (eds) Mammals of South America. Vol. 2. Rodents. University of Chicago Press, Chicago, pp 898–915

Emmons LH, Patton JL (2015c) Echimyidae. *Santamartamys*. In: Patton JL, Pardiñas UFJ, D'Elía G (eds) Mammals of South America. Vol. 2. Rodents. University of Chicago Press, Chicago, pp 928–929

Emmons LH, Patton JL (2015d) Echimyidae. *Toromys*. In: Patton JL, Pardiñas UFJ, D'Elía G (eds) Mammals of South America. Vol. 2. Rodents. University of Chicago Press, Chicago, pp 929–931

Emmons LH, Patton JL, Leite YLR (2015a) Echimyidae. *Dactylomys, Kannabateomys, Olallamys*. In: Patton JL, Pardiñas UFJ, D'Elía G (eds) Mammals of South America. Vol. 2. Rodents. University of Chicago Press, Chicago, pp 877–888

Emmons LH, Leite YLR, Patton JL (2015b) Echimyidae. In: Patton JL, Pardiñas UFJ, D'Elía G (eds) Mammals of South America. Vol. 2. Rodents. University of Chicago Press, Chicago, pp 877–880

Emmons LH, Leite YLR, Patton JL (2015c) Echimyidae. *Echimys*. In: Patton JL, Pardiñas UFJ, D'Elía G (eds) Mammals of South America. Vol. 2. Rodents. University of Chicago Press, Chicago, pp 893–898

Estrada A, Garber PA, Rylands AB et al (2017) Impending extinction crisis of the world's primates: why primates matter. Sci Adv 3(1):e1600946. https://doi.org/10.1126/sciadv.1600946

Fernandez-Duque E, Corley MK, Spence-Alzenberg A (2013) Aotidae. In: Mittermeier RA, Rylands AB, Wilson DE (eds) Handbook of the mammals of the world 3. Primates. Platyrrhini. Lynx Editions, Barcelona, pp 414–431

Ferrari SF (2013) Callicebinae. *Callicebus*. In: Mittermeier RA, Rylands AB, Wilson DE (eds) Handbook of the mammals of the world 3. primates. Platyrrhini. Lynx Editions, Barcelona, pp 259–472

Ferrari SF, Veiga LM, Pinto LP, Marsh LK (2013) Pithecidae. In: Mittermeier RA, Rylands AB, Wilson DE (eds) Handbook of the mammals of the world 3. Primates. Platyrrhini. Lynx Editions, Barcelona, pp 432–457

Ferrari SF, Guedes PG, Figueiredo-Ready WMB, Barnett AA (2014) Reconsidering the taxonomy of the black-faced uacaris, *Cacajao melanocephalus* group (Mammalia:Pitheciidae), from the northern Amazon basin. Zootaxa 3866(3):353–370

Franklin W (2011) Family Camelidae. In: Wilson DE, Mittermeier RH (eds) Handbook of the Mammals of the World. Vol. 2. Hoofed Mammals. Lynx Ediciones, Barcelona, pp 206–246

Gardner AL (2005a) Didelphidae. *Chacodelphys, Cryptonanus, Hyladelphys, Marmosops, Micoureus, Metachirus*. In: Gardner AL (ed) Mammals of South America. Volume 1. Marsupials, xenarthrans, shrews, and bats. University of Chicago Press, Chicago, pp 39–43

Gardner AL (2005b) Paucituberculata. In: Wilson DE, Reeder DM (eds) Mammal species of the world: a taxonomic and geographic reference. The Johns Hopkins University Press, Baltimore, pp 19–20

Gardner AL (2005c) Microbiotheria. In: Wilson DE, Reeder DM (eds) Mammal species of the world: a taxonomic and geographic reference. The Johns Hopkins University Press, Baltimore, p 21

Gardner AL (2005d) Xenarthra. In: Gardner AL (ed) Mammals of South America. Volume 1. Marsupials, xenarthrans, shrews and bats. University of Chicago Press, Chicago, pp 127–128

Gardner AL (2005e) Cingulata. In: Wilson DE, Reeder DM (eds) Mammal species of the world: a taxonomic and geographic reference. The Johns Hopkins University Press, Baltimore, pp 94–103
Gardner AL (2007a) Bradypodidae. *Bradypus*. In: Gardner AL (ed) Mammals of South America. Volume 1. Marsupials, xenarthrans, shrews and bats. University of Chicago Press, Chicago, pp 157–168
Gardner AL (2007b) Cyclopidae. *Cyclopes*. Myrmecophidae. *Myremcophaga*. *Tamandua*. In: Gardner AL (ed) Mammals of South America. Volume 1. Marsupials, xenarthrans, shrews, and bats. University of Chicago Press, Chicago, pp 168–177
Gardner AL (2007c) Mammals of South America, volume 1: marsupials, xenarthrans, shrews and bats. University of Chicago Press, Chicago
Gardner AL (2007d) American marsupials. In: Gardner AL (ed) Mammals of South America. volume 1. marsupials, xenarthrans, shrews, and bats. University of Chicago Press, Chicago, pp 1–117
Gardner AL (2007e) Order Didelphimorphia, Family Didelphidae, Genus *Hyladelphys*. In: Gardner AL (ed) Mammals of South America. Volume 1. Marsupials, xenarthrans, shrews, and bats. University of Chicago Press, Chicago, p 50
Gardner AL (2007f) Cohort Placentalia Owen, 1837. Magnorder Xenarthra Cope, 1889. In: Gardner AL (ed) Mammals of South America. volume 1. marsupials, xenarthrans, shrews, and bats. University of Chicago Press, Chicago, pp 127–128
Gardner AL (2007g) Pilosa. In: Gardner AL (ed) Mammals of South America. Volume 1. Marsupials, xenarthrans, shrews, and bats. University of Chicago Press, Chicago, pp 157–168
Gardner AL (2007h) Pilosa. Suborder Vermilingua. In: Gardner AL (ed) Mammals of South America. volume 1. marsupials, xenarthrans, shrews, and bats. University of Chicago Press, Chicago, pp 168–177
Gardner AL (ed) (2007 [2008]) Mammals of South America. Vol. 1: Marsupials, Xenarthrans, Shrews, and Bats. University of Chicago Press, Chicago.
Gardner AL, Creighton GK (2007a) Didelphidae. *Marmosops*. In: Gardner AL (ed) Mammals of South America. Volume 1. Marsupials, xenarthrans, shrews, and bats. University of Chicago Press, Chicago, pp 61–74
Gardner AL, Creighton GK (2007b) Didelphidae. *Micoureus*. In: Gardner AL (ed) Mammals of South America. Volume 1. Marsupials, xenarthrans, shrews, and bats. University of Chicago Press, Chicago, pp 74–82
Gardner AL, Dagosto M (2007) Didelphidae. *Metachirus*. In: Gardner AL (ed) Mammals of South America. Volume 1. Marsupials, xenarthrans, shrews, and bats. University of Chicago Press, Chicago, pp 35–39
Gardner AL, Naples VL (2007) Megalonychidae. *Choloepus*. In: Gardner AL (ed) Mammals of South America. Volume 1. Marsuials, xenarthrans, shrews, and bats. University of Chicago Press, Chicago, pp 165–168
Garshelis DL (2009) Ursidae *Tremarctos*. In: Wilson DE, Mittermeier RA (eds) Handbook of the mammals of the world. Vol. 1. Carnivores. Lynx Editions, Barcelona, pp 448–497
Glander KE (2013) Alouattinae. *Alouatta*. In: Mittermeier RA, Rylands AB, Wilson DE (eds) Handbook of the mammals of the world 3. Primates. Platyrrhini. Lynx Editions, Barcelona, pp 525–534
Glander KE, Pinto LP (2013) Alouattinae. *Alouatta*, Spix's Howler, *Alouatta discolor*. In: Mittermeier RA, Rylands AB, Wilson DE (eds) Handbook of the mammals of the world 3. Primates. Platyrrhini. Lynx Editions, Barcelona, p 528
Gomez-Laverde M, Voss RS, Pacheco V (2015) Cricetidae. *Handleyomys*. In: Patton JL, Pardiñas UFJ, D'Elía G (eds) Mammals of South America. Vol. 2. Rodents. University of Chicago Press, Chicago, pp 321–323
Gonçales PR, Teta P, Bonvicino CR (2015) Cricetidae. *Holochilus*. In: Patton JL, Pardiñas UFJ, D'Elía G (eds) Mammals of South America. Vol. 2. Rodents. University of Chicago Press, Chicago, pp 325–335

González EM, de Olivwie JA, Teta P (2015a) Cricetidae. *Juliomys*. In: Patton JL, Pardiñas UFJ, D'Elía G (eds) Mammals of South America. Vol. 2. Rodents. University of Chicago Press, Chicago, pp 92–96

González EM, de Oliveira JA, Pardiñas UFJ (2015b) Cricetidae. *Wilfredomys*. In: Patton JL, Pardiñas UFJ, D'Elía G (eds) Mammals of South America. Vol. 2. Rodents. University of Chicago Press, Chicago, pp 105–109

Groves CP (2005) Order Primates. In: Wilson DE, Reeder DM (eds) Mammal Species of the World: A Taxonomic and Geographic Reference, 3rd edn. Johns Hopkins University Press, Baltimore, pp 111–184

Grubb P (2005) Order Artiodactyla. In: Wilson DE, Reeder DM (eds) Mammal Species of the World, Third edn. The Johns Hopkins University Press, Baltimore, pp 637–722

Hafner MS (2015a) Geomyidae. In: Patton JL, Pardiñas UFJ, D'Elía G (eds) Mammals of South America. Vol. 2. Rodents. University of Chicago Press, Chicago, pp 49–51

Hafner MS (2015b) Heteromyidae. In: Patton JL, Pardiñas UFJ, D'Elía G (eds) Mammals of South America. Vol. 2. Rodents. University of Chicago Press, Chicago, pp 52–58

Helgen KM, Pinto CM, Kays R, Helgen LE, Tsuchiya MTN, Quinn A, Wilson DE, Maldonado JE (2013) Taxonomic revision of the olingos (*Bassaricyon*), with a description of a new species, the Olinguito. *ZooKeys* 324:1–83. https://doi.org/10.3897/zookeys.324.5827

Hoffmann RS, Smith AT (2005) Order Lagomorpha. In: Wilson DE, Reeder DM (eds) Mammal species of the world: a taxonomic and geographic reference. The Johns Hopkins University Press, Baltimore, pp 185–211

Honeycutt RL, Rowe DL, Gallardo MH (2003) Molecular systematics of the South American caviomorph rodents; relationships among species and genera in the family Octodontida. Mol Phylogenet Evol 26(3):476–489

Hutterer R (2005) Order Soricomorpha. In: Wilson DE, Reeder DM (eds) Mammal species of the world: a taxonomic and geographic reference. The Johns Hopkins University Press, Baltimore, pp 220–311

Konstant WR, Rylands AB (2013) Subfamily Atelinae, Genus *Ateles*. In: Mittermeier RA, Rylands AB, Wilson DE (eds) Handbook of the mammals of the world 3. Primates. Platyrrhini. Lynx Editions, Barcelona, pp 536–542

Lanzone C, Braun JK (2015) Cricetidae. *Salinomys*. In: Patton JL, Pardiñas UFJ, D'Elía G (eds) Mammals of South America. Vol. 2. Rodents. University of Chicago Press, Chicago, pp 555–556

Lanzone C, Braun JK, Patton JL, Paradiñas UFJ (2015) Cricetidae. *Eligmodontia*. In: Patton JL, Pardiñas UFJ, D'Elía G (eds) Mammals of South America. Vol. 2. Rodents. University of Chicago Press, Chicago, pp 508–522

Larivière S, Jennings AP (2009) Mustelidae. In: Patton JL, Pardiñas UFJ, D'ÉlíaG (eds) Handbook of the mammals of the world. Vol. 1. Carnivores. Lynx Editions, Barcelona, pp 564–656

Leite YLR, Loss AC (2015) Echimyidae. *Phyllomys*. In: Patton JL, Pardiñas UFJ, D'Elía G (eds) Mammals of South America. Vol. 2. Rodents. University of Chicago Press, Chicago, pp 915–928

Luna L (2015) Cricetidae. *Rhagomys*. In: Patton JL, Pardiñas UFJ, D'Elía G (eds) Mammals of South America. Vol. 2. Rodents. University of Chicago Press, Chicago, pp 580–583

Marsh LK (2014) A taxonomic revision of the saki monkeys, *Pithecia* Desmarest, 1804. Neotropical Primates 21(1):1–165

Marsh LK, Ferrari SF (2013) Pitheciinae, *Pithecia* Desmarest, 1804. In: Mittermeier RA, Rylands AB, Wilson DE (eds) Handbook of the mammals of the world 3. Primates. Platyrrhini. Lynx Editions, Barcelona, pp 475–477

Matioli S (2011) Family Cervidae. In: Wilson DE, Mittermeier RH (eds) Handbook of the Mammals of the World. Vol. 2. Hoofed Mammals. Lynx Ediciones, Barcelona, pp 336–343

Medici EP (2011) Tapiridae. In: Wilson DE, Mittermeier RA (eds) Handbook of the mammals of the world. Vol. 2. Hoofed mammals. Lynx Editions, Barcelona, p 182

Mittermeier RA, Rylands AB, Wilson DE (2013) Handbook of the mammals of the world. Vol. 3 Primates. Lynx Ediciones, Barcelona

Molina M, Molinari J (1999) Taxonomy of Venezuelan white-tailed deer (*Odocoileus*, Cervidae, Mammalia), based on cranial and mandibular traits. Canadian Journal of Zoology 77:632–645

Musser GG, Carleton MD (2005) Superfamily Muroidea. In: Wilson DE, Reeder DM (eds) Mammal species of the world: a taxonomic and geographic reference. The Johns Hopkins University Press, Baltimore, pp 894–1531

Myers P, Patton JL (2007) Genus *Lestoros* Oehser, 1934. In: Gardner AL (ed) Mammals of South America. Volume 1. Marsupials, xenarthrans, shrews, and bats. University of Chicago Press, Chicago, pp 124–126

Ortiz PE, Jayat JP (2015) Cricetidae. *Neotomys*. In: Patton JL, Pardiñas UFJ, D'Elía G (eds) Mammals of South America. Vol. 2. Rodents. University of Chicago Press, Chicago, pp 99–101

Pacheco V (2015a) Cricetidae. *Aepemys*. In: Patton JL, Pardiñas UFJ, D'Elía G (eds) Mammals of South America. Vol. 2. Rodents. University of Chicago Press, Chicago, pp 574–577

Pacheco V (2015b) Cricetidae. *Chilomys*. In: Patton JL, Pardiñas UFJ, D'Elía G (eds) Mammals of South America. Vol. 2. Rodents. University of Chicago Press, Chicago, pp 577–580

Pacheco V (2015c) Cricetidae. *Thomasomys*. In: Patton JL, Pardiñas UFJ, D'Elía G (eds) Mammals of South America. Vol. 2. Rodents. University of Chicago Press, Chicago, pp 617–682

Pardiñas UFJ, Bezerra AMR (2015) Cricetidae. *Gyldenstolpia*. In: Patton JL, Pardiñas UFJ, D'Elía G (eds) Mammals of South America. Vol. 2. Rodents. University of Chicago Press, Chicago, pp 222–225

Pardiñas UFJ, Patton JL (2015) Cricetidae. *Calassomys*. In: Patton JL, Pardiñas UFJ, D'Elía G (eds) Mammals of South America. Vol. 2. Rodents. University of Chicago Press, Chicago, pp 479–481

Pardiñas UFJ, Teta P (2015a) Cricetidae. *Deltamys*. In: Patton JL, Pardiñas UFJ, D'Elía G (eds) Mammals of South America. Vol. 2. Rodents. University of Chicago Press, Chicago, pp 219–222

Pardiñas UFJ, Teta P (2015b) Cricetidae. *Podoxymys*. In: Patton JL, Pardiñas UFJ, D'Elía G (eds) Mammals of South America. Vol. 2. Rodents. University of Chicago Press, Chicago, pp 268–269

Pardiñas UFJ, Teta P (2015c) Cricetidae. *Thalpomys*. In: Patton JL, Pardiñas UFJ, D'Elía G (eds) Mammals of South America. Vol. 2. Rodents. University of Chicago Press, Chicago, pp 274–277

Pardiñas UFJ, Teta P, D'Elía G (2015a) Cricetidae. *Abrawayaomys*. In: Patton JL, Pardiñas UFJ, D'Elía G (eds) Mammals of South America. Vol. 2. Rodents. University of Chicago Press, Chicago, pp 73–75

Pardiñas UFJ, Teta P, Alvarado-Serrano GL, Jayat JP, Ortiz PE, Gonçalves PR, de Olivwie JA (2015b) Cricetidae. *Akodon*. In: Patton JL, Pardiñas UFJ, D'Elía G (eds) Mammals of South America. Vol. 2. Rodents. University of Chicago Press, Chicago, pp 144–204

Pardiñas UFJ, D'Elía G, Teta P (2015c) Cricetidae. *Bibimys*. In: Patton JL, Pardiñas UFJ, D'Elía G (eds) Mammals of South America. Vol. 2. Rodents. University of Chicago Press, Chicago, pp 204–208

Pardiñas UFJ, Teta P, Ortiz PE, Jayat JP, Salazar-Bravo J (2015d) Cricetidae. *Necromys*. In: Patton JL, Pardiñas UFJ, D'Elía G (eds) Mammals of South America. Vol. 2. Rodents. University of Chicago Press, Chicago, pp 232–247

Pardiñas UFJ, Galliano A, Teta P (2015e) Cricetidae. *Reithrodon* In: Patton JL, Pardiñas UFJ, D'Elía G (eds) Mammals of South America. Vol. 2. Rodents. University of Chicago Press, Chicago, pp 559–566

Paresque R, Hanson JD (2015) Cricetidae. *Microakodontomys*. In: Patton JL, Pardiñas UFJ, D'Elía G (eds) Mammals of South America. Vol. 2. Rodents. University of Chicago Press, Chicago, pp 354–355

Patterson BD (2007a) Paucituberculata. In: Gardner AL (ed) Mammals of South America. Volume 1. Marsupials, xenarthrans, shrews, and bats. University of Chicago Press, Chicago, pp 119–124

Patterson BD (2007b) Paucituberculata. Caenolestidae. *Rhyncholestes*. In: Gardner AL (ed) Mammals of South America. Volume 1. Marsupials, xenarthrans, shrews, and bats. University of Chicago Press, Chicago, pp 126–127

Patterson BC, Rogers MA (2007) Microbiotheria. *Dromiciops*. In: Gardner AL (ed) Mammals of South America. Volume 1. Marsupials, xenarthrans, shrews, and bats. University of Chicago Press, Chicago, pp 117–119

Patterson BC, Smith MF, Teta P (2015) Cricetidae. *Abrothrix*. In: Patton JL, Pardiñas UFJ, D'Elía G (eds) Mammals of South America. Vol. 2. Rodents. University of Chicago Press, Chicago, pp 109–127

Patton JL (2015a) Sciuridae. In: Patton JL, Pardiñas UFJ, D'Elía G (eds) Mammals of South America. Vol. 2. Rodents. University of Chicago Press, Chicago, pp 3–5

Patton JL (2015b) Cricetidae. *Isthmomys* In: Patton JL, Pardiñas UFJ, D'Elía G (eds) Mammals of South America. Vol. 2. Rodents. University of Chicago Press, Chicago, pp 60-61

Patton JL (2015c) Cricetidae. *Punomys*. In: Patton JL, Pardiñas UFJ, D'Elía G (eds) Mammals of South America. Vol. 2. Rodents. University of Chicago Press, Chicago, pp 103–105

Patton JL (2015d) Cricetidae. *Lenoxus*. In: Patton JL, Pardiñas UFJ, D'Elía G (eds) Mammals of South America. Vol. 2. Rodents. University of Chicago Press, Chicago, pp 231–232

Patton JL (2015e) Cricetidae. *Scolomys*. In: Patton JL, Pardiñas UFJ, D'Elía G (eds) Mammals of South America. Vol. 2. Rodents. University of Chicago Press, Chicago, pp 445–448

Patton JL (2015f) Dinomyidae. *Dinomys*. In: Patton JL, Pardiñas UFJ, D'Elía G (eds) Mammals of South America. Vol. 2. Rodents. University of Chicago Press, Chicago, pp 783–786

Patton JL (2015g) Cuniculidae. *Cuniculus*. In: Patton JL, Pardiñas UFJ, D'Elía G (eds) Mammals of South America. Vol. 2. Rodents. University of Chicago Press, Chicago, pp 726–733

Patton JL (2015h) Echimyidae. *Myocastor*. In: Patton JL, Pardiñas UFJ, D'Elía G (eds) Mammals of South America. Vol. 2. Rodents. University of Chicago Press, Chicago, pp 1020–1022

Patton JL, da Silva MNF (2007) Didelphimorphia, Didelphidae, *Philander*. In: Gardner AL (ed) Mammals of South America. Volume 1. Marsupials, xenarthrans, shrews, and bats. University of Chicago Press, Chicago, pp 27–35

Patton JL, Emmons LH (2015a) Dasyproctidae. *Dasyprocta, Myoprocta*. In: Patton JL, Pardiñas UFJ, D'Elía G (eds) Mammals of South America. Vol. 2. Rodents. University of Chicago Press, Chicago, pp 733–762

Patton JL, Emmons LH (2015b) Abrocomidae. *Abrocoma, Cuscomys*. In: Patton JL, Pardiñas UFJ, D'Elía G (eds) Mammals of South America. Vol. 2. Rodents. University of Chicago Press, Chicago, pp 805–818

Patton JL, Emmons LH (2015c) Echimyidae. *Hoplomys*. In: Patton JL, Pardiñas UFJ, D'Elía G (eds) Mammals of South America. Vol. 2. Rodents. University of Chicago Press, Chicago, pp 940–942

Patton JL, Emmons LH (2015d) Echimyidae. *Mesomys*. In: Patton JL, Pardiñas UFJ, D'Elía G (eds) Mammals of South America. Vol. 2. Rodents. University of Chicago Press, Chicago, pp 943–950

Patton JL, Leite RN (2015) Echimyidae. *Proechimys*. In: Patton JL, Pardiñas UFJ, D'Elía G (eds) Mammals of South America. Vol. 2. Rodents. University of Chicago Press, Chicago, pp 950–988

Patton JL, Pardiñas UFJ, D'Elía G (2015) Mammals of South America, volume 2: rodents. University of Chicago Press, Chicago

Pearson OP (2007) Order Didelphimorphia, Family Didelphidae, Genus *Lestodelphys*. In: Gardner AL (ed) Mammals of South America. Volume 1. Marsupials, xenarthrans, shrews, and bats. University of Chicago Press, Chicago, pp 50–51

Percequillo AR (2015a) Cricetidae. *Phaenomys*. In: Patton JL, Pardiñas UFJ, D'Elía G (eds) Mammals of South America. Vol. 2. Rodents. University of Chicago Press, Chicago, pp 101–103

Percequillo AR (2015b) Cricetidae. *Aegialomys*. In: Patton JL, Pardiñas UFJ, D'Elía G (eds) Mammals of South America. Vol. 2. Rodents. University of Chicago Press, Chicago, pp 293–300

Percequillo AR (2015c) Cricetidae. *Cerradomys*. In: Patton JL, Pardiñas UFJ, D'Elía G (eds) Mammals of South America. Vol. 2. Rodents. University of Chicago Press, Chicago, pp 300–308

Percequillo AR (2015d) Cricetidae. *Eremoryzomys*. In: Patton JL, Pardiñas UFJ, D'Elía G (eds) Mammals of South America. Vol. 2. Rodents. University of Chicago Press, Chicago, pp 310–312

Percequillo AR (2015e) Cricetidae. *Euryoryzomys*. In: Patton JL, Pardiñas UFJ, D'Elía G (eds) Mammals of South America. Vol. 2. Rodents. University of Chicago Press, Chicago, pp 312–321

Percequillo AR (2015f) Cricetidae. *Hylaeamys*. In: Patton JL, Pardiñas UFJ, D'Elía G (eds) Mammals of South America. Vol. 2. Rodents. University of Chicago Press, Chicago, pp 335–346

Percequillo AR (2015g) Cricetidae. *Mindomys*. In: Patton JL, Pardiñas UFJ, D'Elía G (eds) Mammals of South America. Vol. 2. Rodents. University of Chicago Press, Chicago, pp 360–361

Percequillo AR (2015h) Cricetidae. *Nephelomys*. In: Patton JL, Pardiñas UFJ, D'Elía G (eds) Mammals of South America. Vol. 2. Rodents. University of Chicago Press, Chicago, pp 377–390

Percequillo AR (2015i) Cricetidae. *Oreoryzomys*. In: Patton JL, Pardiñas UFJ, D'Elía G (eds) Mammals of South America. Vol. 2. Rodents. University of Chicago Press, Chicago, pp 437–439

Percequillo AR (2015j) Cricetidae. *Oryzomys*. In: Patton JL, Pardiñas UFJ, D'Elía G (eds) Mammals of South America. Vol. 2. Rodents. University of Chicago Press, Chicago, pp 439–443

Percequillo AR (2015k) Cricetidae. *Sooretamys*. In: Patton JL, Pardiñas UFJ, D'Elía G (eds) Mammals of South America. Vol. 2. Rodents. University of Chicago Press, Chicago, pp 451–454

Percequillo AR, Weksler M (2015) Cricetidae. *Drymoreomys*. In: Patton JL, Pardiñas UFJ, D'Elía G (eds) Mammals of South America. Vol. 2. Rodents. University of Chicago Press, Chicago, pp 308–310

Pessôa LM, Tavae WC, Neves ACA, da Silva ALG (2015a) Echimyidae. *Thrichomys*. In: Patton JL, Pardiñas UFJ, D'Elía G (eds) Mammals of South America. Vol. 2. Rodents. University of Chicago Press, Chicago, pp 989–999

Pessôa LM, Tavares WC, de Oliveira JA, Patton JL (2015b) Echimyidae. *Trinomys*. In: Patton JL, Pardiñas UFJ, D'Elía G (eds) Mammals of South America. Vol. 2. Rodents. University of Chicago Press, Chicago, pp 999–1019

Pine RH, Handley CO (2007) Order Didelphimorphia, Family Didelphidae, Genus *Monodelphis*. In: Gardner AL (ed) Mammals of South America. Volume 1. Marsupials, xenarthrans, shrews, and bats. University of Chicago Press, Chicago, pp 82–107

Pinto LP (2013) Callicebinae, *Callicebus*. In: Mittermeier RA, Rylands AB, Wilson DE (eds) Handbook of the mammals of the world 3. Primates. Platyrrhini, vol 460. Lynx Editions, Barcelona, pp 463–472

Pinto LP, Ferrari SF (2013) Pitheciidae. Pitheciinae. *Cacajao*. In: Mittermeier RA, Rylands AB, Wilson DE (eds) Handbook of the mammals of the world 3. Primates. Platyrrhini. Lynx Editions, Barcelona, pp 481–483

Pires Costa L, Leite YLR, Mendes SL, Ditchfield AD (2005) Mammal conservation in Brazil. Conserv Biol 19(3):672–679

Prüfer K, Racimo F, Patterson N, Jay F, et al (2014) The complete genome sequence of a Neandertal from the Altai Mountains. Nature. 2014 Jan 2; 505(7481): 43–49

Ruiz-Garcia M, Alvarez D (2003) RFLP analysis of mtDNA from six platyrrhine genera: phylogenetic inferences. Folia Primatologica 74:59–70. https://doi.org/10.1159/000069998

Ruiz-García M, Pinedo-Castro MO (2009) Molecular systematics and phylogeography of the genus *Lagothrix* (Atelidae, Primates) by means of the mitocondrial COii gene. Folia Primatol 81:109–128. https://doi.org/10.1159/000315070

Ruiz-García M, Castillo MI, Vásquez C, Rodriguez K, Pinedo-Castro M, Shostell J, Leguizamonc N (2010) Molecular phylogenetics and phylogeography of the white-fronted capuchin (*Cebus albifrons*; Cebidae, Primates) by means of mtCOII gene sequences. Mol Phylogenet Evol 57:1049–1061

Ruiz-García M, Pinedo-Castro M, Shostell JM (2014) How many genera and species of woolly monkeys (Atelidae, Platyrrhine, Primates) are there? The first molecular analysis of Lagothrix flavicauda, an endemic Peruvian primate species. Molec Phylog Evol 79:179–198

Ruiz-García M, Pinedo-Castro M, Shostell JM (2014) "*Tapirus kabomani*" a new Amazonian tapir species? The fulminant rise and fall of a "typological species" by means of mitogenomics and craniometrics. J Mammal.

Rylands AB, Mittermeier RA (2013a) Callitrichidae (Marmosets and Tamarins). In: Handbook of the mammals of the world 3. Primates. Platyrrhini. Lynx Editions, Barcelona, pp 262–346

Rylands AB, Mittermeier RA (2013b) Cebidae (Squirrel Monkeys and Capuchins) In: Mittermeier RA, Rylands AB, Wilson DE (eds) Handbook of the mammals of the world 3. Primates. Platyrrhini. Lynx Editions, Barcelona, pp 348–389; 398–404

Rylands AB, Mittermeier RA (2013c) Pitheciidae (Titis, Sakis and Uacaris). In: Mittermeier RA, Rylands AB, Wilson DE (eds) Handbook of the mammals of the world 3. Primates. Platyrrhini. Lynx Editions, Barcelona, pp 432–457

Rylands AB, Mittermeier RA (2013d) Atelidae (Howlers, Spider and Woolly Monkeys and Muriquis). In: Mittermeier RA, Rylands AB, Wilson DE (eds) Handbook of the mammals of the world 3. Primates. Platyrrhini. Lynx Editions, Barcelona, pp 484–523

Rylands AB, Mittermeier RA, Silva JS Jr (2012) Neotropical primates: taxonomy and recently described species and subspecies. Int Zoo Yearb 46:11–24

Salazar-Bravo J (2015a) Cricetidae. *Chinchillula*. In: Patton JL, Pardiñas UFJ, D'Elía G (eds) Mammals of South America. Vol. 2. Rodents. University of Chicago Press, Chicago, pp 78–79

Salazar-Bravo J (2015b) Cricetidae. *Auliscomys*. In: Patton JL, Pardiñas UFJ, D'Elía G (eds) Mammals of South America. Vol. 2. Rodents. University of Chicago Press, Chicago, pp 472–479

Salazar-Bravo J (2015c) Cricetidae. *Calomys*. In: Patton JL, Pardiñas UFJ, D'Elía G (eds) Mammals of South America. Vol. 2. Rodents. University of Chicago Press, Chicago, pp 481–507

Salazar-Bravo J (2015d) Cricetidae. *Galenomys*. In: Patton JL, Pardiñas UFJ, D'Elía G (eds) Mammals of South America. Vol. 2. Rodents. University of Chicago Press, Chicago, pp 522–523

Salazar-Bravo J (2015e) Cricetidae. *Tapecomys*. In: Patton JL, Pardiñas UFJ, D'Elía G (eds) Mammals of South America. Vol. 2. Rodents. University of Chicago Press, Chicago, pp 556–559

Salazar-Bravo J (2015f) Cricetidae. *Sigmodon*. In: Patton JL, Pardiñas UFJ, D'Elía G (eds) Mammals of South America. Vol. 2. Rodents. University of Chicago Press, Chicago, pp 566–571

Salazar-Bravo J, Jayat JP (2015) Cricetidae. *Andinomys*. In: Patton JL, Pardiñas UFJ, D'Elía G (eds) Mammals of South America. Vol. 2. Rodents. University of Chicago Press, Chicago, pp 75–77

Schipper J, Chanson JS, Chiozza F et al (2008) The status of the world´s land and marine mammals: diversity, threat, and knowledge. Science 322:225–230

Sillero-Zubiri C (2009) Canidae (dogs). In: Wilson DE, Mittermeier RA (eds) Handbook of the mammals of the world. Vol. 1. Carnivores. Lynx Editions, Barcelona, pp 352–446

Solari S, Muñoz-Saba Y, Rodríguez-Mahecha JV, Defler TR, Ramírez-Chaves HE, Trujillo F (2013) Riqueza, endemismo y conservación de los mamíferos de Colombia. Mastozoología Neotropical 20(2):301–365

Spotorno AE, Patton JL (2015) Chinchillidae. *Chinchilla, Lagidium, Lagostomus*. In: Patton JL, Pardiñas UFJ, D'Elía G (eds) Mammals of South America. Vol. 2. Rodents. University of Chicago Press, Chicago, pp 762–783

Stein BR, Patton JL (2007a) Didelphidae. Didelphinae. *Chironectes*. In: Gardner AL (ed) Mammals of South America. Volume 1. Marsupials, xenarthrans, shrews, and bats. University of Chicago Press, Chicago, pp 14–17

Stein BR, Patton JL (2007b) Didelphidae, *Lutreolina*. In: Gardner AL (ed) Mammals of South America. Volume 1. Marsupials, xenarthrans, shrews, and bats. University of Chicago Press, Chicago, pp 25–27

Steppen SJ, Ramirez O (2015) Cricetidae. *Phyllotis*. In: Patton JL, Pardiñas UFJ, D'Elía G (eds) Mammals of South America. Vol. 2. Rodents. University of Chicago Press, Chicago, pp 535–555

Sunquist ME, Sunquist FC (2009) Felidae (cats). In: Wilson DE, Mittermeier RA (eds) Handbook of the mammals of the world. Vol. 1. Carnivores. Lynx Editions, Barcelona, pp 50–168

Taber AB, Altrichter M, Beck H, Gongora J (2011) Familia Tayassuidae (peccaries). In: Wilson DE, Mittermeier RA (eds) Handbook of the mammals of the world. Vol. 1. Carnivores. Lynx Editions, Barcelona, pp 292–307

Talebi M (2013) Atelidae, *Brachyteles*. In: Mittermeier RA, Rylands AB, Wilson DE (eds) Handbook of the mammals of the world 3. Primates. Platyrrhini. Lynx Editions, Barcelona, pp 547–549

Teta P, Paradiñas UFJ (2015a) Cricetidae. *Irenomys*. In: Patton JL, Pardiñas UFJ, D'Elía G (eds) Mammals of South America. Vol. 2. Rodents. University of Chicago Press, Chicago, pp 89–92

Teta P, Paradiñas UFJ (2015b) Cricetidae. *Notiomys*. In: Patton JL, Pardiñas UFJ, D'Elía G (eds) Mammals of South America. Vol. 2. Rodents. University of Chicago Press, Chicago, pp 135–138

Teta P, Paradiñas UFJ (2015c) Cricetidae. *Blarinomys*. In: Patton JL, Pardiñas UFJ, D'Elía G (eds) Mammals of South America. Vol. 2. Rodents. University of Chicago Press, Chicago, pp 208–211

Teta P, Paradiñas UFJ, D'Elía G (2015a) Cricetidae. *Chelemys*. In: Patton JL, Pardiñas UFJ, D'Elía G (eds) Mammals of South America. Vol. 2. Rodents. University of Chicago Press, Chicago, pp 127–132

Teta P, Paradiñas UFJ, D'Elía G (2015b) Cricetidae. *Geoxus*. In: Patton JL, Pardiñas UFJ, D'Elía G (eds) Mammals of South America. Vol. 2. Rodents. University of Chicago Press, Chicago, pp 132–135

Teta P, Pardiñas UFJ, D'Elía G (2015c) Cricetidae. *Thaptomys*. In: Patton JL, Pardiñas UFJ, D'Elía G (eds) Mammals of South America. Vol. 2. Rodents. University of Chicago Press, Chicago, pp 277–279

Teta P, Pardiñas UFJ, D'Elía G (2015d) Cricetidae. *Anotomys*. In: Patton JL, Pardiñas UFJ, D'Elía G (eds) Mammals of South America. Vol. 2. Rodents. University of Chicago Press, Chicago, pp 281–282

Teta P, Pardiñas UFJ, D'Elía G (2015e) Cricetidae. *Chibchanomys*. In: Patton JL, Pardiñas UFJ, D'Elía G (eds) Mammals of South America. Vol. 2. Rodents. University of Chicago Press, Chicago, pp 282–284

Teta P, Pardiñas UFJ, D'Elía G (2015f) Cricetidae. *Ichthyomys*. In: Pardiñas UFJ, D'Elía G (eds) Mammals of South America. Vol. 2. Rodents. University of Chicago Press, Chicago, pp 284–287

Teta P, Pardiñas UFJ, D'Elía G (2015g) Cricetidae. *Neusticomys*. In: Pardiñas UFJ, D'Elía G (eds) Mammals of South America. Vol. 2. Rodents. University of Chicago Press, Chicago, pp 287–293

Teta P, Pardiñas UFJ, Salazar-Bravo J, D'Elía G (2015h) Cricetidae. *Loxodontomys*. In: Pardiñas UFJ, D'Elía G (eds) Mammals of South America. Vol. 2. Rodents. University of Chicago Press, Chicago, pp 531–535

Thorington RW Jr, Hoffmann RS (2005) Family Sciuridae. In: Wilson DE, Reader DM (eds) Mammal Species of the World. The John Hopkins University Press, Baltimore, pp 754–818

Timm RM, Patterson BD (2007) Caenolestidae, *Caenolestes*. In: Gardner AL (ed) Mammals of South America. Volume 1. Marsupials, xenarthrans, shrews, and bats. University of Chicago Press, Chicago, pp 120–124

Tognelli MF (2004) Assessing the utility of indicator groups for the conservation of South American terrestrial mammals. Biol Conserv 121:409–417

Tribe CJ (2015) Cricetidae. *Rhipidomys*. In: Gardner AL (ed) Mammals of South America. Volume 1. Marsupials, xenarthrans, shrews, and bats. University of Chicago Press, Chicago, pp 583–617

Verzi DH, Díaz M, Barquez RM (2015a) Octodontidae. *Aconaemys*. In: Gardner AL (ed) Mammals of South America. Volume 1. Marsupials, xenarthrans, shrews, and bats. University of Chicago Press, Chicago, pp 1023–1029

Verzi DH, Díaz M, Barquez RM (2015b) Octodontidae. *Octondontomys*. In: Gardner AL (ed) Mammals of South America. Volume 1. Marsupials, xenarthrans, shrews, and bats. University of Chicago Press, Chicago, pp 1035–1038

Verzi DH, Díaz M, Barquez RM (2015c) Octodontidae. *Octomys.*. In: Gardner AL (ed) Mammals of South America. Volume 1. Marsupials, xenarthrans, shrews, and bats. University of Chicago Press, Chicago, pp 1038–1039

Verzi DH, Díaz M, Barquez RM (2015d) Octodontidae. *Spalacopus*. In: Gardner AL (ed) Mammals of South America. Volume 1. Marsupials, xenarthrans, shrews, and bats. University of Chicago Press, Chicago, pp 1039–1043

Vilela JF, Gonçales PR, de Olivwie JA (2015) Cricetidae. *Brucepattersonius*. In: Gardner AL (ed) Mammals of South America. Volume 1. Marsupials, xenarthrans, shrews, and bats. University of Chicago Press, Chicago, pp 211–219

Voss RA (2015a) Cricetidae. *Delomys*. In: Gardner AL (ed) Mammals of South America. Volume 1. Marsupials, xenarthrans, shrews, and bats. University of Chicago Press, Chicago, pp 79–83

Voss RA (2015b) Cricetidae. *Lundomys*. In: Gardner AL (ed) Mammals of South America. Volume 1. Marsupials, xenarthrans, shrews, and bats. University of Chicago Press, Chicago, pp 346–348

Voss RA (2015c) Cricetidae. *Pseudoryzomys*. In: Gardner AL (ed) Mammals of South America. Volume 1. Marsupials, xenarthrans, shrews, and bats. University of Chicago Press, Chicago, pp 443–445

Voss RA (2015d) Cricetidae. *Zygodontomys*. In: Gardner AL (ed) Mammals of South America. Volume 1. Marsupials, xenarthrans, shrews, and bats. University of Chicago Press, Chicago, pp 460–465

Voss RA (2015e) Erethizontidae. In: Gardner AL (ed) Mammals of South America. Volume 1. Marsupials, xenarthrans, shrews, and bats. University of Chicago Press, Chicago, pp 786–805

Voss RA, Gardner AL, Jansa SA (2007a) Didelphidae. *Chacodelphys*. In: Gardner AL (ed) Mammals of South America. Volume 1. Marsupials, xenarthrans, shrews, and bats. University of Chicago Press, Chicago, pp 39–43

Voss RA, Lunde PW, Jansa SA (2007b) Didelphidae, *Cryptonanus*. In: Gardner AL (ed) Mammals of South America. Volume 1. Marsupials, xenarthrans, shrews, and bats. University of Chicago Press, Chicago, pp 40–43

Voss RS, Helgen KM, Jansa SA (2014) Extraordinary claims require extraordinary evidence: a comment on Cozzuol et al. (2013). J Mammal 95(4):893–898

Weksler M (2015a) Cricetidae. "*Handleyomys*". In: Gardner AL (ed) Mammals of South America. Volume 1. Marsupials, xenarthrans, shrews, and bats. University of Chicago Press, Chicago, pp 323–325

Weksler M (2015b) Cricetidae. *Sigmodontomys*. In: Gardner AL (ed) Mammals of South America. Volume 1. Marsupials, xenarthrans, shrews, and bats. University of Chicago Press, Chicago, pp 449–451

Weksler M (2015c) Cricetidae. *Tanyuromys*. In: Gardner AL (ed) Mammals of South America. Volume 1. Marsupials, xenarthrans, shrews, and bats. University of Chicago Press, Chicago, pp 454–456

Weksler M, Bonvicino CR (2015a) Cricetidae. *Neacomys*. In: Gardner AL (ed) Mammals of South America. Volume 1. Marsupials, xenarthrans, shrews, and bats. University of Chicago Press, Chicago, pp 361–369

Weksler M, Bonvicino CR (2015b) Cricetidae. *Oligoryzomys*. In: Gardner AL (ed) Mammals of South America. Volume 1. Marsupials, xenarthrans, shrews, and bats. University of Chicago Press, Chicago, pp 417–437

Weksler M, Lóss S (2015) Cricetidae. *Melanomys*. In: Gardner AL (ed) Mammals of South America. Volume 1. Marsupials, xenarthrans, shrews, and bats. University of Chicago Press, Chicago, pp 348–354

Weksler M, Valqui M (2015) Cricetidae. *Amphinectomys*. In: Gardner AL (ed) Mammals of South America. Volume 1. Marsupials, xenarthrans, shrews, and bats. University of Chicago Press, Chicago, pp 298–300

Wetzel RM, Garder AL, Redford KH, Eisenberg JF (2007) Order Cingulata Illiger, 1811. In: Gardner AL (ed) Mammals of South America. Volume 1. Marsupials, xenarthrans, shrews, and bats. University of Chicago Press, Chicago, pp 128–157

Wilson DE, Mittermeier RA (eds) (2009) Handbook of the mammals of the world. Vol. 1. Carnivores. Lynx Editions, Barcelona

Wilson DE, Mittermeier RH (2011) Handbook of the Mammals of the World. Vol. 2. Hoofed Mammals. Lynx Edicions, Barcelona

Wilson DE, Reeder DM (2005) Mammal species of the world, 3rd edn. Bucknell Press, Lewisburg PA

Woodman N, Péfaur J (2015) Order Soricomorpha Gregory, 1910. In: Gardner AL (ed) Mammals of South America. Volume 1. Marsupials, xenarthrans, shrews, and bats. University of Chicago Press, Chicago, pp 177–187

Woods CA, Kilpatrick CW (2005) Infraorder Hystricognathi. In: Wilson DE, Reeder DM (eds) Mammal species of the world: a taxonomic and geographic reference. The Johns Hopkins University Press, Baltimore, pp 1538–1600

Wozencraft WC (2005) Order Carnivora. In: Wilson DE, Reeder DM (eds) Mammal Species of the World, Third edn. The Johns Hopkins University Press, Baltimore, pp 532–628

Permissions

1.1 *Megatherium americanum,* old mount (by Juan Bautista Bru de Ramón, described by Cuvier, Georges [1796]). Notice sur le squelette d'une très grande espèce de quadrupède inconnue jusqu'à présent, trouvé au Paraguay et déposé au cabinet d'histoire naturelle de Madrid. Magasin encyclopedique. Public Domain
1.2 Young George Cuvier, "Father of Paleontology," 1769–1832 (from Règne animal—1829 vol I), Public Domain
1.3 Alexander von Humboldt, 1769–1859 (portrait painted by Friedrich Georg Weitsch), Public Domain
1.4 Charles Darwin (1809–1882) (taken about 1784 by Leonard Darwin), Public Domain
1.5 Alcide Dessalines dÓrbigny (1802–1857), Public Domain
1.6 Florentino Ameghino (left) and Carlos Amegino (right) (Verlag Naturhistorisches Museum Wien; E. Buffetaut 2013), Public Domain
1.7 The Borhyaenidae show many features of canines, leading Florentino to consider erroneously this group as ancestral to the dog family (Original by Roman Uchytel)
1.8 Santiago Roth, https://www.ecured.cu/Santiago_Roth, Public Domain
1.9 John Bell Hatcher, Wikipedia, Public Domain
1.10 John Berryman Scott, Biographical memoir of William Berryman Scott 1858–1947, 1948, National Academy of Sciences of the United States of America, Public Domain
1.11 *Pyrotherium* sp. lived in what is now Argentina, during the Early Oligocene. Its body was 3 m long and 1.50 m tall at the shoulders. It has robust legs and a short proboscis and flat, forward-facing tusks (two in the upper jaw, one in the lower one). It has sometimes been seen as a descendent of the Xenungulata (Original by Roman Uchytel).
1.12 Jean Albert Gaudry, Wikipedia, Public Domain

T. Defler, *History of Terrestrial Mammals in South America*, Topics in Geobiology 42, https://doi.org/10.1007/978-3-319-98449-0

1.13 G. G. Simpson. American Museum of Natural History image library, by permission
1.14 Geomagnetic polarity 0–169 Ma since the Middle Jurassic. Dark areas denote periods where the polarity matches today's polarity, while light areas denote periods where that polarity is reversed. (from Wikipedia "Geomagnetic Reversal", Public Domain)
1.15 Fossil evidence for continental drift, United States Geological Survey, Public Domain
1.16 Pangaea and the original position of the continents (Original by Christopher Scolese, Paleomap Project, Permission)

2.1 *Adelobasileus* the most primitive mammal known from the Triassic of 225 million years of Texas (by N. Tamura permission)
2.2 Relation of continents about 200 million years ago, during the time of the first mammals. (238 k) high definition? (Original by Christopher Scotese, Paleomap Project, permission)
2.3 Origin of the first southern tribosphenic australosphenidan mammal was discovered in Argentina in 2001 in Middle Jurassic strata (Original by Diego Casallas-Pabón)
2.4 Continental arrangement at the end of the Jurassic (152 Ma) (by Christopher Scotese Permission, Paleomap Project)
2.5 Recreation of (a) *Asfaltomylos patagonico*, a Mesozoic (168–161 Ma) tribosphenic mammal (related to Monotremes); (b) jaw bone and tribosphenic teeth discovered in Patagonia (drawing based on Rauhut et al. 2002) (Original by Marie Giraud-López, Grupo Evolución y Ecología de Mamíferos Neotropicales, Universidad Nacional de Colombia)
2.6 Photograph of ichnites, open access, Kümmell SB, Frey E (2014) Range of Movement in Ray I of Manus and Pes and the Prehensility of the Autopodia in the Early Permian to Late Cretaceous Non-Anomodont Synapsida. PLoS ONE 9(12): e113911. doi:10. 1371/journal.pone.0113911 CC-BY-4.0, https://creativecommons.org/licenses/by/4.0/legalcode
2.7 Continental positions in Cretaceous 135 Ma (Original by C. Scortese, Paleomap Project)
2.8 *Vincelestes neuquenianus* (Original by Roman Uchytel)
2.9 Continental positions in Cretaceous 94 Ma (Original by C. Scortese, Paleomap Project)
2.10 *Cronopio dentiacutus* (by Guillermo Rougier permission)
2.11 *Obdurodon dicksoni* (Original by Marie Giraud-López, Grupo Evolución y Ecología de Mamíferos Neotropicales, Universidad Nacional de Colombia)
2.12 Biogeographic provinces of North Gondwana and South Gondwana (Original by Diego Casallas-Pabón)

3.1 Map just before Chixulub impact (Original by C. Scotese, Paleomap Project)
3.2 *Pucadelphys andinus* (Original by Kelsey Van Horn)
3.3 *Mayulestes ferox* (Original by Roman Uchytel)

3.4 *Alcidedorbignya* skeleton (de Muizon C 1991. La fauna de mamíferos de Tiupampa (Paleoceno inferior, Formación Santa Lucia), Bolivia. In: Suárez Soruco R (ed) Fósiles y facies de Bolivia. Vol. 1. Vertebrados. Revista Técnica de Yacimientos Petroliferos Fiscales Bolivianos 12/3-4:575-624) by C. de Muizon, Permission
3.5 *Carodnia vieirai* (Original by Roman Uchytel)

4.1 *Sinodelphys szalayi* (Original by Marie Giraud-López, Grupo Evolución y Ecología de Mamíferos Neotropicales, Universidad Nacional de Colombia)
4.2 Skull of *Asiatherium* (oldest fossil marsupial) (by T. E. Williamson, S. L. Brusatte, G. P. Wilson, Wikipedia, CC-BY-3.0 Unported, https://species.wikimedia.org/wiki/Wikispecies:Collaboration_with_ZooKeys_and_PhytoKeys
4.3 *Alphadon* (Original by Marie Giraud-López, Grupo Evolución y Ecología de Mamíferos Neotropicales, Universidad Nacional de Colombia)
4.4 *Peradectes* (Original by Marie Giraud-López, Grupo Evolución y Ecología de Mamíferos Neotropicales, Universidad Nacional de Colombia)
4.5 *Pucadelphys* (Original by Kelsey Van Horn)
4.6 *Argyrolagus* (Original by Marie Giraud-López, Grupo Evolución y Ecología de Mamíferos Neotropicales, Universidad Nacional de Colombia)
4.7 *Dromiciops* (by José Luis Bartheld, Wikipedia, CC-BY-2.0, https://creativecommons.org/licenses/by/2.0/legalcode
4.8 *Mayulestes* (Original by Roman Uchytel)
4.9 *Cladosictis lustratus* (by Charles Knight, in W. B. Scott, 1913, *A History of Land Mammals in the Western Hemisphere*, Public Domain)
4.10 *Borhyaena* (Original by Roman Uchytel)
4.11 *Prothylacinus patagonicus* (by Charles Knight, in W. B. Scott, 1913, *A History of Land Mammals in the Western Hemisphere*, Public Domain)
4.12 *Lycopsis longirostris* (by Ryan Sommer, CC-BY-SA-2.0 Generic, Wikipedia)
4.13 *Dukecynus magnus* (Original by Roman Uchytel)
4.14 *Callistoe vincei* (Original by Roman Uchytel)
4.15 *Arminiheringia* (Original by Roman Uchytel)
4.16 *Proborhyaena gigantea* (Original by Roman Uchytel)
4.17 *Thylacosmilus* attacking *Tapirus* (Original by Roman Uchytel)

5.1 *Meniscotherium* (by Robert Bruce, in W. B. Scott, 1913, *A History of Land Mammals in the Western Hemisphere*, Public Domain)
5.2 *Nesodon imbricatus* (Original by Jorge W. Moreno-Bernal)
5.3 *Thoatherium* (by Charles Knight, 1913, W. B. Scott, *A History of Land Mammals in the Western Hemisphere*, Public Domain)
5.4 *Diadiaphorus majusculus* (Original by Roman Uchytel)
5.5 a. *Paraphysornis* (Original by Roman Uchytel). b. *Kelenken* (Original by Roman Uchytel)
5.6 Macrauchenia (Original by Roman Uchytel)
5.7 *Theosodon garretorum* (by Charles Knight, in W. B. Scott, 1913, *A History of Land Mammals in the Western Hemisphere*, Public Domain)

5.8 Notoungulate relationships (Original by Diego Casallas-Pabón, based on Rose, 1996)
5.9 *Notostylops* (Original by Roman Uchytel)
5.10 *Thomashyuxleya* (Original by Roman Uchytel)
5.11 *Homalodotherium* (Original by Roman Uchytel)
5.12 *Huilatherium* (Original by Restron)
5.13 *Toxodon* (Original by Roman Uchytel)
5.14 *Adinotherium* (Original by Roman Uchytel)
5.15 *Nesodon imbricatus* (Original by Roman Uchytel)
5.16 a. *Campanorco inauguralis*, b. *Coquenia bondi* (from Babot et al. 2017. Mamíferos paleógenos del subtrópico de Argentina: síntesis de estudios estratigráicos cronológicos y taxonómicos. by Permission)
5.17 *Hemihegetotherium trilobus* (by Velizar Simeonovski, by Permission)
5.18 *Astrapotherium magnum* (Original by Roman Uchytel)
5.19 *Astrapotherium magnum* head (by Robert Bruce Horsfall in W. B. Scott, 1913, *A History of Land Mammals in the Western Hemisphere*, Public Domain)
5.20 *Granastrapotherium* (Original by Roman Uchytel)
5.21 *Pyrotherium romeroi* (by Robert Bruce Horsfall in W. B. Scott, 1913, *A History of Land Mammals in the Western Hemisphere*, Public Domain)
5.22 *Pyrotherium romeroi* (Original by Roman Uchytel)
5.23 *Colombitherium tolimense* (Original by Zimices-Julián Bayona)
5.24 *Carodnia vieiri* (Original by Roman Uchytel)

6.1 *Pholidota* (by Piekfrosh, Wikipedia, CC-BY-SA-3.0 Unported license, https://creativecommons.org/licenses/by-sa/3.0/legalcode
6.2 Aardvark (by MontageMan, CC-BY-2.5 Generic License (https://creativecommons.org/licenses/by/2.5/legalcode), Wikipedia
6.3 *Eurotamandua* (Original by Marie Giraud-López, Grupo Evolución y Ecología de Mamíferos Neotropicales, Universidad Nacional de Colombia)
6.4 Phylogeny of Xenarthra (Original by Diego Casallas-Pabón, Applied Biodiversity Foundation)
6.5 Thoracic vertebrae xenarthric (Original by Diego Casallas-Pabón, Applied Biodiversity Foundation)
6.6 Lumbar vertebrae xenarthric (Original by Diego Casallas-Pabón, Applied Biodiversity Foundation)
6.7 Osteoderms (Original by Diego Casallas-Pabón, Applied Biodiversity Foundation)
6.8 Armadillo (by http://www.birdphotos.com, CC-BY-3.0 Unported License), (https://creativecommons.org/licenses/by/3.0/legalcode), Wikipedia
6.9 *Utaetus* (Original by Marie Giraud-López, Grupo Evolución y Ecología de Mamíferos Neotropicales, Universidad Nacional de Colombia)
6.10 *Holmesina* (Original by Roman Uchytel)
6.11 *Peltephilus* (Original by Roman Uchytel)
6.12 *Doedicurus clavicaudatus* (Original by Roman Uchytel)
6.13 *Glyptodon* (Original by Roman Uchytel)

6.14 *Parapropalaehoplophorus* (by Velizar Simeonovski, by Permission)
6.15 *Megatherium americanum* (Original by Roman Uchytel)
6.16 *Glossotherium robustum* (Original by Roman Uchytel)
6.17 *Thalassocnus natans* (Original by Roman Uchytel)

7.1 Hystricognathi skulls (CC-SA-3.0 Unported license, Wikipedia), https://creativecommons.org/licenses/by-sa/3.0/legalcode
7.2 *Canaanimys* rodent (by MJ Orliac, by Permission)
7.3 *Platypittamys brachyodon* (Original by V. Cadena)
7.4 *Cuniculus* (by Hans Hillewaert, CC BY-SA 3.0), https://creativecommons.org/licenses/by-sa/3.0/legalcode
7.5 *Eocardia* skeleton (by Robert Bruce Horsfall—William B Scott (ed) Reports of the Princeton University Expeditions to Patagonia. 1896-1899. 1901-32 [v. 1, 1903]. Public Domain
7.6 *Erithizon* (by Maaatnad, (CC BY-SA 3.0), https://creativecommons.org/licenses/by-sa/3.0/legalcode
7.7 *Chinchilla* (by Thirteen Squared, Public Domain)
7.8 *Diplomys caniceps* (by R. Mintern, 1876, Public Domain)
7.9 *Tympoanoctomys,* an extinct Octodontidae (by F. Ameghino, 1889, Public Domain)
7.10 *Octodon degu* (by Pierre Camateros, CC-BY-SA 3.0), https://creativecommons.org/licenses/by-sa/3.0/legalcode
7.11 *Ctenomys brasiliensis* (in Alcide Dessdalines d´Orbigny) Public Domain
7.12 *Josephoartigasia monesi* (Original by Roman Uchytel)
7.13 Female capybara with young (by Fidel León Darder, CC-BY-SA 3.0 Unported, Wikipedia), https://creativecommons.org/licenses/by-sa/3.0/legalcode

8.1 Comparison of skulls Platyrrhini vs. Catarrhini (Original by Diego Casallas-Pabón)
8.2 Odd drawing of *Tremacebus harringtoni* (by Rusconi, 1935, Public Domain)
8.3 *Killikaike* (Tejedor et al. 2006, by Permission from PNAS)
8.4 *Cebupithecia sarmientoi* (Original by Roman Uchytel) and skull (Original by Diego Casallas-Pabón, Applied Biodiversity Foundation)
8.5 *Cartelles coimbrafilhoi* (Original by Roman Uchytel)

9.1 K/T boundary map (Original by C. R. Scotese, Paleomap Project)
9.2 Middle Eocene 50.2 Ma (Original by C. R. Scotese, Paleomap Project)
9.3 Connection SA to Antarctica (Google Earth Image, US Geological Survey), Public Domain
9.4 Location of Seymour Island (Original by Diego Casallas-Pabón, Applied Biodiversity Foundation)
9.5 Seymour-Marambio Island photo by ACA Genta, CC-BY-SA-4.0 International License, https://creativecommons.org/licenses/by-sa/4.0/legalcode
9.6 Isthmus of Scotia, land bridge between SA and Antarctic (Original by Diego Casallas-Pabón, Applied Biodiversity Foundation)

9.7 Temperature change throughout the Cenozoic (by R. A. Rohde for Global Warming Art, Wikipedia, GFDL Free Documentation License), https://commons.wikimedia.org/wiki/Commons:GNU_Free_Documentation_License,_version_1.2
9.8 *Antarctodolops* sp. (by William Stout, *Frozen in Time,* Permission)
9.9 *Antarctodon sobrali* (Original by Zimices or Julián Bayona)
9.10 *Notiolofos* (Original by Marie Giraud-López, Grupo Evolución y Ecología de Mamíferos Neotropicales, Universidad Nacional de Colombia)10.1 Mega wetlands in middle Miocene (from Hoorn et al. 2010, permission Wiley and Sons)

10.1 Possible extent of mega-wetlands in the middle Miocene (based on Hoorn C, Wesselingh FP, Hovikoski J, Guerrero J (2010) The development of the Amazonian mega-wetland (Miocene; Brazil, Colombia, Peru, Bolivia). In: Hoorn C, Wesselingh F (eds) Amazonia: landscape and species evolution. Wiley-Blackwell, Oxford, pp 123-142) by Diego Casallas Pabón, Applied Biodiversity Foundation)
10.2 *Dukecynus magnus* (Original by Roman Uchytel)
10.3 *Anachlysictis gracilis* (Original by Roman Uchytel)
10.4 *Hondathentes* cazandor (Original by Zimeces or Julián Bayona)
10.5 *Scirraltherium* (Original by Marie Giraud-López, Grupo Evolución y Ecología de Mamíferos Neotropicales, Universidad Nacional de Colombia)
10.6 *Megadolodus* (Original by Zimices or Julián Bayona)
10.7 *Theosodon* (by Charles Knight, 1929, in WB Scott, *A History of Land Mammals in the Western Hemisphere,* Macmillan) Public Domain
10.8 *Miocochilius* (Original by Marie Giraud-López, Grupo Evolución y Ecología de Mamíferos Neotropicales, Universidad Nacional de Colombia)
10.9 *Huilatherium* (Original by Zimices or Julián Bayona)
10.10 *Pericotoxodon* (Original by Marie Giraud-López, Grupo Evolución y Ecología de Mamíferos Neotropicales, Universidad Nacional de Colombia)
10.11 *Granastrapotherium* (Original by Roman Uchytel)
10.12 Skull of *Cebupithecia* (Original by Diego Casallas-Pabón, Applied Biodiversity Foundation)

11.1 Possible extent of mega wetlands in the middle Miocene (based on Hoorn C, Wesselingh FP, Hovikoski J, Guerrero J (2010) The development of the Amazonian mega-wetland (Miocene; Brazil, Colombia, Peru, Bolivia). In: Hoorn C, Wesselingh F (eds) Amazonia: landscape and species evolution. Wiley-Blackwell, Oxford, pp 123-142) by Diego Casallas Pabón, Applied Biodiversity Foundation)

12.1 Caribbean map (CIA World Factbook), Public Domain
12.2 *Megalocnus* (Original by Marie Giraud-López, Grupo Evolución y Ecología de Mamíferos Neotropicales, Universidad Nacional de Colombia)
12.3 Collection sites for Caribbean monkeys (Original by Diego Casallas-Pabón, Applied Biodiversity Foundation)

12.4 *Capromys pilorides* (by Jensflorian, GNU Free Documentation License, Version 1.2), Wikipedia, https://commons.wikimedia.org/wiki/Commons:GNU_Free_Documentation_License,_version_1.2
12.5 *Megalomys desmarestii* (from Gervais, 1855, Histoire naturelle des mammiferes.., vol 1, L. Curmer, Paris), Public Domain
12.6 *Hyrachyus* (by RB Horsfall, 1913, *Land Mammals of the Western Hemisphere*), Public Domain
12.7 *Solenodon* (by Sebaz86556, CC BY-SA 3.0), https://creativecommons.org/licenses/by-sa/3.0/legalcode
12.8 Caribbean latest Eocene–Early Oligocene palaeography (Original by Diego Casallas-Pabón, Applied Biodiversity Foundation)

13.1 Geology of Central America (Original by Diego Casallas-Pabón, Applied Biodiversity Foundation)
13.2 Recurring glacial advances and retreats of savannas and forest vegetation (Original by Diego Casallas-Pabón, Applied Biodiversity Foundation)
13.3 *Cuvieronius hyodon* (Original by Roman Uchytel)
13.4 *Notiomastodon platensis* (Original by Roman Uchytel)
13.5 *Platygonus* (by Charles Knight, 1922, Public Domain)
13.6 *Hemiauchenia* (Original by Roman Uchytel)
13.7 *Palaeolama niedai* (Original by Roman Uchytel)
13.8 *Chapalmalania* (Original by Roman Uchytel)
13.9 *Onohippidion* and *Hippidion* (by Camila Bernardes, Fernando L. Sicuro, Leonardo S. Avilla, and André E.P. Pinheiro, 2012, Rostral reconstruction of South American hippidiform equids: New anatomical and ecomorphological inferences. *Acta Palaeontologica Polonica* 58 (4), 2013: 669-678 doi: https://doi.org/10.4202/app.2011.0107 (CC BY 2.0), https://creativecommons.org/licenses/by/2.0/legalcode
13.10 *Hippidion* (Original by Roman Uchytel)
13.11 *Canis dirus* (Original by Roman Uchytel)
13.12 *Panthera onca mesembrina* (Original by Roman Uchytel)
13.13 Lion and saber-toothed cat (Original by Roman Uchytel)
13.14 *Smilodon populator* (Original by Roman Uchytel)
13.15 *Xenosmilus hodsonae* (Original by Roman Uchytel)
13.16 *Homotherium venezuelensis* (Original by Roman Uchytel)
13.17 *Smilodon gracilis* (Original by Roman Uchytel)
13.18 *Arctotherium angustidens* (Original by Roman Uchytel)
13.19 *Glyptotherium* (by Sergiodlarosa CC-BY-SA-4.0, https://creativecommons.org/licenses/by-sa/4.0/legalcode

14.1 Changes in vegetation during maximum glaciation (Original by Diego Casallas-Pabón, Applied Biodiversity Foundation)
14.2 Changes in land mammal fauna during last nine years (Original by Diego Casallas-Pabón, Applied Biodiversity Foundation)

15.1 *Caluromys philander* (by Moises Silva Lima, CC-BY-SA-2.0 Generic license), https://creativecommons.org/licenses/by/2.0/legalcode
15.2 *Didelphis marsupiales* (by Juan Tello, CC-BY-SA-2.0 Generic license), https://creativecommons.org/licenses/by/2.0/legalcode
15.3 *Marmosa* sp. (by Ramon Campos, CC-BY-SA-2.5 Generic license, Wikipedia), https://creativecommons.org/licenses/by-sa/2.5/legalcode
15.4 *Marmosops incanus* (by Ramon Campos, CC-BY-SA-4.0, https://creativecommons.org/licenses/by-sa/4.0/legalcode CC-BY-SA-2.0 Generic license), https://creativecommons.org/licenses/by/2.0/legalcode 15.6
15.5 *Metachirus nudicaudatus* (by Haplochromis, CC-BY-SA-3.0 Generic license) https://creativecommons.org/licenses/by/3.0/legalcode
15.6 *Monodelphis domestica* (by E. Hill, 2006, *Marsupial Genome Sheds Light on the Evolution of Immunity*. Hill E, PLoS Biology Vol. 4/3/2006 e75 https://doi.org/10.1371/journal.pbio.0040075), (CC BY 4.0), https://creativecommons.org/licenses/by/4.0/legalcode
15.7 *Dromiciops gliroides* (by José Luis Bartheld, CC-BY-SA-2.0 Generic license), https://creativecommons.org/licenses/by/2.0/legalcode
15.8 *Dasypus novemcinctus* (by Jerry Segraves, The copyright holder of this file allows anyone to use it for any purpose, provided that the copyright holder is properly attributed. Redistribution, derivative work, commercial use, and all other use are permitted).
15.9 *Chaetophractus vellerosus* (by Arnaud Boucher, Public Domain)
15.10 *Bradypus variegatus* (by Stefan Laube, Public Domain)
15.11 *Choloepus hoffmanni* (by Masteraah, CC BY-SA 2.0 DE), https://creativecommons.org/licenses/by/2.0/legalcode
15.12 *Cyclopes didactylus* (by Quinten Questel, CC-BY-SA-3.0, Wikipedia), https://creativecommons.org/licenses/by/3.0/legalcode
15.13 *Myrmecophaga tridactyla* (by Marcelo Calzans, CC-BY-SA-2.0 Generic license Wikipedia), https://creativecommons.org/licenses/by/2.0/legalcode
15.14 *Callithrix pygmaea* (by Malene Thyssen, CC-BY-SA-3.0 Generic license Wikipedia), https://creativecommons.org/licenses/by/3.0/legalcode, GNU Free Documentation License
15.15 *Callithrix geoffroyi* (by Mosztics Attila, Public Domain
15.16 *Leontopithecus rosalia* (by Trisha Shears, CC-BY-SA-3.0 Generic license Wikipedia), https://creativecommons.org/licenses/by/3.0/legalcode
15.17 *Saguinus imperador* (by Brocken Inaglory, CC-BY-SA-3.0 Unported, Wikipedia), https://creativecommons.org/licenses/by-sa/3.0/legalcode
15.18 *Saguinus oedipus* (by © Raimond Spekking / CC BY-SA-4.0, (http://creativecommons.org/licenses/by-sa/4.0/)
15.19 *Cebus albifrons* (by Whaldener Endo, CC BY-SA 3.0 Unported (http://creativecommons.org/licenses/by-sa/3.0/)
15.20 *Saimiri sciureus* (by Adrian Pingstone, Public Domain)
15.21 *Aotus zonalis* (by Tom Friedel, CC BY-SA 3.0 Unported), http://creativecommons.org/licenses/by-sa/3.0/
15.22 *Plecturocebus caquetensis* (by Javier Garcia, Public domain)

15.23 *Pithecia hirsuta* (by Thomas Defler, Public domain)
15.24 *Alouatta seniculus* (by Alessandro Catenazzi, CC BY-SA 2.5 (http://creativecommons.org/licenses/by-sa/2.5/)
15.25 *Lagothrix lagothricha* (by Evgenia Kononova, Public Domain)
15.26 *Sylvilagus floridanus* (by HarveyHenkelmann, Public Domain)
15.27 *Cryptotis venezuelensis* (by Marcial Quiroga-Carmona, permission American Society of Mammalogy)
15.28 *Felis wiedii* (by Malene Thyssen, CC BY-SA 3.0 Unported (http://creativecommons.org/licenses/by-sa/3.0/)
15.29 *Panthera onca* (by Charlesjsharp, CC BY-SA 4.0, http://creativecommons.org/licenses/by-sa/4.0/), https://creativecommons.org/licenses/by-sa/4.0/legalcode.de
15.30 *Speothos venaticus* (by Markus Bonnevier, Attribution-ShareAlike 3.0 Unported CC BY-SA 3.0), https://creativecommons.org/licenses/by-sa/3.0/legalcode
15.31 *Chrysocyon brachyurus* (by Sage Ross, CC-BY-SA-3.0 and older versions 2.0 and 1.0), http://creativecommons.org/licenses/by-sa/3.0/), Wikimedia Commons
15.32 *Tremarctos ornatus* (by Silvio Tanaka, CC-BY-SA-2.0 and 1.0 Generic), https://creativecommons.org/licenses/by/2.0/legalcode, Wikimedia Commons
15.33 *Pteronura brasiliensis* (by Renaud d'Avout d'Auerstaedt, CC-BY-SA-2.5), https://creativecommons.org/licenses/by-sa/2.5/legalcode, Wikimedia Commons
15.34 *Conepatus humboldtii* (by Payayita, CC-BY-SA-2.0), https://creativecommons.org/licenses/by-sa/2.0/legalcode, Wikimedia Commons
15.35 *Procyon cancrivorus* (by Szop rakojad.jpg, CC-BY-SA-2.0 Generic and 1.0 Generic), https://creativecommons.org/licenses/by-sa/2.0/legalcode, Wikipedia
15.36 *Tapirus terrestris* (by Jean-Marc Rosier, http://www.rosier.pro, permission)
15.37 *Catagonus wagneri* (by Dave Pape, worldwide public domain)
15.38 *Vicugna vicugna* (by Alexandre Buisse, CC-BY-SA 3.0 Unported, 2.5 Generic, 2.0 Generic, 1.0 Generic), https://creativecommons.org/licenses/by-sa/3.0/legalcode, Wikipedia
15.39 *Blastocerus dichotomus* (by Phillip Capper, CC-BY-SA-2.0 Generic), https://creativecommons.org/licenses/by/2.0/legalcode, Wikimedia Commons
15.40 *Pudu pudu* (by Jaime E. Jimenez, CC-BY-SA-3.0), https://creativecommons.org/licenses/by-sa/3.0/legalcode, Wikimedia Commons
15.41 *Guerlinguetus aestuans* (by Fernando Fernandez da Silva, CC-BY-SA-4.0 International), https://creativecommons.org/licenses/by-sa/4.0/legalcode, Wikipedia
15.42 *Abrothrix sanborni* (by J. Cardenas, CC-BY-2.0), https://creativecommons.org/licenses/by/2.0/legalcode Wikivisual, Wikimedia Commons
15.43 *Rhipidomys mastacalis* (by Leonado Merçon, CC-BY-4.0), https://creativecommons.org/licenses/by/4.0/legalcode, Wikimedia Commons

15.44 *Coendou rothschildi* (by Beatrice Murch, CC-BY-SA-2.0), https://creativecommons.org/licenses/by/2.0/legalcode, Wikimedia Commons
15.45 *Chinchilla lanigera* (by Rumpelstiltzkin, CC-BY-SA-2.5), https://creativecommons.org/licenses/by-sa/2.5/legalcode http://en.wikipedia.org/wiki/Image:CouchChin.jpg
15.46 *Dinomys branickii* (by Benjamin Frable, CC0-1.0 Universal), Wikimedia Commons, https://creativecommons.org/publicdomain/zero/1.0/deed.en
15.47 *Dolichotis patagonum* (by Vassil, Public domain), https://en.wikipedia.org/wiki/Public Domain , Wikimedia Commons
15.48 *Hydrochoerus hydrochaeris* (by Charles J Sharp, CC-BY-SA-4.0), https://creativecommons.org/licenses/by-sa/4.0/legalcode, Wikimedia Commons
15.49 *Dasyprocta leporina* (by Mistvan, CC-BY-SA 4.0, https://en.wikipedia.org/wiki/GNU_Free_Documentation_License, Wikimedia Commons
15.50 *Cuniculus paca* (by Hans Hillewaert CC-BY-SA-3.0 Unported license), https://creativecommons.org/licenses/by-sa/3.0/legalcode
15.51 *Ctenomys flamarioni* (by Cláudio Dias Timm CC-BY-SA-2.9 Generic license), https://creativecommons.org/licenses/by-sa/2.0/legalcode
15.52 *Octodon degus* (by Pierre Camateros CC-BY-SA-3.0 Unreported license), https://creativecommons.org/licenses/by-sa/3.0/legalcode
15.53 *Myocastor coypus* (by Philippe Amelant, CC-BY-SA-3.0 Unreported license) https://commons.wikimedia.org/wiki/Commons:GNU_Free_Documentation_License,_version_1.2
15.54 Comparison of number of species in South American terrestrial mammal orders (https://www.meta-chart.com/pie)

Index

T. Defler, *History of Terrestrial Mammals in South America*, Topics in Geobiology 42, https://doi.org/10.1007/978-3-319-98449-0

D

I

J

T